FARM TRACTOR SYSTEMS

DPs Dominion
Publishing Services
http://www.dominionpublishingstores.yolasite.com

FARM TRACTOR SYSTEMS

FARM TRACTOR SYSTEMS

Maintenance and Operations

Segun R. Bello

*B. Eng (Hons), FUT, Akure, MSc, Ibadan,
MNSE, MNIAE, FSINRHD, R. Engr. (COREN)*

Farm tractor systems
Maintenance &operations

Copyright © 2012 by Segun R. Bello

Federal College of Agriculture Ishiagu, 480001 Nigeria
segemi2002@yahoo.com; bellraph95@yahoo.com
http://www.dominionpublishingstores.yolasite.com
http://www.segzybrap.web.com
+234 8068576763, +234 8062432694

All Rights Reserved

No part of this book may be reproduced, stored in a retrieval system or transmitted, in any form or by any means, electronics, mechanical, photocopying, recording or otherwise, without the prior written permission of the copyright holder.

ISBN-13: 978- 148-102-292-7

First Edition published in January 2013

Printed by Createspace US
7290 Investment Drive
Suite B North Charleston,
SC 29418 USA, www.createspace.com

Dedication

This work is dedicated to

All that impact knowledge with simplicity and those who received it with diligence

FARM TRACTOR SYSTEMS

Acknowledgement

Glory is to God who gave strength to the meek and humble to do exploit. The author wish to express deep appreciations to all staff and students of the Department of Agricultural Engineering Technology, Federal Colleges of Agriculture Ishiagu and Moor Plantation, Ibadan for their contributions during interactive classes and field discussions which has become a major tool in packaging this book. Practical field experiences shared by the following operators; Mr. Lucky Chukwuma, Timothy Ajah, Kalu Ugwunna and Chibuzor Ajah are of immense value in this documentation.

The author acknowledged the contributions of the following referenced organizations, corporations and individuals whose graphics and photographs, some in modified state were used in this work:

Cross Creek Tractor Co. Inc., Massey Ferguson Co. Inc., Ford/New Holland, Allis Chalmer, TigerCat Co., Stewart Instruments, Inc., Same Deutz-Fahr U.K. Ltd., Integrated publishing, National Renewable Energy Laboratory (NREL) Colorado, SAF Service, Germany, Energy Field, Asian Institute of Technology Thailand, SPX Corporation, Owatonna, LRT Germany, University of Putra Malaysia, Serdang, Selangor D.E. Ohio State University and National Safety Council., The Pennsylvania State University, National Safe Tractor and Machinery Operation Program (HOSTA), PCRET.

Despite all the help received from many people, it seems inevitable that there will be some inaccuracies or errors in the text. For these the author accepts responsibility and apologizes in advance for any incorrect statements or impressions given. Should errors be noticed, the author would welcome factual corrections. He would also be happy to receive comments, observations and additional information on any topic, section or statements in any part of the book. This would be particularly useful should any updated or translated edition be planned. Correspondence may be addressed to the author.

My sincere appreciation goes to all who at one point or the other, share my visions and mission. Your encouragements, unflinching supports and faithfulness will forever be acknowledged. The encouragements of Bukola (my wife), and Ayomikun, 'Pelumi, Damilola and Adeola (my children) drove the passion for the realization of this work.

FARM TRACTOR SYSTEMS

Content

Preface xiii
Introduction xv
Book objective xvi
PART 1   FARMPOWER RESOURCES & MAINTENANCE CONCEPTS 1
CHAPTER 1   Farm Power Resources 2
Introduction 2
Farm power sources 2
Practical exercise 21
CHAPTER 2   Decision making in Power Selection 22
Introduction 22
Critical factors in machinery procurement 22
Decision making in machinery procurement 23
Power-implement (machinery) selection 28
Practical exercise 29
CHAPTER 3   Concepts of Maintenance Practices 30
Introduction 30
Machine failure 30
Maintenance 32
Reasons for machinery maintenance 32
Maintenance policy and decision making 34
Types of maintenance 35
Tero-technology and its concept 43
Maintenance functions and probability (maintenanceability) 44
Machinery repair concepts 45
Best maintenance and repairs practice 46
Repairs and maintenance costs 49
Practical exercise 49
CHAPTER 4   Maintenance Workshop and Organization 50
Introduction 50
Workshop information 51
Machine/maintenance workshop layout 51
Workshop organization 53
Keeping records in workshop 54
Fire control in maintenance workshop 55
Practical exercises 57

PART 2	**TRACTOR DIVISIONS AND POWER TRANSMISSION SYSTEMS**..............58

Introduction 59
Tractor divisions 60

CHAPTER 5	**Engines and Engine Systems**	62

Introduction 62
Fundamentals of engine and engine systems 62
Types of engine 63
Internal combustion engine (ICE) 64
Working principles of an IC engine 66
Classification of internal combustion engine 70
Lubricating oil and oil classification 115
Practical exercise 121

CHAPTER 6	**Tractor Power Transmission Systems**	122

Introduction 122
Tractor power drive systems 122
Mechanical transmission (Friction drives) 122
Hydraulic power drives 127
Tractor drive shafts 130
Tractor transmission systems 138
Tractor steering systems 141
Clutch system 146
Brake systems 147
Tractor chassis 150
Practical exercise 151

CHAPTER 7	**Tractor Drive Designs and Differential Systems**	152

Introduction 152
Tractor drive designs 152
Tractor front axle 154
Tractor tyres 155
Tractor differential system 164
Practical exercise 166

PART 3	**ENGINE AND TRACTOR SYSTEMS MAINTENANCE**..............167

Introduction 168

CHAPTER 8	**Engine Tune Up and Overhauling**	169

Introduction 169
Basic requirements for engine maintenance work 169
Planning for engine maintenance 170

Engine tune up	171
Maintenance of petrol engine	178
Maintenance of Diesel (tractor) engine systems	186
Engine overhauling	225
Preparing engine for end of season storage	234
Practical exercise	235
CHAPTER 9 Tractor Maintenance and Servicing Schedule	237
Introduction	237
Maintenance report/record	237
Operation-hour measurement and computation	245
Guide to preparing tractor for maintenance	246
Tractor transmission component maintenance	247
The differential system maintenance	264
The rear axle shaft and housing	266
Shock absorber installation	274
Tyre service and maintenance	274
Greasing and adjustments	284
Tractor service access points	286
Practical exercises	286
PART 4 TRACTOR DRIVING AND OPERATION	287
Introduction	288
CHAPTER 10 Tractor Driving and Operations	289
Introduction	289
Tractor drive control systems	290
Tractor hydraulic controls	295
Tractor instrument controls	296
Tractor driving and operation	304
Nigerian and international highway codes	310
Tractor driving tests	323
Tractor field operations	326
Attaching implements to tractor	328
Tractor gear selection for field operations	337
Draft control on the field	340
Field operation patterns and methods	343
Estimating tractor operating costs and fuel consumption	348
Tractor power and field calculations	351
Practical exercises	354

References 355
Glossary of Terms 360
Appendix 363
Notes 368

Preface

This book provides a link between machine functionality (performance), operations, reliability, and decision making process in the procurement, management and maintenance of farm machinery. Depreciation and functional deviation of a machine from its original state at manufacture could put the life of such machine in danger of breakdown or obsolescence, which is counted a loss to any such organization or the entrepreneur. To avoid such losses, an understanding of machine systems functionality and a well-organized maintenance programme designed to keep, maintain, prevent or restore machine to near original state is required.

Vocational training and entrepreneurship education in Nigeria's tertiary institutions targeted at students to acquire a do-it-yourself skill in maintenance, tracing and repairs of machine faults. A bimodal training programme (maintenance and operation) on tractor system, packaged and presented in this book is all that is required for managerial decision making on maintenance and operation programme for qualitative service delivery.

This book is all you need in emergency breakdown and where there is no mechanic; it offers a quick guide to engine problem troubleshooting. In this way, the enormous costs and valuable time spent on waiting desperately at breakdown points, tracing of faults, annoying breakdowns, unnecessary down time and costly repairs can be adequately reduced.

However, it is important to note that instructions and procedures outlined in this book are simple to read, but putting it to work may not be as simple as it seems. You need training, you need discipline, and you need diligence and tolerance to be able to practice them. An experienced hand (an instructor, operator, or a supervisor) is all you need the first time, thereafter, you will be maintenance compliant! ….. So get started.

This book will go a long way to acquaint students and researchers with the nitty-gritty of tractor systems, operations, repairs and maintenance practices and also provide requisite knowledge and skills for effective tractor system management.

Bello, R. S.
480001, Nigeria

FARM TRACTOR SYSTEMS

Introduction

In Nigeria and elsewhere in the world today, the application of engineering principles to agricultural production has provided basic tools by which inherent drudgeries and low productivities in agricultural operations had been reduced in order to accelerate and enhance productivity. Consequently, these innovations came with attendant problems of utility management, equipment maintenance and repairs, and costs management.

The introduction of tractor vehicles (*tractorization*) into agriculture has grossly increased engineering involvement in agriculture and hence the choice of equipment acquisition, hazards prevention programme and increased maintenance activities. In today's agricultural industry, system automation and remote sensing applications with guided technologies are now being employed in the field of agricultural production. The introduction of mechanical power and advanced technologies into agricultural production system has revolutionized this age-long industry.

Agricultural tractors operate in a poor and most unpleasant environment, and as such must be adequately maintained in order to effectively perform their desired functions. As a result, maintenance and repair programme has become inevitable. The serviceability and reliability of any machine in performing its desired function depends so much on how much of the maintenance practices that are observed in operating such machine.

Thus two broad issues recognized and widely discussed in effective machine operations and utility management which significantly affect their performance are; Operator's comfort/safety and the company's throughputs. Operator's safety is addressed by risk assessment during machine operation and the company's throughputs assessed by how much of maintenance and repairs is carried out at breakdown or before breakdown.

Management decisions related to agricultural machinery could also affect production profits in several ways. Fuel, interest, labour and timeliness of operation are other pertinent factors that contributed to the tractor's productivity and efficiency. To improve productivity and efficiency, it is necessary to have comprehensive information on all aspects of the tractor and implement performance in the field.

This book addressed these problems and provide guide to students on basic practical approach to farm power selection, machinery repairs and operation, to provide them with the necessary theoretical and practical skills for effective entrepreneurship in accordance with Nigeria tertiary institutions' regulatory bodies bearing in mind balance between input and output.

Book objective

This book was thus designed;

1. To help the students understand decision making processes in machinery procurement.
2. To help them understand the basic engine operating systems, tractor operation and maintenance practices
3. To help the students acquire the necessary skills in machinery repairs, maintenance and safety during maintenance operations.
4. To be familiar with the terms associated with fault diagnosis, tractor tests and to know the steps involved in diagnosing and testing an engine and engine component.
5. Be able to prepare tractor system maintenance chart, maintenance records, maintenance schedule etc.
6. To help guide students in tractor field operations.

PART 1

FARM POWER RESOURCES & MAINTENANCE CONCEPTS

Introduction to farm power resources, machinery development, decision making, maintenance practices

CHAPTER 1

Farm Power Resources

Introduction

This chapter describes various sources of generating power on the farm, their availability, applicability, features, and comparative advantages and disadvantages. At the end of this chapter, students are expected to have a good knowledge of various power sources and utilities available within farm settlement and the areas of need for making appropriate mechanization decision.

Farm power sources

It is essential for students to be familiar with the basics of farm power sources for the purpose of evaluation mechanization practices. Different levels of mechanization were defined by the different types of power sources in use. It is quite clear that agriculture has always been mechanized, employing four main sources of power or a combination of two or three of these sources which include: human, animal, mechanical/engine and renewable energy resources.

1. Human power

Traditionally, human power has been the major source of power on the field from the Stone Age. Hand held tools are predominantly used based on operations performed at different phases of crop production on and off the farm. Hand power tools used in farm operations include machete (cutlass), hoes, axe, digger, sickle, knife etc. Load carrying aids for head include baskets, basins, head pan, sacks, wheel barrows etc. other locally made hand push equipment are also available and include hand push weeders, hand wheel hoes, jab planters, etc.

Figure 1-1: Human power employed in farm operations

Human power accounts for the lion's share of work input in overall agricultural production, most especially in the tropical and sub-tropical African countries. The amount of human power-use and intensity has been a source of concern to agricultural engineers owing to its low level of development and natural limitations over machines (Mrema and Mrema, 1993; Comsec, 1990; Anazodo et al., 1989; and Anazodo, 1987). The predominance of human power in the agriculture of a developing country is an important factor to address in dealing with overall economic development of that country. The problem of limited cultivated area of land is not necessarily with the tools used, especially for primary production operations, since efforts have been made to redesign them but this yielded no significant improvements (Makanjuola, 1991 and Odigboh, 1991).

Advantages

a. They are readily available and cheap
b. They are involved in decision making process and therefore reduce the cost of management consultancy
c. Requires no complexity in performing an operation

Disadvantages

a. *Limited area coverage*: Power is the major limitation to increasing the area cultivated by the hand-tool farmers. The toil, drudgery, and severe power constraints on timely field operations, which limit production and increased return capacity, are the inherent characteristics of peasant farmers using hand-tool technology. When you change the technology, you change the farmer's status (Odigboh, 1996).
b. *Labour shortages*: Labour shortage is most crucial during the period of land preparation at the beginning of the rainy season and this may result in untimely operations and limits both area expansion and total food production. This is especially true in cases where the cropping season is short.
c. *Power limitation (human work output):* Only about 25 percent of the energy consumed when handling relatively easy tasks such as pedaling, pushing or pulling is converted to actual human work output. Under more difficult work conditions, the efficiency of converting consumed energy to physical work may be as low as 5 percent or less. This means that, at the maximum continuous energy consumption rate of 0.30 kW and conversion efficiency of 25 percent, the physical power output is approximately 0.075 kW sustained for an 8–10 hour work-day. Naturally, higher rates can be maintained for shorter periods only (Inns, 1992).

2. *Draught animal power*

Because of the limitations of the human power availability on the field, horses, mules, oxen and bullocks became the principal sources of power on the farm. They develop more power than human power for agricultural operations. Because of their availability for use in most stringent conditions, they are often referred to as the beast of burden.

Figure 1-2: Animal draught power

Reduced energy potential is a factor limiting the use of animals for work which is determined by characteristics and working ability of the animal species. A draft animal power (DAP) unit (animal plus equipment) which exerts a tractive force can be compared with a system consisting of a resistant part (equipment) and a power unit (animals).

Nigerian animal traction technology development is characterized by use of limited implements to explore the full potentials of animal traction operation. Most farmers lack animal drawn equipment like ploughs, harrows, planters, weeders and harvesters while Animal traction implements/equipment development and manufacture are left in the hands of local blacksmiths that are constrained by insufficient patronage due to unavailability of raw materials, inadequate workshop facilities and ineffective marketing strategies.

Factors influencing the work achieved by animals

In-depth knowledge of the factors (some can be controlled by farmers, others cannot) influencing the work achieved by animals is required as discussed below

a. *Environmental conditions:* Environmental factors (soil and climate) that define working conditions are uneasy to control. Farmers can improve such conditions but to a small extent. They may prefer to work their animals at the beginning or end of the day, when heat is acceptable. This is profitable in terms of animal capabilities and endurance.
b. *Choice of animals:* The characteristics of animals (breed, species, sex, age, temperament) determine their working abilities. Farmers cannot control these. Their only room for maneuver is in the choice they can make between the various species locally available or affordable.
c. *Use of animals:* Team composition and the choice of implements and harnessing systems depend on the farmer's decisions. They are key factors for transforming the energy accumulated by animals into mechanical power. With current yokes and harnessing, pooling two animals or more in a team results in a reduced efficiency at an individual

level. If the available power is 1 with one animal, it is only 1.85 with two animals, 3.10 with four, and 3.80 with six (Havard and Wanders, 1999). Choosing the more suitable harnessing system, equipment and number of animals depends on local availability and cost, but may rapidly enable significant energy gains.

The characteristics of these animals such as breed, species, sex, age, temperament and environmental factors determine their working abilities. Farmers cannot control these factors thus makes them irrational and unreliable.

Constraints to animal traction may include lack of appropriate implements, limited capital and credit, insufficient animals, animal health problems, inadequate animal nutrition (quantity and/or quality), uncleared fields, farmer traditions, lack of technical knowledge, poor infrastructure and limited marketing possibilities. Most constraints can be overcome when other conditions are favourable and knowledge spreads quickly through informal channels.

Environmental factors such as temperature and humidity affects their performance. Disease attack also limits their performance and range of operation in the tropical and rain forest zones. Technology development made it possible to mechanize their operation to some extent. Team composition and the choice of implements and harnessing systems are the key factors for transforming the energy accumulated by these animals into mechanical power.

Advantages

Advantages of animal draughts power include

a. They are relatively cheap to maintain,
b. They develop more power than human power for agricultural operations.
c. They are cheap to maintain, they have multi-purpose uses, and have self-replacement value.
d. Because of their availability they are used in most stringent conditions

Disadvantages

Constraints to animal traction may include

a. Animal requires continual attention either working or not.
b. They have limited daily working hours, slow, they have high person-to-power ratio.

a. They are naturally slow and are limited to smaller units,
b. Too temperamental and very difficult to control

a. Lack of appropriate implements, limited capital and credit, insufficient animals, animal health problems, inadequate animal nutrition (quantity and/or quality), uncleared fields, farmer traditions,
b. Lack of technical knowledge in harnessing systems,
c. Poor infrastructure and limited marketing possibilities.
d. Subject to adverse weather condition. Environmental factors such as temperature and humidity affects their performance.
e. Disease attack also limits their performance and range of operation in the tropical and rain forest zones.

3. *Mechanical power*

Mechanical power as used on the farm consists of internal combustion engine, the electric motor, and the steam engine sometimes called external combustion engine, the water wheel and windmill. The internal combustion engine and electric motor are the most important. Recently internal combustion engine are being complimented by hydraulic power transmission.

i. *Stationary engines*

Farm engines are generally used for powering stationary machinery or equipment mounted on a person's back. Farm engines are broadly classified as; spark ignition or compression ignition engines, forced-air-cooled *gasoline engine* with a single cylinder using either a two- or four-stroke cycle or a water-cooled, single-cylinder *diesel engine*.

Figure 1-3: Stationary engines

Figure 1-3 show examples of farm engine. There are a few models of small and light weight *air-cooled diesel engines*, which need a high level of engineering technology to develop. The direction of rotation of the output shaft of the engine is counter-clockwise from the farmer facing it, according to the industrial standard.

These engines are used to power crop processing machinery such as digesters, milling machines, etc. and also to power some semi-mounted and mobile machines such as feed mixers, wood chippers and corn shellers. They are either air-cooled or water-cooled engines, petrol or diesel engine. In contrast to electric motor, they produced a lot of smokes, noise and vibrations.

ii. *Power tillers or two-wheel tractors*

Two-wheel tractors are sources of power designed to perform most field operations. Due to the size of such tractors, they become an economic alternative for small scale farming. In addition, two-wheel tractors are much more productive than animal traction and they require less time for attendance and preparation, giving the individual farmer more independence and contact with modern technology.

Figure 1-4: Mitsubishi 8 kW heavy duty power tiller

Two-wheel tractors are widely used for dry land farming, paddy farming, and local transportation with trailer attachments. New special applications have been developed for two-wheel tractors such as drilling holes, cutting wood and moving snow. Also there has been observed a tendency to substitute a diesel powered 2-wheeled tractor for gasoline engines, air-cooled diesel engines of similar speed, for better reduced fuel consumption.

Figure 1-5: Mitsubishi 8 kW heavy duty power tiller

iii. *Tractors*

These are traction machines designed primarily to supply power to agricultural implements and farmstead equipment. A tractor is a self-powered work vehicle, designed for pulling or pushing special machinery or heavy loads over land. The agricultural tractor is a vehicle for off-road and on-road operation, being able to carry, guide, pull and drive implements or machines – moving or stationary – and to pull trailers. Tractors are widely used in agriculture, building construction, road construction, and for specialized service in industrial plants, railway freight stations, and docks.

Figure 1-6: Typical agricultural tractors

Tractors are widely used in agriculture, horticulture, building construction, farm road construction, and for specialized services in industrial plants, railway freight stations and docks. They are the major source of power available on the farm these days because of their versatility and ease of operation, though very expensive to procure and maintain. Tractors suited for agricultural and horticultural works are generally of three types.

- ✓ *Agricultural tractors:* An agricultural tractor propels itself and provides a force in the direction of travel to enable attached soil engaging and other agricultural implements to perform their intended function according to standard are large, heavy-duty tractors suited for commercial farming.
- ✓ *Utility tractors* are smaller, less powerful or both than agricultural tractors, but heavy duty and usually sufficient for private farms and small commercial farming operations.
- ✓ *Compact tractors* (some manufacturers refer to these as sub-compact or compact-utility tractors). They are smaller, less powerful or both than utility tractors and suited for small, private farms. They offer more functional capability than lawn-and-garden tractors and ATVs, and they can be acquired with necessary features, such as a PTO and three-point hitch, which most lawn-and-garden tractors lack.

iv. *Truck/haulage tractors*

These machines provide energy for hauling, distribution of goods and inputs between and within farms. Movement of goods and services to and from point of processing, storage and market is achieved by truck engines. On-farm and off-farm activities are largely done by engaging hauling to get farm utilities and harvested products to processing centers and market.

Figure 1-7: A log haulage truck

Advantages

a. They have multi-purpose use, operate in harder conditions than animals or humans, operate in both wet and dry conditions,
b. They are capable of producing continuous power for a long period under favourable operating conditions
c. They develop more power than human power for agricultural operations.
d. Because of their availability they are used in most stringent conditions

Disadvantages

a. They produce exhaust fume that is not environmentally friendly
b. They generate a lot of noise thus constituting pollution menace
c. High operator fatigue, although ride-on versions are now available.
d. The initial cost of acquisition and operating them are very high
e. Requires a lot of maintenance and thus expensive to run

4. *Electric power source*

a. *Fuel cells (batteries)*

Fuel cells are electrochemical devices that convert the chemical energy of a fuel directly and very efficiently into electricity (DC) and heat, thus doing away with combustion. A fuel cell consists of an electrolyte sandwiched between two electrodes. Oxygen passes over one electrode and hydrogen over the other, and they react electrochemically to generate electricity, water, and heat.

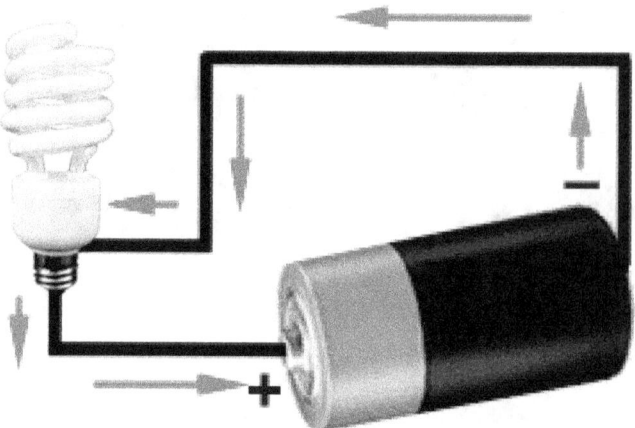

Figure 1-8: Energy from fuel cell

Different types of batteries

Different types of batteries use different types of chemicals and chemical reactions. Some of the more common types of batteries are:

i. *Alkaline battery* - Used in Duracell and Energizer and other alkaline batteries. The electrodes are zinc and manganese-oxide. The electrolyte is an alkaline paste.
ii. *Lead-acid battery* -These are used in automobiles. The electrodes are made of lead and lead-oxide with a strong acid as the electrolyte.
iii. *Lithium battery* - These batteries are used in cameras for the flash bulb. They are made with lithium, lithium-iodide and lead-iodide. They can supply surges of electricity for the flash.
iv. *Lithium-ion battery* - These batteries are found in laptop computers, cell phones and other high-use portable equipment.
v. *Nickel-cadmium* or NiCad battery - The electrodes are nickel-hydroxide and cadmium. The electrolyte is potassium-hydroxide.
vi. *Zinc-carbon battery or standard carbon battery* - Zinc and carbon are used in all regular or standard AA, C and D dry-cell batteries. The electrodes are made of zinc and carbon, with a paste of acidic materials between them serving as the electrolyte.

b. *Induction motor*

Electricity generated by electric motor is used to power farm machines. This source of energy use has the advantage of producing low noise level, smooth operation and clean energy. No environmental pollution of any kind. The major challenge of this energy source is the unreliability of electric power source. In the developing nations, the use of this energy source is widely limited and unreliable.

Figure 1-9: An electric powered wood chipper

c. *Generators*

These are engines coupled to magnetic coils for electricity generation and distribution for farm operations, farm house lighting programmes and processing of agricultural products. Electricity is essential in the farm for powering process plants, farm electrification, farm lighting, refrigeration, hatchery powering and other essential utilities that provide comfort for human, animals and machines alike.

Figure 1-10: A generating plant

5. *Renewable energy sources*

All energy sources mentioned above have an impact on the environment. Concerns about the greenhouse effect and global warming, air pollution, and energy security have led to increasing interest and more development in renewable energy sources such as solar, wind, geothermal, wave power and nuclear energy. Each of these energy sources are discussed below.

i. *Solar energy*

This is the process of conversion of sunlight into heat in the oven or electrical energy in solar cells for further production activities in the farm. This energy conversion process involves three major steps;

FARM TRACTOR SYSTEMS

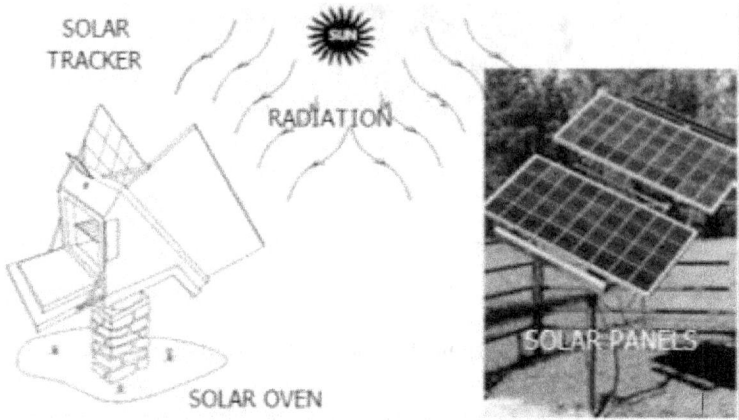

Figure 1-11: A solar tracking system

- ✓ Absorption of the sunlight by solar cell (heat source);
- ✓ Heating up of the thermocouple junction or black surface thus obtaining a temperature difference; and
- ✓ The transfer of these thermoelectric potentials or temperature difference for outside application in the form of electric current or heat.

Figure 1-12: Solar dryer system; Courtesy PCRET

Solar thermal energy is used for heating water for industrial and domestic purposes, cooking/heating, drying/timber seasoning, distillation, electricity/power generation, cooling, refrigeration, cold storage. Solar energy can also be used to meet our electricity requirements. Solar power technologies, from individual home systems to large-scale concentrating solar power systems, have the potential to help meet growing energy needs and provide diversity and reliability in energy supplies.

ii. *Wind energy*

Wind power is generated by the harnessing of wind with turbines to produce mechanical power or electricity. Wind turbines, like windmills, are mounted on a tower to capture the most energy.

Figure 1-13: Wind mills in a maize farm

At 100 feet (30 meters) or more aboveground, they can take advantage of the faster and less turbulent wind. Turbines catch the wind's energy with their propeller-like blades. The wind turns the blades of a windmill-like machine. The rotating blades turn the shaft to which they are attached. The turning shaft typically can either power a pump or turn a generator which produces electricity.

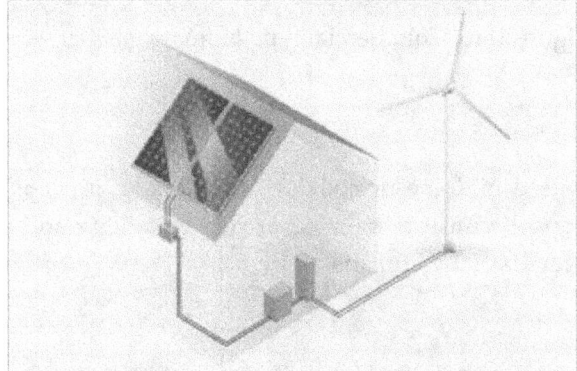
Figure 1-14: Wind mills in home energy system

The main components of a wind turbine system (Figure 1-15) include the rotor, generator, tower, and storage devices. Usually, two or three blades are mounted on a shaft to form a *rotor*. The rotor consists of a hub that connects the rotor to the turbine system. The blade can spin in either horizontal or vertical axis to the ground.

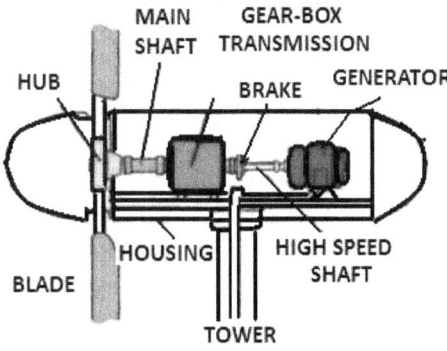
Figure 1-15: Components of wind mill

iii. Biomass energy

Biomass is a renewable energy resource derived from the carbonaceous waste of various human and natural activities. It is derived from numerous sources, including the by-products from the timber industry, agricultural crops, raw material from the forest, major parts of household waste and wood.

Figure 1-16: Commercial type biomass gasifier stove

iv. Solid fuels

Solid fuels are widely used in domestic cooking and include firewood and charcoal, organic wastes, and energy crops. Because of its widespread availability and ease of use, fuel wood was the first fuel used regularly by humans.

Figure 1-17: Cooking with firewood

In addition to the primary use of these fuels for cooking, they are essential in developing countries like Nigeria for many small to medium-scale trades and industries such as brick making, textile manufacture, baking, and brewing. In rural communities, fuel wood and charcoal can have important other uses such as crop drying or curing.

v. Charcoal

Charcoal is produced by the pyrolysis (burning) of wood, wood wastes, agricultural and food processing by-products. The principal worldwide use for charcoal is for preparing food, and the preferred form is in lumps because of its ease of handling and ignition.

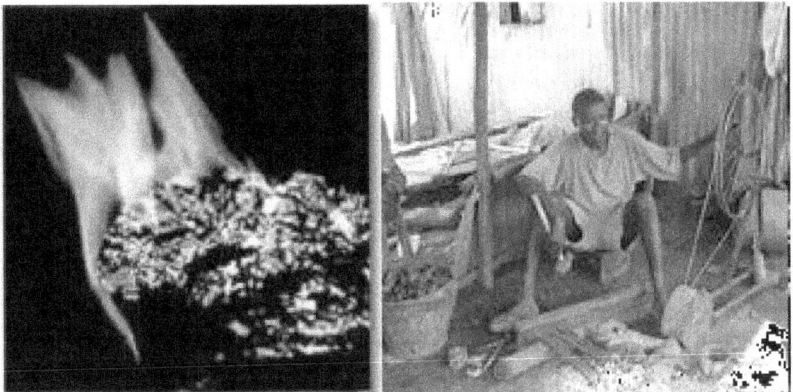
Figure 1-18: Charcoal used in blacksmith bellow

vi. Briquettes

Briquettes are composed of fine charcoal plus a binder and sometimes filler. Wood ash, coal or petroleum solids, sawdust, or calcium carbonate (in the form of ground limestone, chalk, or ground sea shells) are sometimes used as fillers, ostensibly to provide a slower burn, but they are also significantly cheaper than charcoal.

Figure 1-19: Rice husk briquettes

vii. Forestry residues

Forestry residues are usually classified as either primary (e.g. pulp, paper, lumber) or secondary (e.g., furniture, composite boards, and wooden handles) manufacturing operations. The various types of residues can include wood flour, sawdust, shavings, sander dust, pole and post peelings, chip rejects, end cuts, slabs, flawed dimension lumber, and other residues.

Figure 1-20: Processed wood chip

Availability: Although pulp and paper mills generate large quantities of wood residues (i.e., bark, fines, and unusable chips), this waste is typically used in boilers and is not available domestically.

viii. Agricultural crop residues

Agricultural crop residues—materials left on the land after harvest of the economically valuable part of the crop—may be collected during harvest of the crop or left for later harvest. It includes orchard and vineyard pruning and field residues from horticultural and agronomic crops such as rice bran, corn stalks, stovers, etc.

ix. Fossil fuels (petrochemicals)

Fossil fuels are produced from plant and animal remains converts by natural processes over millions of years into a form which we can use as a source of energy. Fossil fuels such as coal, oil and natural gas are the main source of our energy, used to power our power stations and vehicles.

Figure 1-21: Fossil fuel

x. Geothermal energy

Heat from the earth core can be drawn either as hot water or steam reservoirs to be used on both large and small scales. A plant can use the hot water or steam to drive generators and produce electricity. Other applications apply the heat produced from geothermal directly to various uses in buildings, roads, agriculture, and industrial plants. Still others use the heat directly from the ground to provide heating and cooling in homes and other buildings.

Figure 1-22: Generating hot water from earth surface

Other geothermal resources exist miles beneath the earth's surface in the hot rock and magma there. In the future, these resources may also be useful as sources of heat and energy.

Other applications apply the heat produced from geothermal plant directly to supply various uses in buildings, roads, agriculture, and industrial plants. Still others use the heat directly from the ground to provide heating and cooling in homes and other buildings.

xi. Hydropower (hydel) energy

This is energy derived from water sources. The energy wave in the flowing water is used to produce electricity by creating a reservoir or basin behind a barrage and then passing tidal waters through turbines in the barrage to generate electricity.

Figure 1-23: Hydel (hydroelectric) power dam

How hydro power works: The water stored in the dam is released through pipes that run through turbines that turn generating dynamos.

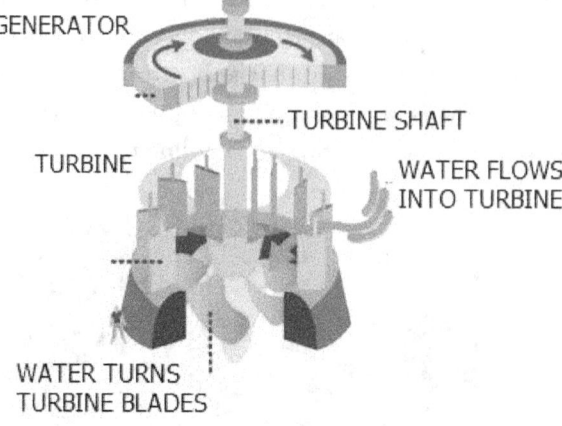

Figure 1-24: Hydroelectric power generator

Hydro power is one of the best, cheapest, and cleanest sources of energy, although, with big dams, there are many environmental and social problems such as concerns about impact of dams on environments and river flows, vulnerability to variations in rainfall, and large set up costs. Small dams are, however, free from these problems.

xii. *Nuclear energy - fission and fusion*

Another major form of energy is the nuclear energy, the energy that is trapped inside an atom. An atom's nucleus can be split apart. When this is done, a tremendous amount of energy is released *(nuclear fission)*. The energy is both heat and light energy. This energy, when let out slowly, can be harnessed to generate electricity. When it is let out all at once, it can make a tremendous explosion in an atomic bomb.

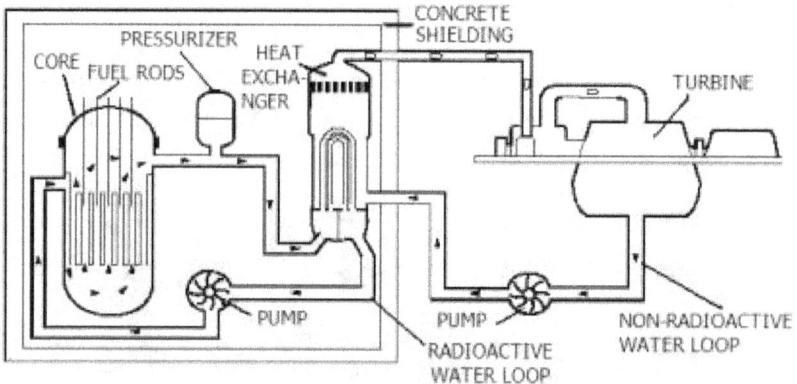

Figure 1-25: Nuclear power plant; (*Courtesy* Nuclear Institute)

Nuclear fusion: Another form of nuclear energy is called fusion. Fusion means joining smaller nuclei (the plural of nucleus) to make a larger nucleus. The sun uses nuclear fusion of hydrogen atoms into helium atoms. This gives off heat and light and other radiation.

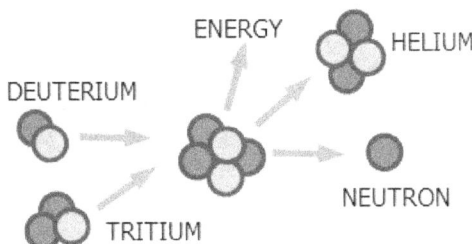

Figure 1-26: Energy fusion (UCLA, Berkeley)

In Figure 1-26, two types of hydrogen atoms, deuterium and tritium, combined to make a helium atom and an extra particle called a neutron. Also given off in this fusion reaction is energy! Scientists have been working on controlling nuclear fusion for a long time, trying to make a fusion reactor to produce electricity. But they have been having trouble controlling the reaction in a contained space. Nuclear fusion creates less radioactive material than fission, and its supply of fuel can last longer than the sun.

xiii. Ocean energy

There are three basic ways to tap the ocean for its energy. We can use the ocean's waves, we can use the ocean's high and low tides, or we can use temperature differences in the water. Let us take a look at each.

Wave energy: Kinetic energy (movement) exists in the moving waves of the ocean which can be used to power a turbine. The wave rises into a chamber (Figure 1-27) which forces the air out of the chamber.

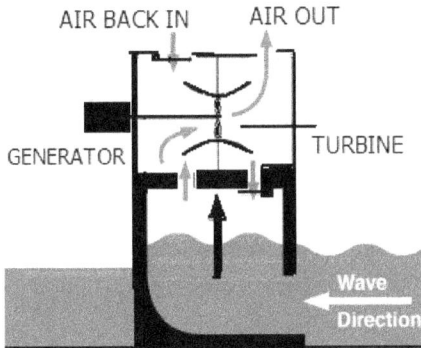

Figure 1-27: Wave energy generation

The moving air spins a turbine which can turn a generator. When the wave goes down, air flows through the turbine back into the chamber through doors that are normally closed. This is only one type of wave-energy system. Others actually use the up and down motion of the wave to power a piston that moves up and down inside a cylinder. That piston can also turn a generator. Most wave-energy systems are very small. But, they can be used to power a warning buoy or a small light house.

Tidal energy: Another form of ocean energy is called tidal energy. When tides come into the shore, they can be trapped in reservoirs behind dams. Then when the tide drops, the water behind the dam can be let out just like in a regular hydroelectric power plant. In order for tidal energy to work well, you need large increases in tides. An increase of at least 16 feet between low tide and high tide is needed.

Figure 1-28: Tidal wave in the ocean

The world's ocean may eventually provide us with energy to power our homes and farms. Right now, there are very few ocean energy power plants and most are fairly small.

Advantages of renewable energy resources

a. They are environmentally friendly.
b. Because of their availability for use in most stringent conditions, these sources of energy are renewable and there is no danger of depletion. These recur in nature and are inexhaustible.
c. The power plants based on renewable sources of energy do not have any fuel cost and hence negligible running cost.
d. Renewable are more site specific and are used for local processing and application. There is no need for transmission and distribution of power.
e. Renewables have low energy density and more or less there is no pollution or ecological balance problem.
f. Most of the devices and plants used with the renewables are simple in design and construction which are made from local materials, local skills and by local people. The use of renewable energy can help to save foreign exchange and generate local employment.
g. The rural areas and remote villages can be better served with locally available renewable sources of energy. There will be huge savings from transporting fuels or transmitting electricity from long distances.

Disadvantages of renewable energy resources

a. Low energy density of renewable sources of energy need large sizes of plant resulting in increased cost of delivered energy.
b. Intermittency and lack of dependability are the main disadvantages of renewable energy sources.

c. Low energy density also results in lower operating temperatures and hence low efficiencies.
d. Although renewables are essentially free, there is definite cost effectiveness associated with its conversion and utilization.
e. Much of the construction materials used for renewable energy devices are themselves very energy intensive.
f. The low efficiency of these plants can result in large heat rejections and hence thermal pollution.
g. The renewable energy plants use larger land masses.

Practical exercise

1. Identify the various power sources in your institution farm or in a selected farm settlement. Provide all necessary details to fully describe each of these power sources. Discuss the mode of utilization of each identified sources.
2. Carry out a feasibility study of the potentials for utilization of renewable energy sources as alternative energy source in your institution. Make recommendations on how to harness such potential for power optimization. Report your findings in your workbook.
3. Identification and classification of farm tractor. The purpose of this exercise is to enable you compare features of two tractor models of tractor found in your farm shop for the purpose of standardization.

CHAPTER 2

Decision making in Power Selection

Introduction

The introduction of mechanical power and advanced technologies into agricultural production system has revolutionized this age-long industry. In today's agricultural industry, the conventional agriculture has expanded in scope to such an extent that system automation and remote sensing applications with guided technologies are now being employed in agricultural production. This process referred to as **motorization** comprises of engine-powered machinery for traditional agriculture involving the use of machines (simple or complex) as power sources.

The introduction of tractor vehicles (*tractorization*) into agriculture has grossly increased engineering involvement in agriculture and hence the choice of equipment acquisition, hazards prevention programme and increased maintenance activities. Based on these reasons, acquisition of farm power resources has become a serious engineering business with diverse critical factors. Each of these factors has to be addressed to enable investors make choices in planning and execution of agricultural operation.

Critical factors in machinery procurement

Machinery procurement is all about decision making in the acquisition and management of agricultural utilities. In as much as you know the cost of power in the farm and size of your undertaken, you just might not need a tractor. Higher fuel prices will lead to higher machinery costs that agricultural producers will have to absorb in the short run whether they are doing the operations themselves or hiring somebody else to do it. Therefore you must consider the following initial critical steps before embarking on decision making process:

Step one: identify your needs

Size of farm: Consider how much acreage you will put to use each year, how rough the terrain is (e.g. sloped or flat, heavy or light, sandy soil), and how much time you have to do the work. If you have heavier soils or otherwise more mechanically stressful conditions to deal with, then you might be served with a heavier-duty tractor.

Type of operation and fuel price: Consider what sorts of jobs you expect to accomplish with your tractor. Take into account the heaviest anticipated usage because if you have to continually operate your tractor at peak output, it will age the machine. In if you anticipate fuel prices will remain high in the future, more management decisions are required to lower this cost either by reduced tillage, purchase machinery that is more fuel efficient per unit of use, change crop rotation, etc.

Choice of implement: Consider practical issues, such as what kinds of implements you will need and whether they will attach to the front or the back of the tractor. If you will have to transport the tractor to and from your products, also be sure to consider whether your vehicle is up to the job.

Time of operation: Consider how much time you have to do your tractor work. If your time is limited, you may need a more-powerful tractor so you can work at a faster pace.

When conclusions on these considerations have been made, then proceed to step 2.

Step two: find the right tractor

Once you have completed the list of your needs using the criteria enumerated above, you only have three other variables to consider in finding the right tractor for your operation. These are;

- *Power rating*: What horsepower tractor do I need?
- *Size* of tractor, what model of tractor and desirable features such as power steering, remote hydraulic control etc.
- The *cost* of your choice?

In order to find solutions to these variables, the decision making process is required.

Decision making in machinery procurement

The following decision making processes should then be followed in procuring equipment if profit is your concern.

1. The *first* stage include decision making stages where the issue of ownership, lease or hiring must be made,
2. *Secondly*, other decisions based on choice of equipment- used or new, what machinery to buy etc.
3. *Thirdly*, equipment specifications, power ratings and other attachments or accessories are necessary and
4. *Finally*, safety devices that are available must be checked.

FARM TRACTOR SYSTEMS

Stage 1: Decision making process

In decision making over the procurement of tractor and equipment for farm enterprise there are three most common choices to choose from alternative to purchasing machinery based on economy of scale. These alternatives are *equipment leasing, equipment rental* and *hiring*.

Decision of ownership of implement: In large hectares of land, it is advisable to buy equipment but on small hectares of land or small holdings, it is better to hire, considering the cost of equipment. When purchase consideration has been made on those implements or machinery that will suite the size of farm holding or power of the available machine, timeliness of particular farm operation may be disadvantaged to get machine for hire at the appropriate time needed.

When you made up your choice of purchase of a machinery or tractor, either used or new, you should consider the risks the tractor or machinery may introduce at your workplace. After considering these risks you should ensure that the health and safety design features of the chosen tractor or machinery could control such risks.

For instance, the recent trend towards less tillage brought about by advances in farm chemicals usage, especially herbicides, has sharply increased the availability and interest in self-propelled crop sprayers that can be used for both pre- and post-plant treatments.

Machinery investment decisions are inherently complex because they involve time and monetary issues. A few examples of time issues regarding machinery are

1. That machinery depreciates over time;
2. Tax depreciation and market depreciation typically occur at different rates;
3. Repairs tend to increase as a machine ages; and
4. As machines age they become less dependable (more prone to breakdowns), leading to owner concerns about timeliness.

Such potentially important considerations are left to the users to assess.

In terms of economic analysis, machinery ownership and operating costs are often classified into the following categories:

1. Interest;
2. Depreciation;
3. Repair and maintenance;
4. Labour;
5. Fuel and lubrication; and
6. Property taxes, insurance, and shelter (TIS).

Although the timing of tax depreciation impacts on overall costs and profitability, the depreciation of interest will be market depreciation. Market depreciation is the change in machine market value over time, which represents a real loss in asset value. Although based on prevailing lender interest rates, the interest cost to be considered most important here is the opportunity interest, rather than the interest associated with an actual loan arising from an owner's financing decision. That is, because equity could be invested elsewhere, it is considered to bear interest just as debt does.

Stage 2: Choice of buying new machinery

The choice of buying a new tractor is hinged on the size of farm land to be cultivated, the amount of use the tractor should be put into use, ownership factors etc. As a guide, machinery (tractor and equipment) buyers should seek a tractor that has the following features for more efficient and easy operation: more power, power steering, power brakes, three point hitches, remote hydraulic cylinders, a variety of transmission power take off (PTO) drives, hydraulic seat, air conditioned, remote control system and actuators (infrared sensors etc), enclosure for operation protection and comfort, four-wheel drive etc.

Stage 3: Choice of buying used farm machine

An influx of used machinery into the country (Nigeria) has grossly increased the number of junk stores in most establishments and farm sites due to non-consideration of some basic requirements that will ensure a selection of a good machine. It is absurd to know that some imported machines have never been put to use once since procurement. A clean and freshly painted machine can cover up a multitude of problems.

The following processes will help you make a wise choice of machine.

1. Know the seller's reputation in reconditioning, service, use and operation of farm machinery.
2. Check with the last owner of the machine (the reason for selling) and ask for the maintenance records (If available).
3. Check with the dealership on availability of parts and service.
4. It is a good idea to know the seller personally and know the equipment that he is selling. Is the machine popular among other farmers?
5. Test drive the unit before the sale date and make some extensive checks on the following units before purchases; breaks, transmission, clutch, steering, air intake, exhaust, and check all gauges very carefully.
6. Check tractor horsepower, fuel and oil consumption rate. While the tractor is under load, check exhaust smoke, oil leaks, blow-by and any unusual noises.
7. Check hydraulic oil pressure and oil flow with a pressure gauge.
8. Remove the air cleaner element and use flashlight to check the seals for air by-pass, intake hoses and connections for leaks.

9. Drain off a sample of hydraulic and transmission oil in a clear glass; hold it up to the light and check for filings and contamination.
10. Check cooling system for scale, rust and contamination.
11. Check tyres for breaks, cuts, wrinkles and buckles in the sidewalls.
12. Check grease fittings for evidence that they have had contact with a grease gun. Dirt or grease that has been painted over can be a clue.
13. Jack up a front wheel and rotate, checking for side clearance in bearing and seals.
14. Your first impression of a machine is important. Look for caked-on dirt and oil leaks, cracks, new welds, rust, faded paint, loose nuts, cracked belts or hoses.
15. The hour meter does not always tell the whole story about the value of the machine.
16. Check for excessive clearance and vibration in components such as in bearings, pulleys, shafts, chains, bushings and nonalignment of pulleys and belts.

No mechanical device is going to run perfectly, but if you take time and do your homework carefully, there is no reason you cannot select a good used piece of equipment. Always remember that the only way you can justify machinery is for it to work effectively and efficiently.

Stage 4: Choice of farm equipment

You can make your choice of tractor and accessories based on some selected features.

a. *Tractor type*: Four main types of tractors are recommended for agricultural production from which your choice can be made: Crawler or track laying, Standard, Row Crop and Market garden tractors. Each has special features in its design, which make them suitable for particular farm operation.
b. *Engine power*: Approximately one-third of the energy used for agricultural production is used by tractors in such operations as tillage, planting, cultivating and harvesting. The engine should hence be capable of giving high torque or twist at low speeds.
c. *Fuel economy:* The cost of fuel use in machine operation is becoming high, thus consideration should be given to selecting engines with low fuel consumption. Fuel economy factor is always included in the owner's manual accompanying tractors.
d. *Combination of reliability, long life and low maintenance cost* is also a key factor in choice of machinery. Selected machine must be the one that is readily adaptable to local soil and weather conditions, has a long-life pedigree, availability of replacement parts and low maintenance cost
e. *Accessibility to all moving parts and adjustme*n**ts.** For the purpose of maintenance and repairs, machine must not be too complex, compact and components must not be too hidden. For instance, filters should be easily accessible etc.
f. *Transmission*: Select a tractor with as many forward speeds as possible. The highest gear for tractors on steel wheels or tracks should be about 5mph, but for those on pneumatic tyres a much higher gear giving some 1.2-25mph is essential for a rapid transport work.

g. *Wheel requirement*: A tractor should be capable of transmitting a high percentage of the power developed by the engine to the drawbar.
h. *Tractor performance criteria:* Maximum drawbar power is normally the most useful performance criterion for farm tractors. The fuel consumption is also an important criterion that can be used to indicate directly or indirectly the efficiency of the tractor.

Stage 5: Choice of health and safety features

Health and safety features: Such tractor should have the following safety features

a. Roll-over protective structures (ROPS) and/or Falling-object protective structures (FOPS) factory fitted;
b. Factory designed and fitted safeguards;
c. Adequate ventilation if a cabin is fitted;
d. Easy access to and exit from the tractor;
e. The positioning of the exhaust outlet to direct gases away from the operator;
f. Adequate for task and terrain for which tractor or machinery is purchased;
g. Adequate noise control. Where noise cannot be reduced sufficiently at source, hearing protection equipment should be supplied to the operator;
h. The location of switches and levers within easy reach of the operator to avoid repetitive injury risks and to reduce the risk of the wrong lever or control being used;
i. A well-sprung, adjustable seat and seat belt for tractors with cabin; and
j. Control of exposure to ultraviolet radiation (harsh weather), e.g. by provision of shade or canopy.

Stage 6: Environmental considerations

The following machine and environmental factors are equally important in making a decision to buy a tractor in ensuring optimum level of mechanization:

a. *Soil-implement and tractor interaction* – The nature of soil and the speed of operation has profound effect on tractor implement performance and the travel reduction considerably. Maximum permissible forward speed is related to such factors as: Nature of operation, condition of field and amount of power available. All these factors should be taken into consideration before procuring a tractor.
b. *Crop-implement suitability*: Implement chosen must be such as to produce optimum growth condition for the selected crop. There must be a relationship existing within one operation and the other.
c. *Appropriate power-implement match*: This implies that there exists an optimum power implement match for every available power from the power source. Factors affecting this include nature of soil-soil types, rocky outcrops i.e. The remnants of rock materials after weathering.

d. *Selection of operation*: Equipment procurement are determined by operations to be mechanized which were affected by some factors such as:

   i. Type of crop to be mechanized
   ii. Availability of some specific machinery for specific operation
   iii. Weather condition of the area to be mechanized
   iv. Topography of the field area.

e. *Selection of implement size:* The following factors should be considered in choosing the size of machine to buy:
   i. Difference in cost between large and small machine
   ii. Amount of use that will be made of the machine each year
   iii. The amount and cost of available labour
   iv. The financial position of the buyer.

Power-implement (machinery) selection

Putting together an ideal machinery system is not quite easy. Equipment that works best one year may not work well the next because of changes in weather conditions or crop production practices and other associated problems. Because many of these variables are unpredictable, the goal of the good machinery manager should be to have a system that is flexible enough to adapt to a range of weather and crop conditions while minimizing long-running costs and production risks. To meet these goals several fundamental questions must be answered.

Machine performance: First, each piece of machinery must perform reliably under a variety of field conditions or it is a poor investment regardless of its cost. Hence, the performance of a machine often depends on the skill of the operator, or on weather and soil conditions. Nevertheless, differences among machines can be evaluated through field trials, research reports and personal experience.

Machinery costs: Once a particular type of tillage, planting, weed control, or harvesting machine has been selected, the question of how to minimize machinery costs must be answered. Machinery that is too large for a particular farming situation will cause machinery ownership costs to be unnecessarily high over the long run.

Operating costs: Operating costs include; costs of fuel, lubricants and repairs. Operating costs per hectare change very little as machinery size is increased or decreased. Using larger machinery consumes more fuel and lubricants per hour, but this is essentially offset by the fact that more hectares are covered per hour. This is equally true of repair costs.

Labour cost: As machinery capacity increases, the number of hours required to complete field operations over a given area naturally declines. This factor is a trade-off for the initial machine cost. Decision on labour cost is a critical one and must be handled relative to timeliness of operation, production cost and effectiveness of labour management.

Timeliness costs: In many cases, crop yields and quality are affected by the dates of planting and harvesting. This represents a "hidden" cost associated with farm machinery. The value of these yield losses is commonly referred to as "timeliness costs". When labour is adequately managed, man-hour or man-day considerations effectively managed, timeliness will be reduced.

Matching tractor power and implement size

For tillage and planting implements the size of the machine that can be used is often limited by the size of the available tractor. The horsepower needed to pull a certain implement depends on the width of the implement, the ground speed, draft requirement, and soil condition. The general formula for estimating the required horsepower measured at the power take-off (PTO) is:

$$PTO\ hp = Width\ (feet) \times speed\ (mph) \times draft\ (lb./ft.) \times soil\ factor$$

Practical exercise

1. You have been hired as a consultant by Greenfield Horticultural Farms, Ishiagu a subsidiary of Greenfield Group of Industries, to help develop a proposal for the procurement of machinery for the setting up a 100 ha horticultural farm to produce variety of fruits and vegetables and to process them into fruit juice and canned products. Write up a proposal to advise the management on machinery procurement.
2. Estimate the number of field required to plant and harvest 10 hectares. Prepare a worksheet.

CHAPTER 3

Concepts of Maintenance Practices

Introduction

Maintenance can be defined as the practice of keeping of equipment, machine system or object in form or shape in its original state as much as possible. It is a process of retarding or correcting deterioration. It is also a means of achieving optimum value in equipment in order to perform its desired and designed functions.

Thus maintenance is protecting a machine so that it does not break down or wear out quickly. Note that maintenance is not repairing a machine after it breaks or when it stops work or deteriorated.

The objective of this chapter is to enable the students know and be familiar with machine failure, causes of such failure, various aspects of maintenance, maintenance policy and how and when it is most appropriate to carry out a particular type of maintenance and repairs.

Machine failure

Assets such as engines, tractor and machinery are put into practical use and service to fulfill some specific functions. As a result, failure and deterioration of some of its components is inevitable. Farm machine starts to deteriorate as soon as they are manufactured. Some of the deterioration is due to exposure to air, sunlight, and moisture. Such deterioration could be retarded by storing the machines in a dark shelter. But the most significant deterioration is from the use of the machine.

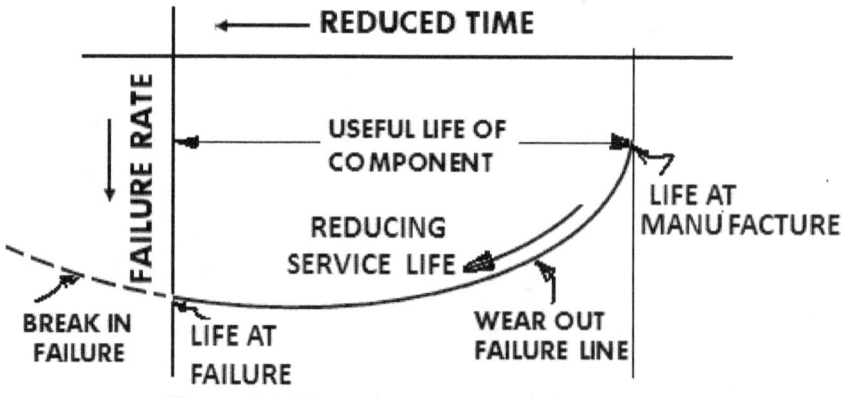

Figure 3-1: Typical component failure curve

Figure 3-1 shows the behaviour of a typical machine component or system under constant usage. At manufacture, the component has a full life t_o; the failure rate is lower, $1/\lambda$ (1/sec) than after a time t, at which the component has gradually worn out after repairs along the failure line (Life at failure). There is a gradual reduction in service life and increased tendency of failure rate beyond the time t when the component eventually fails (break in failure). At this point, the component can no longer be repaired but must be replaced.

Major causes of machine failure

Machine failure or breakdown does not happen automatically, just like the way accident happens, they were caused. Machine and equipment failure are largely traceable to the followings:

1. *Design features*: There are occasions where aesthetics in design become a trade-off for structural designs, thus such machine fails to meet the required and desired purpose. Machine design failures do actually happens and equally depreciation due to age and usage are factors. This factor has no control.
2. *Age*: This factor can only be controlled to some extent when the designed service life is exceeded, depreciation sets in leading to an increase in service and repair costs, frequent part failure etc.
3. *Environmental factor:* When machine operates in unpleasant weather and heavy or sticky soil, there is the tendency of machine breakdown or over-bearing of some components, thereby increasing the frequency of component breakdown.
4. *Lack of good maintenance practice:* This factor is interrelated to human factor problems, but differs slightly in that knowledge of good maintenance practice will save the machine from maintenance neglect.
5. *Human factor*: The single most prevalent factor for machine breakdown and failure is largely traceable to the *attitude* of the stakeholders at all levels toward maintenance. Negligence of small unpleasant noises and creaking equally leads to major breakdowns.

Attitude, how?

The operators' poor attitude and lack of commitment to his work, equipment maintenance engineer/supervisor's quick response, the contractor and management's interest in timely procurement of genuine replacement/spare parts will jointly affect the continuous operation of the system and equipment breakdown.

Attitude, why?

The life of any machine depends so much on the *attitude of the operator* towards its maintenance, and quick fault alert when noticed. The operator become nonchalant when neglected in the process of fault-correction and decision making. In as much as they are more acquainted to such machine, their opinion is required in decision making.

The *equipment engineer or supervisor's quick response* to maintenance schedule, fair fault diagnosis, true assessment / recommendations on component repair or replacement part, sourcing and certification of replacement component will help to reduce sudden equipment breakdown, annoying downtime and economic waste of production time.

The contractor and management's interest in timely procurement of genuine replacement parts will enhance mean-time-to-repair (MTTR) and mean-time-to-failure (MTTF) of component. Some individuals take such advantage in making extra money by procuring refurbished parts "Tokunbo or Belgium" at the expense of genuine parts for replacement.

Maintenance

Maintenance is defined as the combination of all technical, administrative and managerial actions during the life cycle of an item intended to retain or restore it to, a near-original state in which it can perform the required function (function or a combination of functions which are considered necessary to provide a given service). The old concept of maintenance was to attend a machine after it has failed or stopped functioning. But now some more concepts have been identified and applied in the realm of *maintenance*. It is no more ' work after failure".

Maintenance is generally recognized as the single largest controllable cost factor in any production assembly when handled appropriately. But up till now, maintenance issues have presented a challenge to most management which has made it the most expensive production factor in mostly governmental organizations. A visit to any agricultural machinery workshop anywhere in Nigeria would make the management (employer) to reevaluate their maintenance strategies in order to operate at profit.

For instance, regular maintenance is one of the prerequisites for a long lasting and reliable engine performance. The recommended engine servicing procedure for a particular system and engine is contained in the Owners Handbook for each engine model. *Additional servicing* is often necessary and depends on the type of operation the engine is subjected to.

Caution! It must be stressed at this point that if you are to carry out your own servicing, make sure you have the proper tools and perhaps attend short course training on maintenance.

> *Mistakes made due to lack of skill or knowledge are often more expensive than employing a reputable mechanic*!

Reasons for machinery maintenance

Thus the maintenance of these assets is necessary for the following reasons:

1. To preserve the service life of such machine.
2. To enhance machine performance at peak load i.e. to get the most of your machine.

3. To conserve cost in unnecessary procurement of new machine thereby increasing the overhead cost.
4. To restore machine to or near its original state at manufacture.
5. To increase output and profit
6. To provide the operator some comfort while in operation.

Maintenance management system (MMS)

Maintenance management can be defined as all the managerial activities that determine the maintenance objectives or targets priorities assigned and accepted by the management and maintenance department, strategies, and responsibilities and implement them by means such as maintenance planning, control and supervision, including economical aspects in the organization.

Objectives of the maintenance management system (MMS)

a. Optimize the use of available funds, personnel, and facilities and equipment through effective maintenance management methods.
b. Provide accurate data for maintenance and construction program decision-making.
c. Systematically identify maintenance needs and deficiencies and capital improvement needs at all field stations.
d. Determine the unfunded maintenance backlog for the service.
e. Establish field station, regional, and national maintenance and construction project priorities.
f. Enable preparation of service maintenance and construction budget requests using systematic, standardized procedures.
g. Monitor and document corrective actions, project expenditures, and accomplishments.
h. Conduct comprehensive condition assessments of all service real property and personal property.

Good maintenance practice

Good maintenance practices are essential for efficient operation of all types of machinery. Efforts spent in this area of farm management is more than repaid by consistent and reliable operation of machinery, reduced fuel consumption and costs, extended machine life among others. Maintenance of farm machinery is complicated by the usage pattern of machine for short but intense period of activity, followed by a long periods of non-use or storage.

To effectively safeguard the service life of your engine and guarantee performance, you must protect your machine from the following enemies;

1. *Wear* (grease and oil are used to protect machines from wears),
2. *Dirt* (Filters are used to catch and hold dirt before it gets inside and damages parts),

3. *Heat* (the cooling and lubrication systems protects the machine from heat).

Objectives of good maintenance practices

Objectives of a good maintenance practice include the followings:

1. To intervene before failure occurs
2. To carry out maintenance only when required and not when necessary.
3. To reduce number of failure and shutdown times.
4. To reduce maintenance cost and breakdown cost due to production lost.
5. Increase service life of equipment.
6. Reduction in inventory cost / effective inventory control.

Maintenance policy and decision making

Any farm organization that must have a free flow of production must have a well defined maintenance policy and instrument of implementation. It is important to note that maintenance policy is an inseparable part of corporate social responsibility (of any such organization) involved in production activity.

Maintenance policy

Maintenance policy is an organizational programme designed to be followed in selecting appropriate maintenance practice and priorities (Figure 3-2). It is always a concern for the decision makers to determine which type of maintenance, what type of tools and the safety equipment that are most appropriate for optimum machine performance.

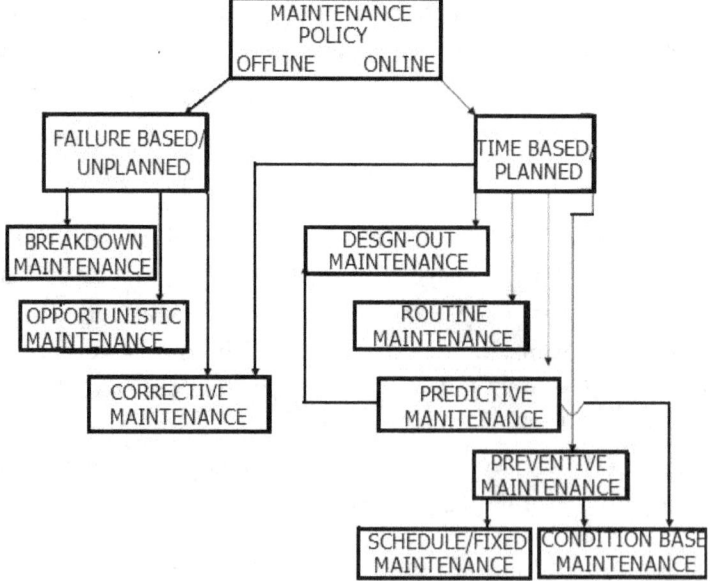

Figure 3-2: Block diagram for maintenance policies

Decision makers therefore, need to take into account the following maintenance priorities:

1. The *needs* (interests) of their business,
2. *Recommendations* from the original equipment manufacturer,
3. Their own *experience* and that of other users of similar equipment, and
4. Online or offline *information* on available machine conditions.

Types of maintenance

Many types of maintenance techniques have been used in machinery and equipment maintenance and reconditioning and are classified as follows: *reactive maintenance* (corrective or break down maintenance), *conventional maintenance* (which include routine maintenance, preventive maintenance, default type, discard type, offline and online type) and *proactive maintenance*, etc.

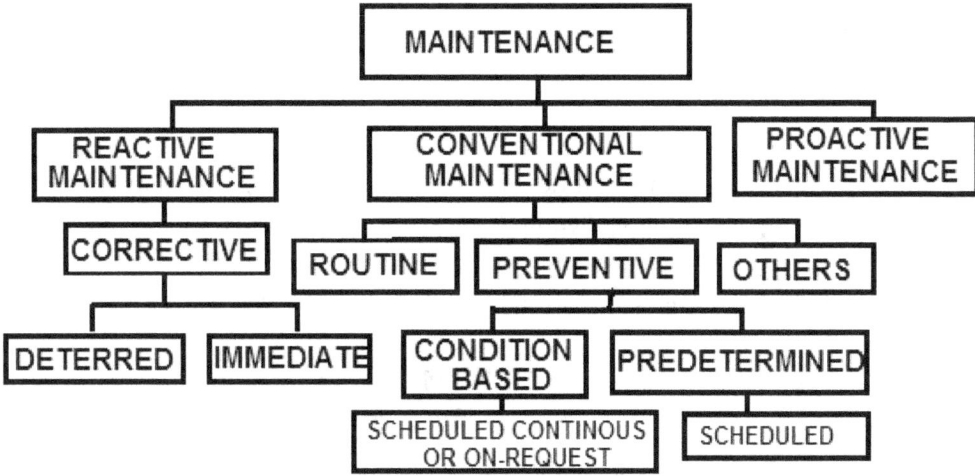

Figure 3-3: A typical maintenance system (modified)

Traditionally, maintenance is performed in either time-based-approach on fixed intervals, called preventive maintenance, or by failure-based-approach termed corrective maintenance (Figure 3-3). For some equipment and faults, corrective maintenance action must be performed immediately, while for others the maintenance action can be deferred in time, all depending on the equipments function.

The different types of maintenance are discussed as follows;

1. Reactive maintenance

Reactive maintenance is basically the "run it till it breaks" maintenance mode. No actions or efforts are taken to maintain the equipment as the designer originally intended it to ensure design life is reached.

Advantages to reactive maintenance can be viewed as a double-edged sword. If we are dealing with new equipment, we can expect minimal incidents of failure. If our maintenance program is purely reactive, we will not expend manpower cost or incur capital cost until something breaks. Reactive maintenance is further categorized as corrective/curative and failure-based/breakdown maintenance.

a. *Corrective /curative maintenance*

These are maintenance carried out when a system has finally breaks down i.e. repairs is to be done on the engine system. Thus it is included in both planned and unplanned maintenance.

b. *Failure based /breakdown maintenance*

Failure based or breakdown maintenance is carried out on the basis of failure occurrence in machine. A little lubrication and minor adjustments are done in this system. It is applied basically when

i. Number of equipments are few
ii. Equipment is very simple and repair does not call for specialist or special tools /tackles.
iii. Where sudden stoppage /failure of equipment will not cause severe financial loss in terms of delivery commitment or further damage to other equipment /components.
iv. Where sudden failure will not cause severe safety or environmental hazards.

A disadvantage of this system is that it cannot be applied to where equipments are many. It is worked out to eliminate the offline jobs of lower order priority and reduce repetitive breakdown.

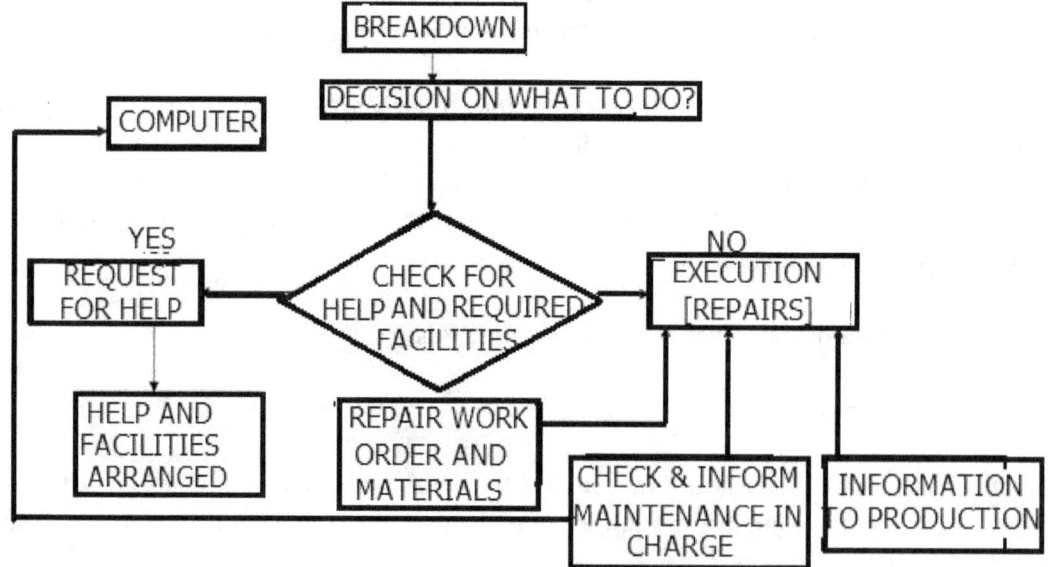

Figure 3-4: Breakdown maintenance block diagram

Advantages of reactive maintenance

- Low cost involvement
- Less staff required.

Disadvantages of reactive maintenance

- Increased cost due to unplanned downtime of equipment
- Increased labor cost, especially if overtime is needed.
- Cost involved with repair or replacement of equipment.
- Possible secondary equipment or process damage from equipment failure.
- Inefficient use of staff resources

2. *Conventional/ traditional maintenance*

Conventional or traditional maintenance practices include emergency maintenance, routine maintenance and preventive maintenance.

a. *Emergency maintenance*

This is the aspect of maintenance, which is necessary to put machine and equipment in good working condition immediately to avoid serious consequences, for instance cleaning of distributor cap in the electrical system of an engine. The machine can still function but when not attended to, can cause major breakdown in the system.

b. *Routine maintenance*

Routine maintenance is the simplest form of planned maintenance but very essential. As the name implies, it is carried out at regular intervals. It involves periodic check of relevant areas. The frequency of such checks ranges between hourly, daily, weekly and monthly or as recommend by the manufacturers.

Routine maintenance reduces fuel bills and extends equipment life. Readings obtained from such checks could be collated in a maintenance record over a long period to give a behaviour history of the equipment. Examples are washing and cleaning, filing of distributor cap, change of oil, topping of battery electrolyte, lubrication, inspection and minor adjustments of pressure, flow, tightness etc.

Routine maintenance reduces fuel bills and extends the equipment life. Good maintenance practices are essential for efficient operation of all types of farm machinery. Effort spent in this area of farm management is more than repaid by consistent, reliable operation of machinery, reduced fuel bills and extended equipment life. Maintenance of farm machinery is complicated by the usage pattern of short periods of intense activity, followed by periods of non-use or storage.

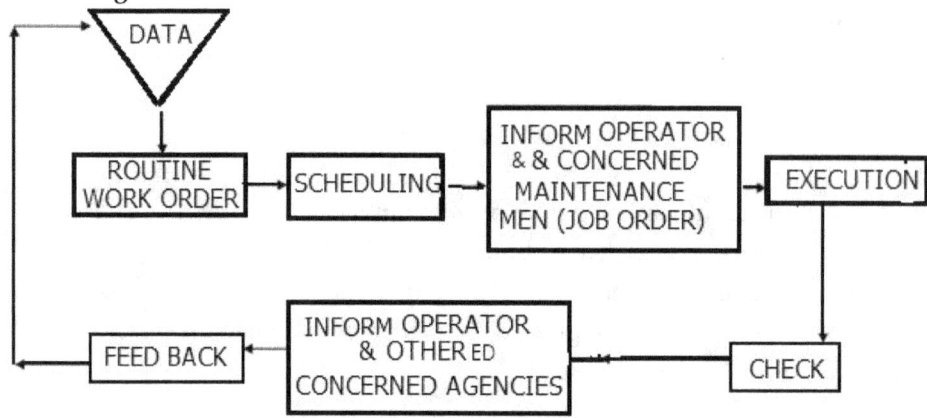

Figure 3-5: Routine maintenance block diagram

During the "standing" or non-use periods chemical interactions between metals and fluids can cause more damage than normal wear and tear from active usage. This must be considered in planning machinery maintenance and the following suggestions are worthy of consideration.

i. Follow manufacturers' instructions for all settings, adjustments, maintenance instructions, operating requirements and long term storage.
ii. Follow manufacturers' recommendations on safety aspects of operation and repair. Maintain all safety equipment as installed or recommended by manufacturers.
iii. Do not overload equipment, or operate at higher speeds than manufacturer recommends.
iv. Do not add counterweights to equipment to increase load capacity unless authorized by manufacturer. Store equipment in clean and dry conditions.
v. Remove all vegetation such as grass, hay, crops and crop residue from equipment before storage periods. Decomposing vegetable matter causes corrosion to metal surfaces. This is particularly important where surfaces are polished from usage.
vi. Keep all cutting edges sharp and clean. Sharp cutters require less power and reduce overall load on equipment. Cracked or damaged cutting edges are also easier to detect on clean equipment.
vii. Replace these items at end of season rather than at season commencement.
viii. Inspect machinery at end of season or harvest.
ix. Repair and adjust as required. Carry out maintenance work without pressure between seasons.

c. *Preventive maintenance*

This is one of the oldest and traditional methods of maintenance. It is used mostly along with corrective maintenance and condition-based maintenance (diagnostic maintenance).

Preventive maintenance is a planned maintenance of plants resulting from periodic inspection in order to minimize the breakdowns and depreciation rates. This includes the followings: servicing; adjusting; operating; repairing and caring for agricultural machines so as to prevent unnecessary wear out of parts, and keep time loss due to breakdown to a minimum.

Preventive maintenance can be defined as actions performed on a time- or machine-run-based schedule that detect, preclude, or mitigate failure of a component or system with the aim of sustaining or extending its useful life through failure control to an acceptable level. Once the unit is placed into full operation, a Preventative Maintenance Program should begin. This program should include regular inspection set up on a periodic basis. Preventive Maintenance is time-based. Activities such as changing lubricant are based on time, like calendar time or equipment run time. For example, most people change the oil in their vehicles every 3,000 to 5,000 miles traveled. This is effectively basing the oil change needs on equipment run time. No concern is given to the actual condition and performance capability of the oil. It is changed because it is time.

For instance the preventive maintenance program for conveyor or any belt system should include a general inspection of:

a. Drives and V-belt drives for wear on belts and proper tension.
b. Roller chains of chain drive for lubrication and proper tension.
c. Loose, worn and/or damaged coupling bolts.
d. Ensuring that guards are all in place and properly installed.
e. Re-lubrication schedule.
f. Other items to be routinely inspected:
g. Screw flighting and hanger bearings for possible damage and/or wear.
h. Operating temperature, signs of wear (noise) and lubrication.
i. Oil in gearbox.

Advantages of conventional maintenance

- Cost effective in many capital intensive processes.
- Flexibility allows for the adjustment of maintenance periodicity.
- Increased component life cycle.
- Energy savings.
- Reduced equipment or process failure.
- Estimated 12% to 18% cost savings over reactive maintenance program.

Disadvantages of conventional maintenance

- Catastrophic failures still likely to occur.
- Labour intensive.
- Includes performance of unneeded maintenance.

- Potential for incidental damage to components in conducting unneeded maintenance.

Preventive maintenance is divided into two categories; Condition Based Maintenance (CBM) and Predetermined Maintenance.

i. Predetermined (predictive) maintenance systems

Predetermined maintenance is scheduled in time. Predictive maintenance can be defined as measures taken to detect from the onset any degradation in mechanism, thereby allowing causal elimination or control prior to any significant deterioration in the component physical state. Results indicate current and future functional capability.

Basically, predictive maintenance differs from preventive maintenance by basing maintenance need on the actual condition of the machine rather than on some preset schedule.

You will recall that this methodology would be analogous to a preventive maintenance task. If, on the other hand, the operator of the car discounted the vehicle run time and had the oil analyzed at some periodicity to determine its actual condition and lubrication properties, he/she may be able to extend the oil change until the vehicle had traveled 10, 000 miles. This is the fundamental difference between predictive maintenance and preventive maintenance, whereby predictive maintenance is used to define needed maintenance task based on quantified material/equipment condition.

The advantages of predictive maintenance are many. A well-orchestrated predictive maintenance program will all but eliminate catastrophic equipment failures. We will be able to schedule maintenance activities to minimize or delete overtime cost. We will be able to minimize inventory and order parts, as required, well ahead of time to support the downstream maintenance needs. We can optimize the operation of the equipment, saving energy cost and increasing plant reliability.

Advantages of predetermined maintenance

- Increased component operational life/availability.
- Allows for preemptive corrective actions.
- Decrease in equipment or process downtime.
- Decrease in costs for parts and labor.
- Better product quality.
- Improved worker and environmental safety.
- Improved worker moral.
- Energy savings.
- Estimated 8% to 12% cost savings over preventive maintenance program.

Disadvantages of predetermined maintenance

- Increased investment in diagnostic equipment.
- Increased investment in staff training.
- Savings potential not readily seen by management.

ii. *Condition based maintenance systems (CBMS)*

The condition based Maintenance can have dynamic or on request intervals. Condition based maintenance (CBM) is defined as maintenance actions based on actual condition obtained from in-situ measurement (Mitchell, 1998). The main point being that the material condition is assessed under operation with the intention of making conclusions as regards to whether it is in need of maintenance or not and if so at what time does the maintenance actions needed to be executed and not to suffer a breakdown or malfunction.

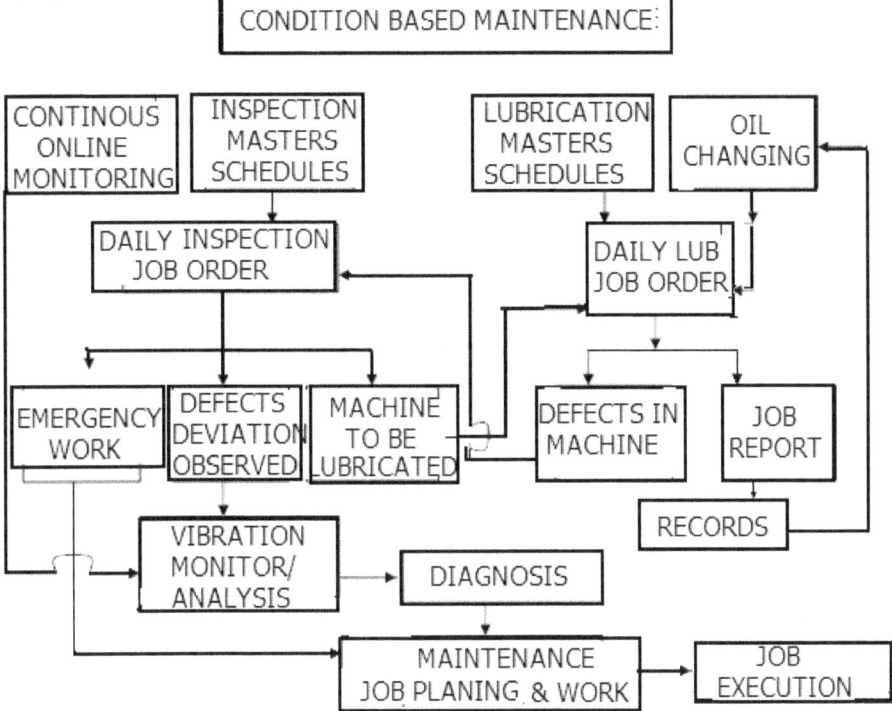

Figure 3-6: Condition based maintenance block diagram

d. *Opportunistic maintenance*

This maintenance arises when an equipment or system is taken down for maintenance of one or two components, the opportunity can be utilized for maintaining other wearing out components. In multi component systems with several failing components, it is often required to use opportunistic maintenance. This would be probably economically. It is also useful in non-monitored components.

3. *Proactive maintenance*

Proactive maintenance is that type of maintenance that employs corrective actions aimed at solving the problem of machine failure from the source. It is designed to extend the life of mechanical machinery as opposed to:

a. Making repairs when nothing is broken,
b. Accommodating failure as routine and normal, and
c. Preempting crisis failure maintenance - all of which are characteristics of the predictive/preventive disciplines.

Generally, maintenance is considered to be the largest single uncontrollable expenditure in any establishment and its cost often exceeds annual net profit. The problem of costly maintenance has truly reached a serious and unimaginable level, but as it has been found out, maintenance costs can be cut drastically by establishing a "proactive" line of maintenance.

When it comes to the life of any particular machine, cleanliness counts. the accepted methods currently being used to combat machine damage are based on either detecting the warning signs of failure once they have already begun (predictive) or regular maintenance according to a schedule rather than the machine's true condition (preventive).

None of the maintenance type has previously taken a closer look on machine damage by concentrating on the causes instead of the symptoms except proactive maintenance. Proactive maintenance is the single most important means of achieving savings unsurpassed by conventional maintenance. Imagine being able to pinpoint and eliminate a disease long before any symptoms occur in your body. It would save you money spent on doctor's bills and keep you out of the hospital in the long run. This is the advantage of proactive maintenance over predictive maintenance.

While effective to a degree, neither preventive nor predictive maintenance is geared to detect the most common and serious source of failure- contamination. Therefore, the first logical step to proactive maintenance is the implementation of a strict contamination control program for lubrication fluids, hydraulic fluids, gear oils, and transmission fluids.

4. *Other maintenance programmes*

a. *Design–out maintenance*

This is a design oriented curative measure aimed at rectifying a design defect originated from improper method of installation or poor material choice etc. Design-Out Maintenance requires a strong maintenance design interface so that maintenance engineer works in close cooperation with design engineer. Where most of the maintenance concepts aimed at minimizing the number of failures or effects of failures, design-out maintenance aims at eliminating the cause of maintenance. It is more suitable for item/equipment of high maintenance cost. The choice to be made is between the cost of redesign and cost of recurring maintenance.

b. Reliability centered maintenance (RCM)

Reliability centered maintenance is a process used to determine the maintenance requirements of any physical asset in its operating context. Basically, RCM methodology deals with some key issues not dealt with by other maintenance programs. It recognizes that all equipment in a facility is not of equal importance to either the process or facility safety. It recognizes that equipment design and operation differs and that different equipment will have a higher probability to undergo failures from different degradation mechanisms than others. It also approaches the structuring of a maintenance program recognizing that a facility does not have unlimited financial and personnel resources and that the use of both need to be prioritized and optimized.

In a nutshell, RCM is a systematic approach to evaluate a facility's equipment and resources to best mate the two and result in a high degree of facility reliability and cost-effectiveness. RCM is highly reliant on predictive maintenance but also recognizes that maintenance activities on equipment that is inexpensive and unimportant to facility reliability may best be left to a reactive maintenance approach.

Advantages

- Can be the most efficient maintenance program.
- Lower costs by eliminating unnecessary maintenance or overhauls.
- Minimize frequency of overhauls.
- Reduced probability of sudden equipment failures.
- Able to focus maintenance activities on critical components.
- Increased component reliability.
- Incorporates root cause analysis.

Disadvantages

- Can have significant startup cost, training, equipment, etc.
- Savings potential not readily seen by management.

Tero-technology and its concept

Terotechnology, a word derived from the Greek root word "tero" or "I care", refer to the study of the costs associated with an asset throughout its life cycle - from acquisition to disposal. The goals of this approach are to reduce the different costs incurred at the various stages of the asset's life and to develop methods that will help extend the asset's life span.

Terotechnology is an interdisciplinary field, joining mechanical, electrical, diagnostic, signal processing, and maintenance experts. Student gains sufficient knowledge to support maintenance of complex mechanical systems.

Terotechnology uses tools such as net present value, internal rate of return and discounted cash flow in an attempt to minimize the costs associated with the asset in the future. These costs can include engineering, maintenance, wages payable to operate the equipment, operating costs and even disposal costs.

Practice of terotechnology is a continuous cycle that begins with the design and selection of the required item, follows through with its installation, commissioning, operation, and maintenance until the item's removal and disposal and then restarts with its replacement. For example, let's say an oil company is attempting to map out the costs of an offshore oil platform. They would use terotechnology to map out the exact costs associated with assembly, transportation, maintenance and dismantling of the platform, and finally a calculation of salvage value.

Comparing major types of maintenance

In maintenance practice, the trend should be focused to cost-saving and a maintenance program that targets the root causes of machine wear and failure. In this way, Proactive Maintenance is more beneficial than the traditional maintenance methods because of cost saving.

With *preventive* maintenance approach, maintenance is performed in order to prevent equipment breakdown

With *corrective* maintenance approach, maintenance is performed after a breakdown or an obvious fault has occurred.

With *proactive* maintenance approach, maintenance is performed to save cost on machine maintenance every year.

Maintenance functions and probability (maintenanceability)

The following time related functions are defined with respect to maintenance and repairs;

1. *Mean time to repair (MTTR) an assembly*: This includes both the time elapsing before it is realized that there has been a failure, which will be very short for self-revealing faults and the time for maintenance personnel to be deployed and achieve their repair mission. Good design can reduce MTTR and it is also influenced by maintenance procedure.
2. *Mean time to failure (MTTF):* This is the time elapsed for repairs to be done after which failure will occur. Good design can increase MTTF.

3. *Mean time between failures (MTBF):* This is the measure of the recurrence of failure in a system. It is a system mean time between two successive failures and also a factor of MTTF and MTTR i.e.

$$MTBF = MTTF + MTTR \quad \ldots\ldots\ldots\ldots\ldots\ldots\ldots.3.1$$

4. *Reliability (r):* Reliability is defined as the probability of survival of any system or machine component(s) throughout the time the machine is in operation.

$$R(t) = \exp(-\lambda t) \quad \ldots\ldots\ldots\ldots\ldots\ldots\ldots..3.2$$

Where

λ = Constant failure rate in unit of per (1/sec) i.e. Failure/ time

5. *Availability:* The concept of availability is the fraction of time for which a machine or component(s) can perform its desired/designed functions.

$$Availability = \frac{MTTF}{MTTF + MTTR} = \frac{MTTF}{MTBF} \quad \ldots\ldots\ldots\ldots\ldots\ldots..3.3$$

Maintenance probability (maintenanceability)

Maintenanceability is expressed as the probability that a device that has failed will be restored to operational effectiveness within a given period of time when the maintenance action is performed in accordance with prescribed procedures. With an unrealistic but ideal assumption of a constant repair rate, the situation can be analyzed with reference to reliability and failure.

Maintenanceability is mathematically expressed as;

$$Maintenanceability\ (in\ time\ t) = 1 - e^{-t/MTTR} \quad \ldots\ldots\ldots\ldots\ldots\ldots.3.4$$

Where
MTTR = Mean Time To Repair. This is the time taken to make a repair.
e = Exponential relation
t = time expressed in seconds

Machinery repair concepts

Machinery repairs is the application of maintenance services, including fault location / troubleshooting, removal/installation, and disassembly/assembly procedures and maintenance actions to identify troubles and restore serviceability to an item by correcting specific utility damage, fault, malfunction, or failure of a machine part, a subassembly, module (component or assembly), or system. There are two types of repairs applicable to machinery reconditioning; Minor repairs and major repairs.

Minor repairs: This includes jobs performed on an engine component such as;

1. Adjustments made on governing systems
2. Brake pad replacement
3. Carburetor and injection system setting

Major repairs: Major repairs depend on the extent of damage. This include such jobs as

1. Replacing sleeves or re-boring cylinders (Sleeves are linings that are put inside the cylinder within which the piston slides)
2. Replacing the timing gears
3. Installing new pistons
4. Replacing crankshafts and main journal bearings
5. Replacing valve guides and inserts and
6. Any major works on the gears or final drive.
7. Replacing valve guides and inserts, and any major works on gears and final drive.

Best maintenance and repairs practice

Best maintenance repair practices is a concept in maintenance culture where equipment are kept operating at optimum (peak machine reliability) level to achieve increased productivity and capacity through reduced maintenance costs. The concept might just become a mandatory requirement for the future success of an organization in today's economy.

Repair costs are those expenditures for parts and labour to install replacement parts after a failure and to recondition renewable parts as a result of mechanical wear. Two examples of best maintenance practice are as described below:

Practical example 1: *Consider a maintenance task of* lubrication of bearing; the *required best practices* to be performed include the followings

1. Clean the bearing fittings.
2. Clean end of grease gun.
3. Lubricate with proper amount of grease and the right type of lubricant.
4. Lubricate within variance of time frequency.

Probability of future failures: The tendency that the bearing will fail when best practices are not carried out against when carried out isabout20 vs. 1 in %.

Practical example 2: Consider a maintenance task: of maintaining a hydraulic component. The *required best practices* include the followings:

1. Hydraulic fluid must be input into the hydraulic reservoir utilizing a filter pumping system only. Filters must be rated to meet the needs of the component reliability and not equipment manufacturer's specification.
2. Filters must be changed on a timed basis and based on filter condition.
3. Oil samples must be taken on a set frequency and all particles should be trended in order to understand the condition and wear of the hydraulic unit.

Consequences for not following best practices: Premature failure or unknown hydraulic pump life. Length of equipment breakdown causes lost in production.

Probability of future failures: The tendency that the pump will fail when best practices is not carried out against when carried out is 30 vs. 1 %

Reasons for deviating from best maintenance practice

A few of the most common reasons why an organization or mechanics do not follow best maintenance repair practices include:

a. The maintenance workforce lacks either the discipline or direction to follow best maintenance repair practices.
b. Management is either not supportive, and/or does not understand the consequences of not following the best practices (real understanding must involve a knowledge of how much money is lost to the bottom line).
c. Maintenance being a totally reactive process is not seen in the right perspective of its definition, which is to protect, preserve, and prevent a system from decline.
d. Maintenance personnel do not have the requisite skills.

Fault tracking process

Performance skill: malfunctions diagnoses and reporting

Significant production capacity and financial losses in most organizations have been traceable to maintenance and repair issues. In order to assess such situations, the following 3-steps approach should be taken to trace out and correct faults to avoid component malfunctioning and a subsequent major breakdown.

Step 1: A repetitive equipment failure is an indication of a problem. For instance, when a component such as electrical fuse flashes off repeatedly, it is an indication of a major fault which could result from either wire bridging, overload or local heating of a particular component.

Step 2: Identify the source of the problem. Track or trace the line to identify the source and causes of the problem. This could be a combination of issues including the problem of not following "Best Maintenance Repair Practices". A sequential course of action should be taken as stated below:

a. *Maintenance skill level* – Carry out a skill assessment (written and performance based) of personnel to evaluate whether skill levels are adequate to meet "Best Maintenance Repair Practices" for your specific organization
b. *Maintenance culture* – Provide training to all maintenance crew relative to a change in maintenance strategy and how it will impact them individually (i.e., increase in profit for the plant, less overtime resulting from fewer equipment breakdowns, etc.). Track and measure the changes and display the results to everyone
c. *Maintenance strategy* – Develop a plan to introduce a proactive maintenance model with preventive or planned maintenance at the top of planned priorities. This will provide more time for performing maintenance utilizing the "Best Maintenance Repair Practices".

Step 3: Implement the changes needed to move toward following "Best Maintenance Repair Practices" and measure the financial gains. Several of the reasons why implementing a program of change, such as the one discussed above, can be doomed to fail include:

a. Lack of discipline on the operator and direction by the supervisor
b. Lack of management (employer) commitment and accountability
c. Changes in operational direction and modalities, slowed down momentum due to indecision
d. Lack of an adequately skilled workforce and lack of retraining, and
e. No specific maintenance action plan to guide the effort to close the skill gaps analysis

Fault tracking and assessment procedure

Given typical machinery and a description of problem symptoms or actual symptoms of poor performance, the technician will troubleshoot such symptoms, identify vehicle malfunctions, and report problem. The following steps are essential to fault tracing:

- Check each component and vehicle system. Identify vehicle systems or components that are functioning properly, identify those that are in imminent danger of failing, or functioning improperly.
- Identify and interpret symptoms of malfunction.
- Match symptom to possible list of problems.
- Describe symptoms of improper operation completely and accurately to maintenance personnel.
- Correct problems within jurisdiction.

- Avoid attempting to perform maintenance for which driver is unqualified.
- Properly report breakdowns occurring en route.
- Properly complete a machinery condition report (MCR).
- Fix problems within jurisdiction of operator, as described by company policy and regulation.

Repairs and maintenance costs

Operating costs (also called variable costs) for any particular machine or machinery include repairs and maintenance cost, fuel cost or bill, lubrication cost, and operator (labour) costs. Repair costs occur because of routine maintenance, wear and tear, and accidents. Repair costs for a particular type of machine vary widely from one geographic location to another because of varying soil types, rocks, varying terrain, climate, and other conditions. Within a local area, repair costs vary from farm to farm because of different management policies and operator skill.

The best data for estimating repair costs are the operator's own records of past repair expenses. Good records indicate whether a machine had above or below average repair costs and when major overhauls may be needed. They also will provide information about the operator's maintenance program and mechanical ability. Without such data, repair costs could only be estimated from average experience.

Practical exercise

1. When next you visit a farm shop, ask the maintenance engineer and the operator for a sample of the record chart of past repairs and expenses made on a typical machine. Make comments and draw conclusions based on the recorded data.
2. Visit a technician or service manager for hints on good preventive maintenance. Ask why it is important. What happens that causes wear or damage? Report on what you discovered.
3. Visit an implement dealer. Interview the dealer technician or service manager for hints on good preventive maintenance. Ask why it is important. What are the costs? What happens that causes wear or damage? Report on what you discovered.
4. Carry out a routine maintenance check on a farm tractor system

CHAPTER 4

Maintenance Workshop and Organization

Introduction

Maintenance workshop should be designed to meet the requirements of carrying out machining, repairs and maintenance operations. In the same way, an array of workshop tools and equipment are required. The extent of equipment in a maintenance workshop varies considerably from one place to another; these vary from the very basic hand tools and few machines required, to complex digital test equipment, required by highly mechanized workshops and organizations.

However, safety remains a concern in these workshops and has increased in recent times. People working in maintenance workshop are exposed to serious risk or fatal injury and illnesses associated with hazards due to the misuse of virtually every item of equipment. Adopting safe work practices is imperative when carrying out equipment maintenance. Up to 20 percent of the registered farm injuries are reportedly caused by maintenance works.

Causes of hazards associated with work in the workshop have been largely traceable to:

1. Poor workshop design and layout (inadequate machine space, equipment not properly arranged and installed),
2. Fire outbreak due to fuel storage,
3. Unprotected contact with electricity (electrocution),
4. Using power hoists (crushing),
5. Using power and hand tools (cut, bruise, pinch etc.),
6. Grinding without proper guides (stone cut),
7. Exposure to welding arc light (eye damage and skin burn), and
8. Oxyacetylene or gas cylinder explosion.

The types of injury common in the workshop range from serious injury requiring hospitalization and work down time, to *"nuisance injury"* that stops work for a short time, or makes work slower and reduces productivity. Extreme cases of such injury could lead to death. In order to appreciate safety in workshops and better service delivery, the following issues must be addressed:

Workshop information

The maintenance workshops are available to all engineering students, staff and faculty workers. Everyone is mandated to read the safety handout and pass a safety test before using the tools in the shop. Because farm/machine shops are communal areas used by so many people, it is important to keep the shop clean and orderly. This means that every user must clean the machines and work areas they use, and put away all tools and material before leaving the shop. Disregarding shop rules, working unsafely or leaving a mess will result in suspension of shop privileges. These rules apply to the entire workshop area including the student project workroom. Hawking of any kind is not allowed within workshop area.

Sign-in book

All farm shop users, most especially students, must sign in before being giving access to work! The sign in book should be conspicuously displayed on the tool storage cabinet across from the tool board in the shop.

Machine/maintenance workshop layout

The effect of workshop layout does not only impact on productivity but on health and safety of the workers. Employers should consider the following issues relating to safety and workshop layout:

1. How safe the designated raw material and delivery areas are?
2. Are the tools and replacement parts stored in an easily accessible area? Or close to the job site?
3. Are there linear workflows through the production line?
4. Is the work area clear of unfinished work?
5. How well are the blind spots reduced throughout the workshop?

Answers to these questions are addressed by the following considerations.

Workshop building construction

Machine workshops should be constructed of fireproof materials such as concrete blocks, concrete, steel and non-asbestos fiber cement sheeting. Conditions leading to dampness or condensation should be avoided. Solid walls are best as they can provide support for tool boards, shelves and anchorage for benches.

Various components of the maintenance workshop that needs special consideration in design include: Doors, floors, inspection pit, machine space, work benches etc. each of these components are discussed as below.

FARM TRACTOR SYSTEMS

Figure 4-1: A typical tractor maintenance workshop

Workshop doors

The workshop entrance should be high and wide enough to allow entry of modern large tractors and equipment. Entrance doors need to be at least 3.8m high and 4.5m wide. Because of the size of these doors, double sliding types are the best. Never use swinging doors - they could be too dangerous in wind. A sub-door with adequate headroom should be provided to avoid having to open the main doors for access. Access to the workshop should be limited to technical staff at work. The workshop should always be locked when unattended.

Figure 4-2: A Workshop entrance

Floor

A wooden float finish is ideal; but a concrete floor with rough surface is preferred to have grip. Do not use a steel float or power float as this will lead to a very slippery floor after oil spillage.

FARM TRACTOR SYSTEMS

Inspection pit

This should be well lighted, tiled and the steps kept free of oil. The pit outer edge should be railed to prevent vehicle falling into the pit. Entrance to the pit should be guided to prevent unauthorized access and children fall.

Machine space

You need to have at least 2m workspace around each installed machine. In practice this leads to the minimum building size of 10m x 10m and, if working with combines, you will need a 10m x 15m building.

Workshop organization

Service and maintenance tasks in a poorly organized workshop can lead to serious injury. The farm workshop and the field are the primary locations where repair operations and construction are completed.

Make sure your farm shop is part of a farm safety solution, not a problem.

- Organize your workshop so that every item and equipment has a designated place. Make sure items are secure so they will not fall on anyone.

Figure 4-3: A well arranged service table

- Clean walkways to reduce trips and falls. Obstructions and sharp edged tools such as anvil, metal plates etc are quite dangerous and could cause injury.
- When working on agricultural equipment, make sure that the equipment is turned off, ensure all rotating parts have stopped moving, and safety locks are put in place.
- Keep all guards and shields in place on power equipment.
- Use hand tools only for their intended purpose.
- Equip your shop with Ground Fault Circuit Interrupters to help prevent electrical shock.

Figure 4-4: Organized component display table

- Make sure your shop is well lit. If the shop is heated, ensure it is properly vented and that flammable liquids are kept out of the shop area.
- Wear Personal Protective Equipment (PPE) when performing repair jobs. Standard PPE for a farm shop include leather gloves, chemical-resistant gloves, safety glasses, face shields, earplugs or muffs, steel-toed boots, respirators, hard hats, protective aprons and

Keeping records in workshop

Records should be kept current and retained for the management of the life of the plant. It must be accessible for examination if requested by an inspector from the Work Health Authority. Each unit within the workshop should have a record of the following documentations:

- Unique machine number-identification tag/number;
- Date and results of inspections and maintenance;
- Information on major repairs carried out;
- Results of tests on safety devices;
- Results of risk assessments carried out on plant
- Information, instruction and training which has been given to employees on the use of the plant and the associated risks.

Induction and training

Both supervisors and workers should be formally trained to have the required knowledge for the application of safe practices which are involved in the use of their machinery. This is particularly important for new employees or inexperienced people undergoing training. Areas of training and education should include:

- Machinery safety procedures, including emergency procedures;
- Correct and safe way of operating machinery;

- Knowledge and understanding of possible dangers;
- The purpose and function of safety;
- Reporting of faults including guard defects;
- Possible need for wearing and care of PPE;
- Need for good house and safe workstation keeping;
- Other statutory requirements.

Fire control in maintenance workshop

Fire can be a serious hazard in workshops. As a precautionary measure, *flammable materials* such as fuels, oil, paints, thinners or grease should not be stored in the workshop. These materials are best stored in isolated buildings or shed not close to utility buildings or workshop but must be easily accessible to the workshop.

Checking for fire hazards before starting work each day will reduce the chance of costly fire to equipment. Some types of machinery found in the farm shop can present a special type of fire hazard. All equipment including elevators, conveyers and augers need to be checked before and during their use. Preventive maintenance is the key to preventing many of the fires which occur on farm equipment. Good preventive maintenance not only prolongs equipment life but also reduces fire hazards.

Fire equipment

The following fire protection and prevention materials must as a matter of necessity be provided in all standard workshops.

Fire buckets

This is essentially a bucket, several of which is filled with sand and strategically placed, easily accessible and conspicuously displayed- not hidden, within the workshop areas. The bucket must be painted in the correct colour *code of red* and well labeled with white paint (Figure 4-5). At the incidence of fire, the bucket should be lifted and aimed at the source of fire and not the flame.

Figure 4-5: Sand buckets

FARM TRACTOR SYSTEMS

Fire extinguishers

Fire extinguishers are chemical compounds in either liquid or powder form compressed in bottles or cylinders which when released form an envelope over fire causing extinction. Fire extinguishers must be easily accessible and effective. Choose either a dry powder and/or a carbon dioxide type for the workshop as they are suitable for use on electrical fires and burning liquids.

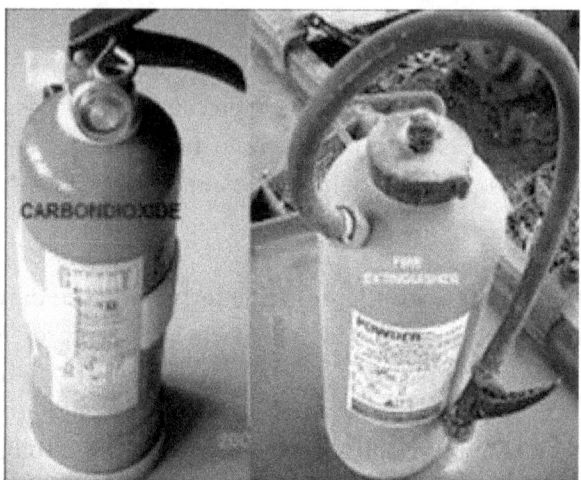

Figure 4-6: Fire extinguisher

All fire extinguishers are marked with letter(s) and or symbols that tell you the kinds of fires they can put out. The letters and symbols are explained in the table below (including suggestions on how to remember them).

Figure 4-7: Fire extinguisher decals

If it becomes necessary to use a fire extinguisher on occasion of fire outbreak, follow these steps:

- THINK and not be panicky;
- Lift the extinguisher and pull the safety pin;
- Aim the nozzle at the base of the fire;
- Slowly squeeze the handle, and
- Sweep the spray from side to side.

DO NOT discharge the entire fire extinguisher initially. Check to see if the fire is out. If not, you will have more extinguishing agent to use. Always keep yourself between the fire and a path to safety.

Fire blankets

In addition, the workshop should be equipped with fire blankets to provide covering for rescuer during fire incidence.

Figure 4-8: Fire blanket (right) and casing (left)

Practical exercises

1. Agricultural Engineering department in your Institution is facing full accreditation by the regulatory body such as NBTE or NUC after the initial resources inspection. One of their recommendations is that your workshops did not meet the required layout standards. Inspect the workshops, single out areas of need and write out how you will solve such problems
2. Carry out further workshop organization in your metal and maintenance workshops according to standards. Report your findings appropriately.
3. Do the following:

   a. List ten safety devices in a well-equipped farm shop and explain the function of each.
   b. Demonstrate proper safety apparel and equipment to be worn and used when operating a grinder, wire-brush wheel, welder, or drill.
   c. Draw a plan showing a well-equipped farm shop. Point out mandatory safety devices and features in the shop.

4. Do TWO of the following:

   a. Replace the handle of any tool found on the farm.
   b. Build a tool rack with storage for nails, bolts, nuts, and washers.
   c. Properly grind the mushroom head off of a chisel or punch.
   d. Correctly grind or file a screwdriver tip.

PART 2

TRACTOR DIVISIONS AND POWER TRANSMISSION SYSTEMS

Tractor divisions, components and functions

Introduction

A tractor is a self-powered work vehicle, designed for the purposes of producing and making power available for various operations on the field. Evolution of the tractor has brought about changes in farm size and technology and has also proved man to be an efficient producer controlling power rather than being the source of power.

Development of new tractors has progressed considerably. Today's agricultural tractors have several built in features such as listed below to enhance more efficient and ease of operation;

1. More power development,
2. Hydraulic controlled power steering and power brakes,
3. Incorporated three point hitches,
4. Remote hydraulic cylinders,
5. Power take off drives,
6. Hydraulic controlled seat ,
7. Installed air conditioner,
8. Installed integrated remote control system and actuators (infrared sensors etc.),
9. Enclosed cab for operator's protection and comfort among others

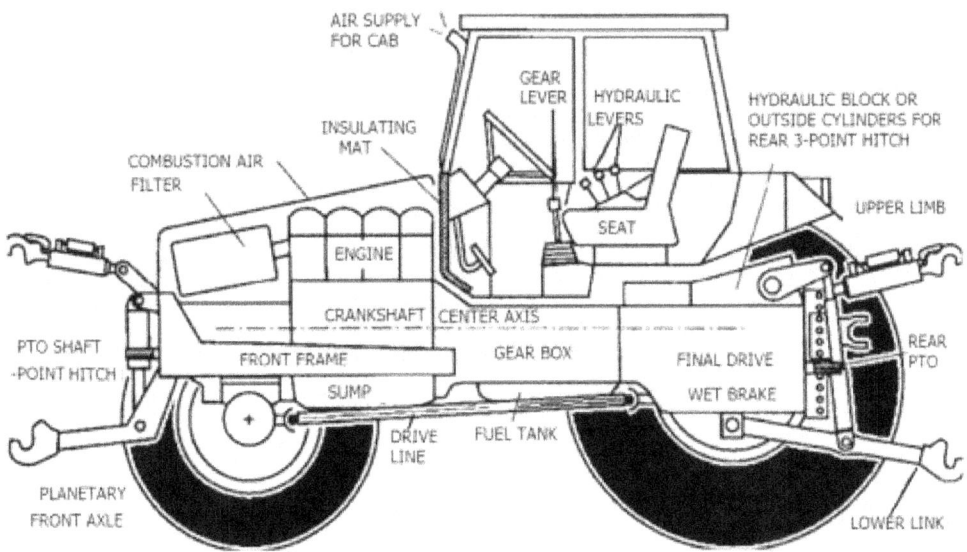

Features of today's tractors

The power capability of the modern tractor has led to higher productivity with a significantly reduced workforce. In recent years there has been a trend from four-wheeled to three-wheeled vehicles, where a single, central front wheel can operate more successfully among crops planted or cultivated in rows

Areas of application

Tractors are designed for one or several of the following applications

1. Pulling or pushing special machinery-moving or stationary (in either off-road or on-road operation),
2. Hauling of heavy loads over land.
3. Widely used in agriculture, building construction, road construction, and
4. For specialized services in industrial plants, railway freight stations, and docks.
5. Other areas of applications also include landform and land clearing development operations.

The place of tractor in achieving the set goals of mechanization has been very significant.

Tractor divisions

The tractor (agricultural tractor) is generally divided into three major segments;

1. The engine,
2. The transmission
3. The differentials and
4. The block design chassis

Each of these segments/divisions comprises of the following systems and components which together constitute the tractor systems.

1. *The engine*

The engine as an integral segment of the tractor comprises of four major functional divisions

   a. The power train,
   b. The stationary parts,
   c. The operating systems and
   d. The auxiliary systems.

The *stationary* and the *power train systems* are essentially the basic engine components such as the engine block, piston, crankshaft etc. The *operating systems* are the various system units with specific functions that works together to enable power generation in engine and such systems include the valve system, air intake system, fuel injection/supply system etc. The *auxiliary systems* are complimentary systems added to engine to improve engine performance and operation. Such system include the following systems: cooling systems, lubrication system, ignition system, exhaust system etc.

The engine's primary function is t produce power (known as brake power or output power) at the crankshaft which is delivered to flywheel.

2. *The tractor power transmission systems*

The tractor power transmission systems are in three categories;

a. The mechanical (friction drives) transmission systems such as: gear systems, belt and pulley, chain and sprocket,
b. The hydraulic and hydrostatic systems such as used in steering systems, brake systems and clutch systems and
c. The drive shaft system which include PTO shaft, drive shaft etc.

The power delivered to the flywheel is transmitted through the transmission systems to the differentials, the final drives, the hydraulic control, the PTO and the drawbar.

3. *The differentials*

The tractor differential is essentially a specially designed and unique feature of the final drive system. Its special feature includes the specially designed final drive wheel, the PTO drive and gear systems for the rear of the tractor.

4. *The block design chassis*

Chassis is a metal frame on which other components, systems or assemblies are mounted. The tractor design did not incorporate a chassis but a well-known *block chassis design* utilizing the engine and transmission housing for support instead of a frame.

The description, function and operation of each of these major tractor divisions will be discussed in subsequent chapters in this part.

CHAPTER 5

Engines and Engine Systems

Introduction

An engine is a mechanical system which transforms heat energy into mechanical energy using fuel. An engine is a machine that makes energy more usable. Engines usually turn heat energy into motion. The useful energy produced by the engine is to be delivered to the crankshaft and the flywheel for vehicular movement, power generation and conversion to other form of energy for utilization. The heat generated is dissipated or converted to useful forms of energy as turbo charging and heating of vehicle interior.

Learning objectives

Students are required to be familiar with the structural composition of various components of the internal combustion engine (ICE) and their functions for the purpose of maintenance, fault tracking and repairs. Therefore students are expected to be familiar and be able to distinguish between an Otto cycle (petrol/gasoline) and Diesel engines when they see one. This chapter thus describes the various engine components, functions and their locations within the engine.

Fundamentals of engine and engine systems

Engine design goals

Over the years, there have been diverse improvements in engine design. In the market, there are varying engine shapes, outlook and form for the purposes of aesthetics, performance and economy. Engine design goals are therefore aimed at improving the economy of petrol/diesel engines.

Typically, other purposes for engine design include:

1. To reduce the engine mass/power ratio (kg/kW)
2. To reduce engine maintenance and repair costs, and
3. To considerably extend the engine life. The life of an engine is the number of hours at which no more than 10 percent of the engines in a given group have failed.

Types of engine

Engines are generally categorized based on their mode of operation, type of fuel burn and design features. Four major types have been identified as follows:

1. *The external combustion engine (ECE)*

The external combustion engine usually called *EC engine* uses steam from a boiler to generate power in an engine. Such engines are usually inefficient, thus, they are no longer in commercial use except for experimental purposes. EC engines work by turning heat energy into motion. Some of the earliest engines ran on steam power, like steam locomotive. Steam traction engines were used in the early tractors, of the USA around 1880 in small quantities. Later on, steam engines were replaced by internal combustion engines between 1890 and 1910.

Figure 5-1: Typical steam engine (1865)

2. *Internal combustion engine (ICE)*

The internal combustion engine called *ic engines* uses the expansive force of burnt gases from an enclosed space called combustion chamber to generate output power in the form of motion called output power for other uses. Examples of IC engine are the diesel and petrol engines.

Figure 5-2: Internal combustion engines

The *diesel engine* is a compression ignition kind of engine in which air is taken in and compressed to high temperatures and fuel is injected for combustion at very high pressures.

The *petrol engine* otherwise called spark ignition engine takes in fuel mixture (air + petrol) and a spark plug ignite the mixture to produce power.

3. *Gas turbine or LP gas engine*

A simple-cycle gas turbine includes a compressor that pumps compressed air into a combustion chamber. Gas turbine engine employs gas flow as the working medium by which heat energy is transformed into mechanical energy. Gas is produced in the engine by the combustion of natural gas, kerosene and diesel oil and stationary nozzles discharge jets of this gas against the blades of a turbine wheel. The impulse force of the jets causes the shaft to turn.

4. *Hybrid engines*

This is a special form of IC engine that uses new technology capable of burning two types of fuel to produce power. Hybrid engine uses such fuels as gas, alcohol, and petrol for combustion.

Internal combustion engine is the most widely used technology in the manufacturing of engine and typically, the diesel engine is the most suited engine for agricultural tractors because of inherent power output advantage, low repair costs among other factors. Owing to these factors, the internal combustion engine and particularly diesel engine will b our focus in subsequent chapter and discussions.

Internal combustion engine (ICE)

Internal combustion are widely classified into two based on the mode of operations and fuel they burn. These classes are the *Otto cycle* and the *diesel engines*; so named after their inventors. The Otto cycle engine is also known as the petrol or spark ignition engine. There are two main types of petrol engines, *reciprocating engines* and *rotary engines*. The reciprocating engines have pistons that move up and down or back and forth. The rotary engine, also known as a Wankel engine, uses a device called rotor to produce rotary motion directly instead of pistons.

The Diesel engine burns a heavier hydrocarbon fuel called diesel to generate more power and is thus preferred in heavy duty applications such as agricultural tillage, ridging etc. These engines vary in their mode of burning fuel as will be explained later.

Comparing the Otto cycle and Diesel engine

The essential parts of Otto-cycle (petrol) and Diesel engines are the same. The only difference between them is in the mode of burning fuel. The petrol engine uses spark ignition mechanism (plugs) to ignite/burn fuel while the Diesel engine uses fuel injection mechanism to inject fuel into a highly compressed hot air in the combustion chamber. The mixing of fuel takes place outside the combustion chamber in petrol engine while it takes place in the combustion chamber in diesel engines. The trade-off factor between the two systems is the power delivery.

Engine selection criteria

Selection of an engine for a specific application involves many factors. CI engines are a good choice in situations where the maintenance and repair infrastructure is limited. SI engines require more frequent tune ups; CI engines can operate for long periods with minimal maintenance beyond regular oil changes and maintenance of fuel cleanliness.

Selection of air-cooled CI engines eliminates the maintenance required for liquid cooling systems. Engines used to drive steady loads, such as those imposed by water pumping, require minimal torque reserve. Conversely, high torque reserve is important in engines used in agricultural tractors, since loads on such tractors can vary widely within a given field. Good fuel economy is obtained by operating the engine with governor control but loaded close to governor's maximum.

Two-stroke and four-stroke cycle engine

Engines; either diesel or petrol undergo four basic processes in generating power through the burning of fuel as follows.
Process 1: *Intake* of fuel mixture (fuel + air) or air only
Process 2: *Compression* of fuel mixture or air
Process 3: *Ignition* of the compressed fuel mixture or hot air and
Process 4: *Exhaust* of waste or burnt gasses from the combustion chamber

Each of these processes is referred to as a *stroke*. A stroke is completed when the crankshaft travels through 180 degrees (1/2 of a revolution) from the starting point. Most engines operate on either two-stroke or four-stroke cycles operation as discussed below.

Two-stroke cycle engine: A two-stroke cycle engine combines the exhaust and intake steps near the end of the power stroke. Two-stroke cycle engines are less fuel-efficient and generate less power than four-stroke cycle engines and are simpler and cheaper to build.

A two-stroke cycle engine is used where low cost is required, such as in a power lawn mower. It delivers more power for a given weight and sizes than does a four-stroke cycle engine. Each cylinder in a two-stroke cycle engine produces a power stroke for every turn of the crankshaft. But in a four-stroke cycle engine, a cylinder produces a power stroke on every other turn.

Four-stroke cycle engine: A four-stroke cycle engine has intake, compression, power, and exhaust strokes taken place independently as the power train complete two revolutions.

Working principles of an IC engine

The combustion chamber consists of a cylinder, usually fixed, in which a close-fitting *piston* slides. The reciprocating motion of the piston varies the volume of the chamber between the inner face of the piston head and the closed end of the cylinder.

Figure 5-3: Cut-away views of engine

The outer face of the piston (the hollow) is attached to a crankshaft by a connecting rod. The crankshaft transforms or converts the reciprocating motion of the piston into rotary motion. In multi-cylindered engines, the crankshaft has one offset portion, called a crankpin, for each connecting rod so that the power from each cylinder is applied to the crankshaft at the appropriate point in its rotation. Crankshafts have heavy flywheels and counterweights, which by their inertia minimize irregularity in the motion of the shaft.

Engine operation

When a piston travels up to the *top dead center* (TDC) and return to the *bottom dead center* (BDC) a *cycle of operation* is complete. Thus two stroke engines complete a cycle in 1 revolution of the crankshaft while the four-stroke engine completes a cycle in 2 revolutions.

Engine configuration and linear measurements

The size of an engine cylinder is indicated in terms of the size of the cylinder bore and length of stroke. The *bore* is the inside diameter of the cylinder while the *stroke* is the distance between top dead center (TDC) and bottom dead center(BDC)(Figure 5-4).

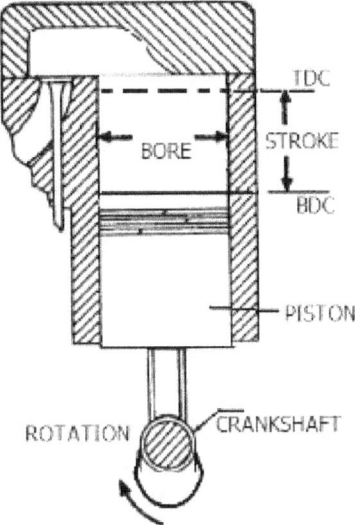

Figure 5-4: Engine configurations

In describing an engine by specification, the bore is always mentioned first, then the stroke. For example, a '3 ½ by 4' cylinder means that the cylinder bore, or diameter, is 3 ½ inches and the length of the stroke is 4 inches. These measurements are used to calculate piston displacement.

Piston displacement: Piston displacement is the volume of space that the piston displaces, as it moves from one end of the stroke to the other. Thus the piston displacement in a 3 ½ -inch by 4-inch cylinder would be the area of a3 ½ inch circle multiplied by the length of the stroke between the TDC and BDC at the specific point.

$$D = \pi r^2 x\, l_s \quad \dots\dots\dots\dots\dots\dots\dots\dots\dots .5.1$$

Where
D= Piston displacement (inches)
r = Radius of the cylinder
l_s = Displaced volume

Engine volume determination: The volume of an engine is determined by the product of the surface area of piston head/cylinder and the length of stroke. The relationship for calculating the area of a circle is πr^2, where r is the radius (one half of the diameter) of the circle. The surface area of the piston head is represented by the area of the circle. With S being the length of stroke, the formula for calculating the volume (V) of the engine is given as:

$$V = \pi r^2 S \ldots\ldots\ldots\ldots 5.2$$

The total displacement of an engine is found by multiplying the volume of one cylinder by the total number of cylinders.

$$V = \pi r^2 n S \ldots\ldots\ldots\ldots 5.3$$

Where

n = Total number of cylinders
S = Length of the stroke between the TDC and BDC at specific point

Engine test and performance

As an engineer or a technician/mechanic, you must know the various ways by which engine and engine performance are measured. An engine performance may be measured in terms of cylinder diameter, piston stroke, and number of cylinders. It may also be measured performance-wise by the torque and horsepower it develops or by its efficiency. An engine that does more work per minute than another is more powerful. The work capacity of an engine is measured in horsepower (hp). A number of devices are used to measure the hp of an engine. The most common device is the dynamometer.

Engine dynamometer: An engine dynamometer (Figure 5-5) maybe used to bench test an engine that has been removed from a vehicle. If the engine does not develop the recommended horsepower and torque, further adjustments must be made and/or repairs on the engine may be required.

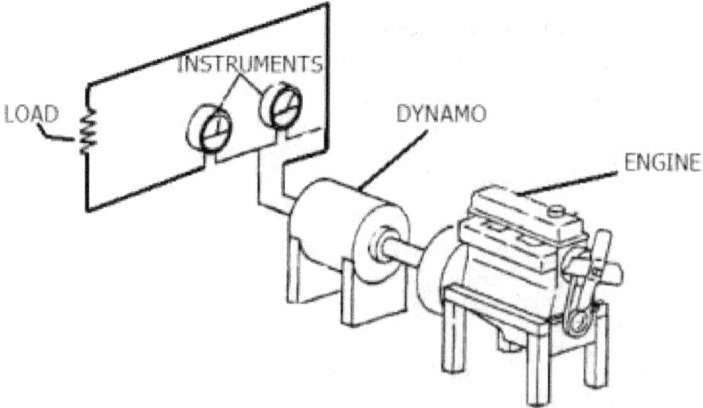

Figure 5-5: Engine dynamometer

Another device that measures the actual usable horsepower of an engine is the prony brake (Figure 5-6). Though it is rarely used today, but is simple to understand. It is useful for learning the concept of horsepower-measuring tools.

Prony brake: The prony brake consists of a flywheel surrounded by a large braking device. One end of the arm is attached to the braking device, while the other end exerts pressure on the scale. In operation, the engine is attached to, and drives, the flywheel. The braking device wrap around the flywheel is tightened until the engine is slowed to a predetermined revolution per minute (rpm).

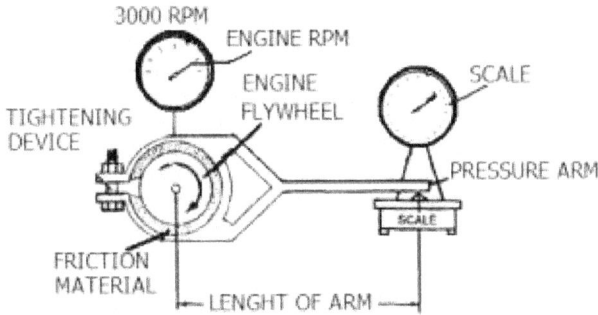

Figure 5-6: The prony brake dynamometer

As the braking device slows the engine, the arm attached to it exerts pressure on a scale. Based on the reading at the scale and engine rpm, a brake horsepower value is calculated by using the following formula:

$$B_p = \frac{6.28 \, x \, l \, x \, n \, x \, r}{33{,}000} (hp) \quad \ldots \ldots \ldots \ldots \ldots \ldots \ldots 5.4$$

Where
B_p = Brake power (hp)
l = length of arm in ft
n = engine rpm
r = scale reading in lb
6.28 & 33,000 = conversion factors

It must be noted that 6.28 and 33,000 are constants in the formula, meaning they never change. For example, a given engine exerts a force of 300 pounds on a scale through a 2-foot-long arm when the brake device holds the speed of the engine at 3,000 rpm. By using the formula, calculate the brake horsepower as follows:

To calculate $B_{p,}$ in hp and kW

Given that
B_p = Brake power (hp)
l = length of arm = 2ft long
n = engine rpm = 3000 rpm
r = scale reading = 300 lb
6.28 & 33,000 = conversion factors
Utilizing equation 5.4

$$B_p = \frac{6.28 \, x \, l \, x \, n \, x \, r}{33{,}000} \, (hp)$$

$$B_p = \frac{6.28 \times 2 \times 3000 \times 300}{33{,}000} = 343.55 \, (hp)$$

To calculate power in kilowatt

$$B_p = \frac{2 \, x \pi x l x f x n}{60{,}000} \, (kW)$$

Where
L = Prony brake length of arm in meter = 2ft (0.6096m) note 1ft = 0.3048m
F = Force in Newton = 300lb (136.077kg) note 1lb = 0.4535kg (1335.55N)
n = Revolutions per minute = 3,000.

$$B_p = \frac{2 \, x \pi x \, 0.6096 \times 1335.55 \times 3000}{60{,}000} = 255.4 \, kW$$

Alternatively, $1hp = 0.746 (kW)$

$$\therefore 342.55 hp = 342.55 \times 0.746 kW = 255.4 \, kW$$

Classification of internal combustion engine

Internal combustion engines are classified based on the following features:

a. *Type of fuel burn* e.g. Diesel, petrol or gas engines.
b. *Fuel and air supply:* Engines are also classified as carbureted or as fuel-injected engines. Because combustion depends upon both air and fuel, an engine may also be classified as either supercharged or turbocharged. See intercooler and turbocharger for more details.
c. *Cycle/stroke of operation:* Most engines operate on either a two-stroke or a four-stroke cycle and are thus classified.
d. *Type of compression:* Engines are classified either as high-compression engine (a compression ratio of 10 to 1) or as low-compression engine (may have a ratio of 8 to 1). High-compression engines burn petrol more efficiently than do low-compression engines. But high-compression engines require high-octane petrol.
e. *Cylinder arrangement:* Engines are also classified by the number and arrangement of cylinders. The most common types include in-line cylinders, V- engine, radial engine, and horizontal opposed engines. Radial engines used in aircrafts have an odd number of cylinders, such as 3, 5, 7, or 9. Most other engines have an even number of cylinders-4, 6, 8, or 12.

f. *Cooling system:* Engines are classified as either air-cooled or water cooled. Most automotive petrol engines are water cooled while aircraft engines are air cooled to reduce weight.
g. *Valve arrangement*: The majority of internal combustion engines also are classified according to the position and arrangement of the intake and exhaust valves, whether the valves are located in the cylinder head or cylinder block.

Internal combustion engine divisions

Engine parts are mutually constrained (restricted) to convert the expansive force generated in the cylinder into rotational motion at the flywheel. Engine components/parts are functionally divided into the following four divisions and will be the basis for our descriptions.

1. *The power train*: This part receives, exert and transmit the motion forces
2. *The stationary parts*: These parts constrain and supports all moving parts
3. *The governing system*: These parts function as a timer for the operating system, programmes the sequence of operation and enhance engine performance
4. *The auxiliary parts*: All other accessories that help enhance engine performance such as exhaust, and ignition systems.

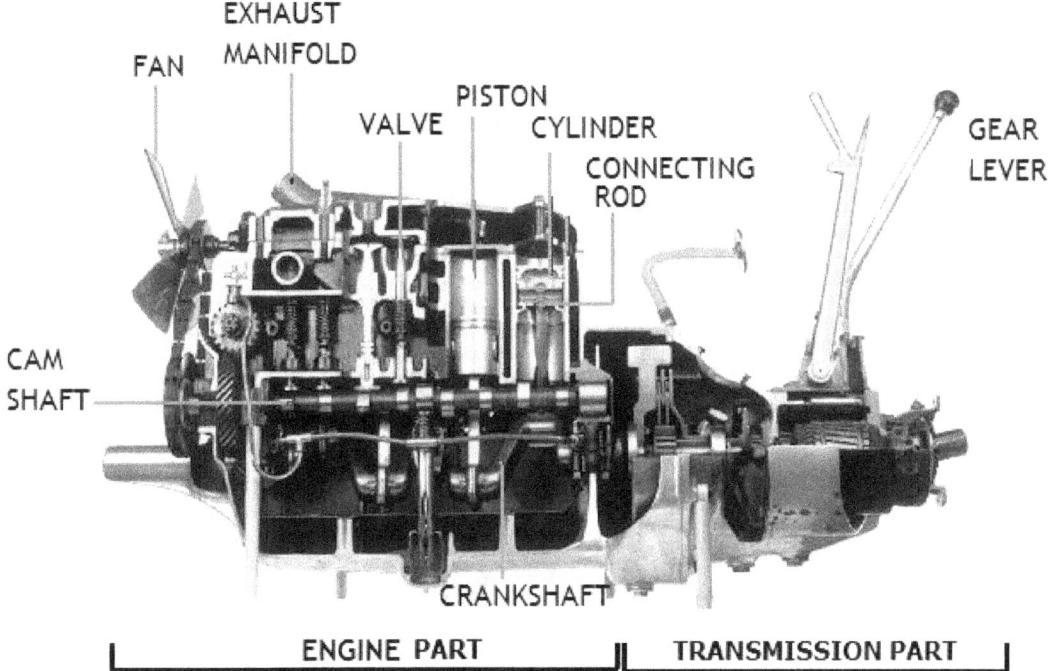

Figure 5-7: Transverse section of engine& transmission

1. **Power train components of engine**

The component parts of the power train in the engine and their functions are described below:

Piston

This part is directly exposed to the expanding gases and when forced outward (downward movement), transmits its motion to the connecting rod. The top of the piston is known as the *head* while the side is refers to as the *skirt*.

Figure 5-8: Piston and the piston assembly

Piston rings

These are made of cast iron that fit into the grooves on the piston. Their purpose is to form a gas-tight combustion chamber at all positions of the piston. The top solid rings close to the piston head are called compression rings and are subjected to high pressures. The bottom rings, known as oil rings, has slots with springs for the purpose of proper lubrication.

Figure 5-9: Piston rings

Piston pin or wristpin

This is a hollow case-hardened steel pin grounded to a smooth polish, serves to fasten the piston to the upper end of the connecting rod. It can either be classified as stationary, oscillating or floating.

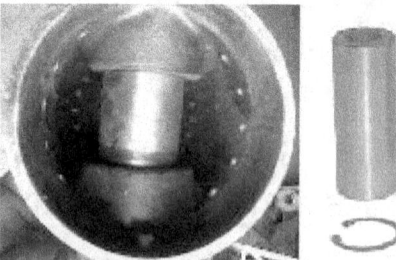

Figure 5-10: Underside of piston showing piston pin

Connecting rod

This connects the piston to the crankshaft. The upper part of the rod has a bearing made of bronze bushing while the lower part (bearing) is split into two halves, one end fix to the rod while the other end fits into the bearing cap. It converts rectilinear motion of the piston constrained by the cylinder walls to rotational motion on the crankshaft.

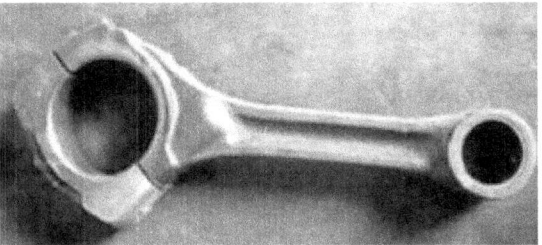

Figure 5-11: Connecting rod

Crankshaft

This is the next link in the power train. The rectilinear motion of the piston is transmitted to rotational motion in the crankshaft. The crankshaft is supported by two or more main bearings in the crankcase. The part forming the crank arm is known as the cheek. The crankpin is at the outer end of the crank arm and forms the journals for the bearing at the lower end of the connecting rod.

Figure 5-12: Ford/Newholland crankshafts

The front end of the shaft is machined for timing gear and may be shaped to accommodate a starting crank. The gear end usually carries a drilled flange for mounting the flywheel.

Flywheel

This is the final component in the power train link. The primary purpose of the flywheel is to reduce speed variation to some acceptable determined by its use.

Figure 5-13: Engine flywheel and ring

The minor purpose is to provide energy when the clutch is engaged under load. The flywheel stores energy during the power strokes and returns the energy during the idle strokes, producing a more uniform rotation.

2. **Stationary components of engine**

The stationary parts are the constraining parts for the power train. These include: Cylinder block, cylinder head, crankcase, base, engine supports, manifolds and other parts.

Cylinder block

The cylinder block is made of cast iron, cast aluminum, or semi-steel and is that part, which confines the expanding gases and forms the combustion chamber. The engine block may be short or long according to designs (Figure 5-14).

Figure 5-14: Different types of engine blocks

Other designs include rope style crankshaft seal (with sleeves), lip style crankshaft seal (with sleeves), short block rope seal crankshaft with flat top pistons for indirect and direct injection, short block 4-cylinder lip seal (front mount balancer) etc. The cylinder may be cast singly, in pairs or as three, four, six or eight on block.

Cylinder head

The cylinder head houses the valves and forms a cover to the cylinder. Part of the combustion chamber is carved into the cylinder head. There are holes known as water jackets and oil galleries machined or cast into the cylinder head , matching those of the engine block and through which lubrication oil and water coolant circulated to the top cylinder and valve cam assembly.

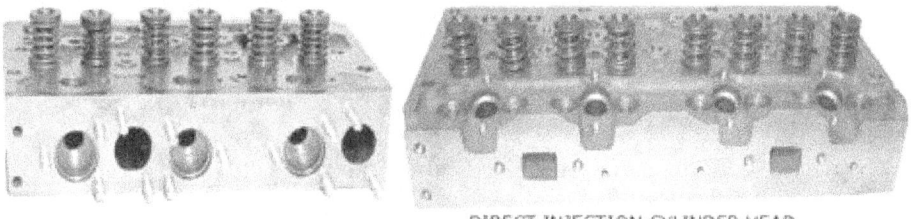

Figure 5-15: Cylinder heads

Combustion chamber

The combustion chamber is a small depression made between the piston head and the cylinder head. These depressions when meshed together form the combustion chamber. The size and shape of the combustion chamber has a deciding effect upon the proper mixing of the fuel and air, and also upon the fuel detonation. In general, the combustion chamber is designed to create greater flame propagation velocity thereby decreasing the combustion time.

Types of combustion chamber

In diesel engines, ignition could be delayed until the fuel-to-air ratio is in the combustible range, thus different designs of combustion chamber were employed to improve fuel ignition and include:

- *Direct injection or open-chamber*: This design employs a concave piston head-mixing of the fuel and air is aided by an induction-produced air swirl or by a movement of air from the outer rim of the piston toward center of the piston commonly known as squish.

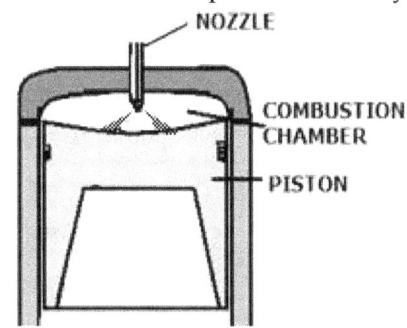

Figure 5-16: Direct combustion chamber

- *Pre-combustion chamber:* This is sometimes a part of the injection-nozzle, or may be part of the cylinder head. The entire fuel charge is injected into the pre-combustion chamber, which contains 25-40% of the clearance volume.

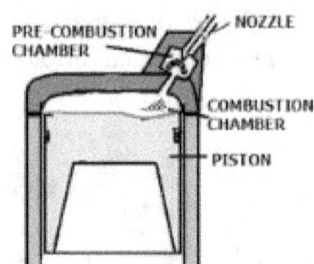

Figure 5-17: Pre-combustion combustion chamber

Compare to the open combustion chamber, it has the following advantages;

1. Lower fuel-injection pressure requirement
2. Has ability to use a wider range of fuels.

Disadvantage includes higher specific fuel consumption.

- *Swirl combustion chamber*: This is sometimes called Richard turbulence chamber. The burning fuel-air mixture is caused to swirl in circular motion, thereby improving the mixing and subsequent combustion in the main combustion chamber. When combustion begins, a reversal of flow occurs and the burning gases stream out of the chamber into the cylinder. The swirl chamber contains 50-90% of the compressed volume when the piston is on TDC.

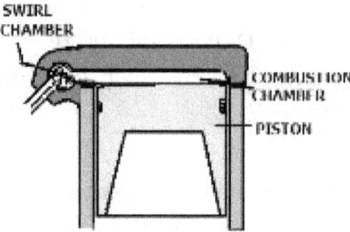

Figure 5-18: Swirl combustion chamber

- *Auxiliary chamber:* Also called air cell or energy cell, is an open chamber with a small cell far from the injection nozzle. When the cell is located on top of the piston, it is called air cell and when located directly across the cylinder from the injection nozzle, it is called energy cell.

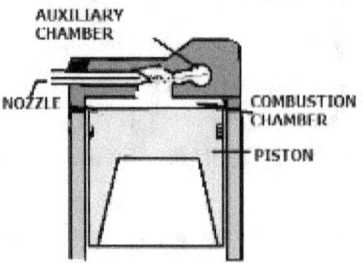

Figure 5-19: Auxiliary combustion chamber

In operation, about 60% of the fuel from the injection nozzle is directed into that auxiliary chamber. The blast from the fuel ignition in the auxiliary chamber will then be directed against the remainder of the fuel being sprayed from the nozzle. Duration of combustion is longer for this type of design, resulting in lower peak pressures at the expense of slightly higher fuel consumption.

Engine balancer

Engine balancers removes or minimize the unbalanced forces resulting from the piston side thrust, unbalanced centrifugal forces from rotating parts etc by using balancing weights or through the design of the crankshaft. Because of the unbalance, it is desirable to eliminate this unbalance by adding counterbalance weights driven by the crankshaft. The weights are arranged so that one side is heavier than the remainder of the weight. Balancers could be front mount or mid mount as shown in the Figure 5-20.

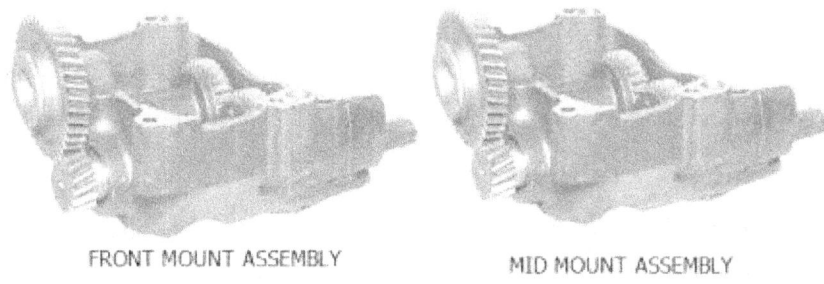

Figure 5-20: The engine balancer

Crankcase

The Crankcase serves the purpose of supporting the crankshaft, mounting the cylinders, housing for the running parts, and forming a reservoir for lubricating oil.

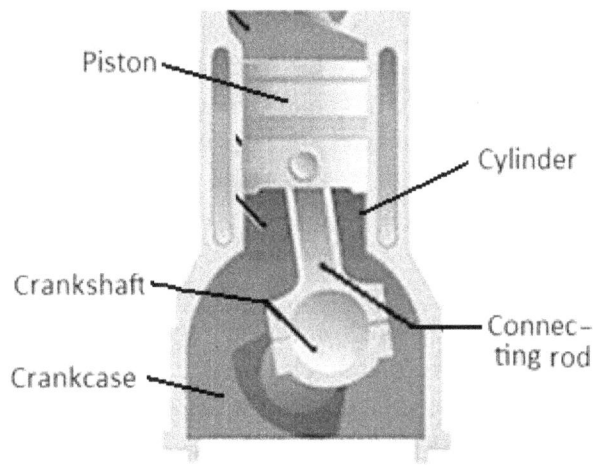

Figure 5-21: Engine crankcase

Oil pan/sump

The oil pan/sump forms the lower part of the crankcase and is made of cast iron, cast aluminum or pressed steel. It forms a reservoir for the oil. The oil pan is attached to the crankcase with bolts. The contacting surface of the crankcase and the pan is sealed with a fiber gasket. Any cut in this gasket causes oil leakage from the sump.

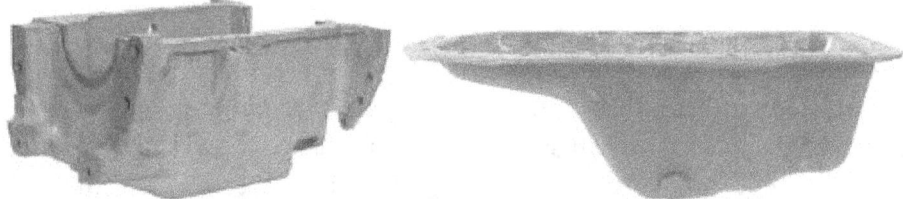

Figure 5-22: Oil pans

Manifolds

The manifold is the channel through which fuel mixture and exhaust gases travel within the engine. The engine manifold is of two types; inlet and exhaust manifolds. Each is described as follows:

The inlet manifold is attached to the side of the cylinder head or block and serves to conduct air mixture into the cylinders. It is made of cast iron or aluminum. The inlet or intake manifold connects the air inlet system with the combustion chamber.

Figure 5-23: Intake manifold

The exhaust manifold attached to the side of the cylinder head or blocks. It is the passageway for the discharge of waste gases from the combustion chamber. Exhaust from each cylinder is collected into a single pipe which is linked to the exhaust pipeline which serves to conduct the burnt gases away from the engine.

Gaskets

These are light sheet materials formed from either asbestos, fiber or reinforced fiber with holes carved to match surfaces it is intended to be fixed. It provides a joint and seal for two mating metal surfaces. For example, the top gasket has slots that match those on the engine block and the cylinder head for the passage of oil and coolant. The gasket has functions of conducting heat away from the metal block and also to prevent pressure leakage from the combustion chamber.

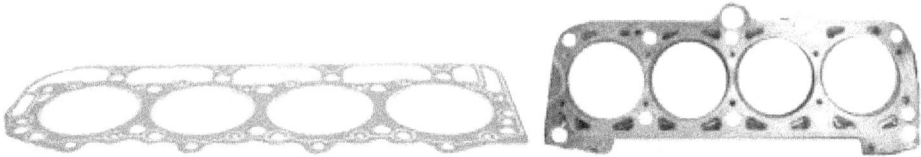

Figure 5-24: 4-cylinder head gasket

3. Engine operating systems

The operating systems of an engine are those individual systems that function together to enhance efficient engine operation. These systems includes the air-intake system, fuel supply system, carburetion and injection systems, valve system, lubrication system, cooling system, ignition system and exhaust system.

b. Air intake system

The air intake system traps unclean air, passes it through some filtering elements and delivers clean air into the combustion chamber. Air entering into the combustion chamber must be kept clean of dust, dirt and any other forms of contamination that could impair efficient burning of fuel mixture. To ensure clean air supply, the air is passed through series of specially designed filtering elements ranging from simple system to more complex designs in different engines.

Problems associated with air intake systems

The filtering elements in the air intake system easily get clogged over a period of time due to the dusty environment they operate. As a result of this clogging, problems associated with air intake system include poor engine performance, incomplete combustion and pre-ignition which could lead to knocking sound etc. To prevent such problems, the air intake system must be well maintained and kept clean at every service.

Types of air intake systems

The air intake system of the tractor consists of the precleaner and air filters. Each system has distinct ways of filtering air that enters into them before entering the engine. Each system is discussed below.

Precleaner

Most farm engines, especially tractors and articulated vehicles operate under the most unpleasant, severe dust condition. Thus air available for intake is often contaminated with dust and particles which when taken in unclean could cause pre-ignition, detonation and poor engine performance. Hence this air must be thoroughly cleaned before supply into the combustion chamber.

FARM TRACTOR SYSTEMS

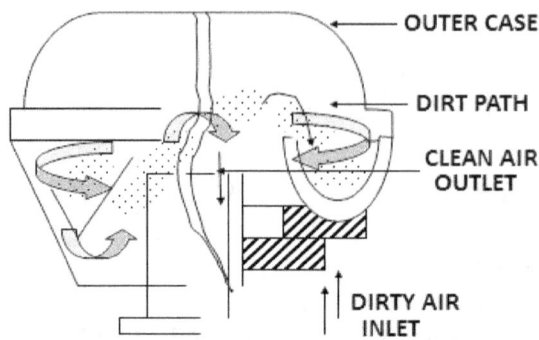

Figure 5-25: Air flow mechanism in pre-cleaners

Precleaner is used under severe dust conditions to protect the air cleaner by reducing the load on it. Vanes fitted into the inlet induce a rotary motion to the air stream as air enters the inlet at high speed, thus generating a centrifugal force.

The centrifugal force causes the heavier dusts and other foreign matter to be thrown into the space between the inner and outer shells. The pre-cleaned air then passes into the air cleaner oil bath. The clean air passes through the clean air outlet. The pre-cleaner is often found mounted on the inlet tube, in tractors above the radiator hood panel assembly.

Figure 5-26: MF precleaner

Air filter

Air filter used in engine systems are of two types; the oil bath type and dry type filters. The *oil type filters* are commonly used in tractor engines because of the dusty environment in which they operate. However, recent tractors now uses dry filter elements with improved performance in dusty condition. The dry type filter is commonly used in small engines such as lawn mowers and auto cycles.

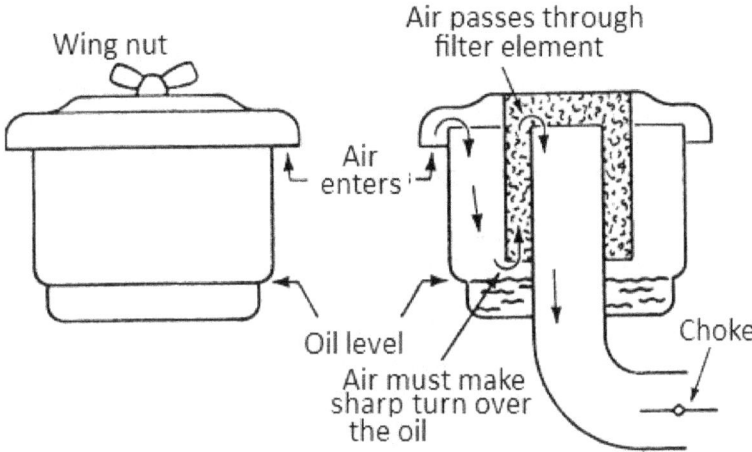

Figure 5-27: Oil bath air cleaner

Oil bath air cleaner: Oil bath air cleaners, shown in Figure 5-27, Courtesy of Briggs & Stratton Corp., are commonly used on engines that perform regularly in dirty conditions, because of their ability to collect larger amounts of dirt particles, and thus are particularly used in tractor engines. Incoming air is forced through a sharp turn directly above an oil reservoir which captures any dirt or particles that may accompany it. Cleaning this type of air cleaner simply involves washing out the metal bowl with a solvent, drying the bowl, and refilling the bath with fresh oil.

Dry type air filter: The dry type filter utilizes a dry paper element, (Figure 5-28 Courtesy of Briggs & Stratton Corp.) or polyurethane foam (Figure 5-29).

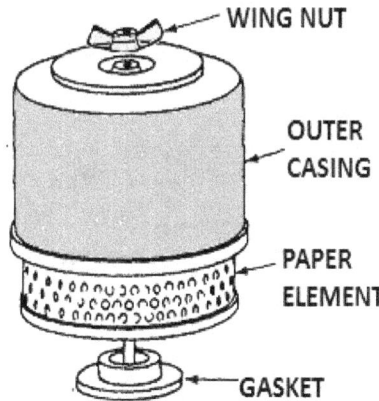

Figure 5-28: Dry paper type element

The paper element type draws dirty air through its porous surface which allows it to catch any particles that may accompany the air flow. Cleaning the filter is achieved by lightly tapping it or by blowing the dirt away from the inside-out with low pressure air. Do not soak or wash a dry paper element. Also, a dry paper element should never be coated with oil. The oil easily traps dirt that makes it difficult to blow it off.

Polyurethane foam elements: Foam elements are popular types of air cleaners used in small IC engines. The foam, (Figure 5-29 Briggs & Stratton Corp.), is saturated with oil which allows it to catch all of the dirt that flows through it. They are particularly used in small engines. They are more durable than the dry type filters. The durability and reliability of this type of filter results from their ability to be washed and re-used.

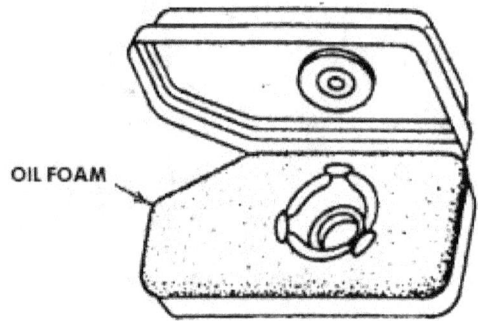

Figure 5-29: Polyurethane foam type element

These filters are cleaned by washing them in warm soapy water, and are then allowed to air dry after all of the excess water is squeezed out. After drying, the filter should be saturated with fresh oil and then replaced.

Dual-element air cleaners: Some air cleaners are considered to be dual-element cleaners because they combine two of the filtering materials mentioned previously. A very common dual element cleaner, (Figure 5-30 Briggs & Stratton Corp.), is the combination of dry paper and polyurethane foam without an oil coating. Oil bath dual-element air cleaners may also incorporate polyurethane foam, but the foam is usually coated with oil.

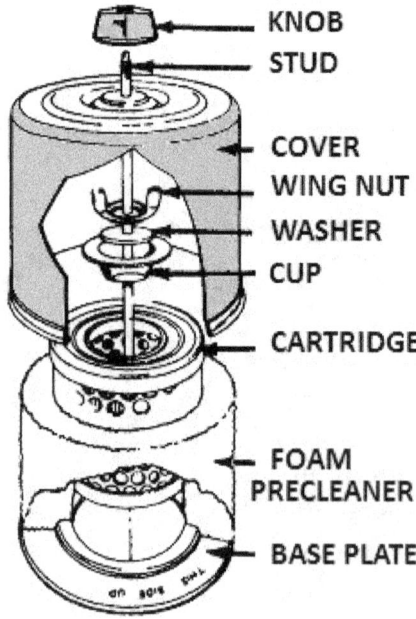

Figure 5-30: Dual element air cleaner

FARM TRACTOR SYSTEMS

b. *Fuel supply system*

Generally, the fuel supply system of an internal-combustion engine (petrol or diesel) consists of a tank, a filter or filters, a fuel lift pump, fuel lines, and a device for vaporizing or atomizing the liquid fuel (carburetor or injection pump).

Fuel tank

The fuel tank comes in different shapes and designs depending on tractor engine design. Fuel flow from the tank could either be by gravity flow or flow supported by a fuel pump (forced flow). Gravity flow is common in single cylinder engines and 2-stroke engine while forced flow system is used in automobiles with their engines below the engine level.

Figure 5-31: Fuel tanks

Fuel filters

The fuel filters removes dirt and other particles that could cause blockage from the fuel being supplied to the combustion chamber. The primary function of the filters is in dirt remover. Filters come in different design and shapes depending of the type of fuel used and manufacturer. For instance, the tractor and other heavy duty engines have two set of filters; the primary filter and the secondary filters. The primary filter is directly connected to the fuel lift pump while the secondary filter is connected to the injection pump. Other light duty engine such as automobiles has single filters.

Figure 5-32: Fuel filters

FARM TRACTOR SYSTEMS

Types of fuel supply systems

The fuel supply system in a petrol engine consists of a sediment bowl, a fuel pump fuel lines and strainers, a carburetor, an intake and exhaust manifold. The fuel is metered in required proportion into the combustion chamber in two separate processes termed *carburetion* and *injection systems*.

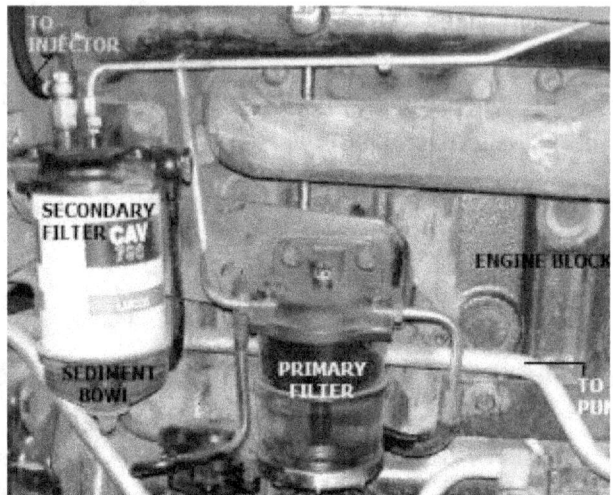

Figure 5-33: Fuel supply system for MF 435

i. *Fuel carburetion system*

This is a process of mixing air and fuel in a device called carburetor and supplying or metering it in required quantity into the combustion chamber for ignition. In most recent automobiles, the use of carburetion system is being gradually replaced with highly automated electronically controlled systems. Carburetion systems are being gradually replaced by hi-tech injection systems because of fuel economy and enhanced engine performance. Carburetion systems are still being used in small Otto cycle/ spark ignition engines.

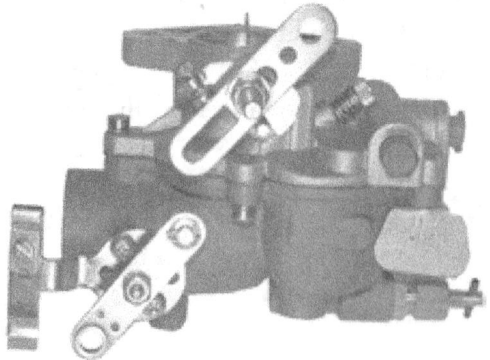

Figure 5-34: Fuel carburetor

The chief job of the carburetor is to mix fuel and air for delivery into the combustion chamber. The air/fuel ratio for petrol engine is *15 parts air to 1 part fuel* by mass. The basic principle of carburetion involves the confining of a small amount of fuel in a bowl, and a nozzle/jet which dispensed fuel into the stream of air.

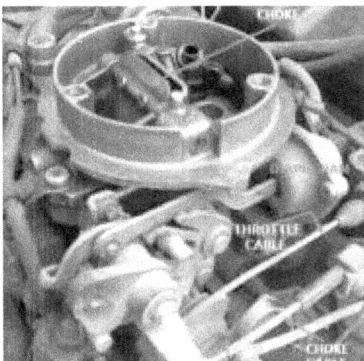

Figure 5-35: An electro-mechanical carburetor

A vacuum is created in the nozzle/jet which causes the atmospheric pressure to push liquid into the nozzle where it is discharged with air in a fine spray. The amount of Fuel: Air mixture determines fuel combustion in the chamber and energy/power output.

Less air and more petrol mixture gives a *rich mixture* e.g. 10:1 ratio. Too rich a mixture results in wasted fuel and in piston and ring wear due to the washing off of lubrication oil, dilution and increased carbon formation.

More air and less petrol give a *lean mixture* e.g. ratio 20 to 1 or 20:1. Too lean a mixture results in poor economy because of loss of power, acceleration and a tendency to burn valves and spark plug fouling.

Components of carburetor

A simple carburetor consists of the following parts:

Fuel bowl which received and hold small quantity of fuel from the tank at a given time. Dirty tank or fuel could lead to dirt settlement in the fuel bowl or water sedimentation could be as a result of condensation in tank, filter or dispensed fuel.

Figure 5-36: Fuel bowl

Float/cork acting upon a needle valve at the fuel inlet controls the fuel level in the fuel bowl. When the correct level of fuel enters the bowl from the supply; the float lifts the needle valve against its seat which shuts off the supply by blocking the entry.

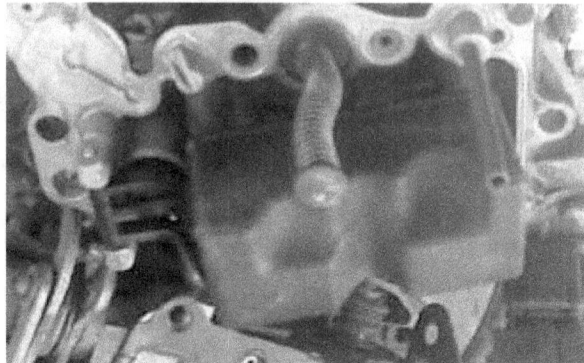

Figure 5-37: Floating cork

Throttle valve: This controls engine speed by regulating the fuel / air mixture discharge from the carburetor to the combustion chamber of the engine.

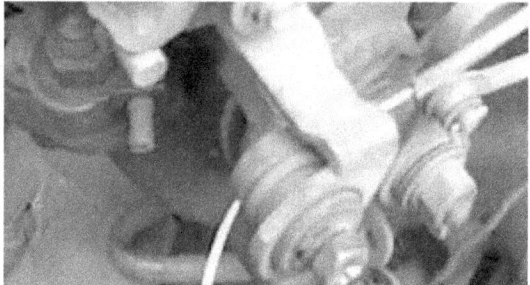

Figure 5-38: Throttle control and cable

Choke valve: This is used when starting the engine in a cold weather condition. It is located in the carburetor air intake path ahead of the venturi. When in a closed condition, more fuel will be drawn from the bowl creating a richer mixture. As soon as the engine starts, the choke valve is opened sufficiently to maintain operation. The choke valve is opened fully when the engine's temperature becomes warm enough and the fuel vapourizes more easily.

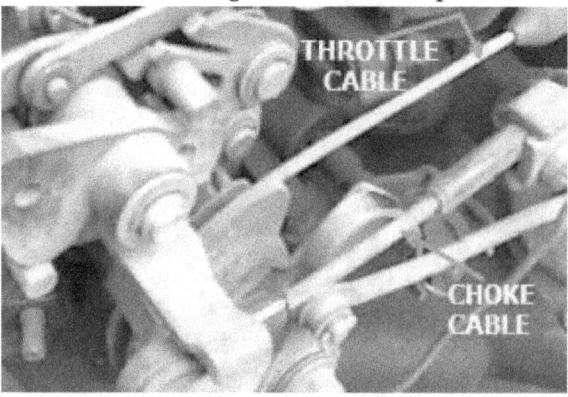

Figure 5-39: Choke system

Venturi: This is a mechanical device designed to vary the pressure difference in the carburetor.

ii. *Fuel injection system*

The fuel-injection system replaces the carburetor in most new automobiles and heavy duty diesel engines to provide a more efficient fuel delivery system and high power performance. Electronic sensors respond to varying engine speeds and driving conditions by changing the ratio of fuel to air. The sensors send a fine mist of fuel from the fuel supply through a fuel-injection nozzle into a combustion chamber, where it is ignited by compressed hot air. The mixture of fuel and hot air triggers ignition.

Figure 5-40: Fuel injection pump

Components of fuel injection system

The components of a fuel-injection system include injectors that atomize (spray a dust of tiny droplets) fuel, a pump that delivers fuel from a storage tank to the injector, a variety of sensors, and a computer, generally called an electronic control module (ECM) or brain box.

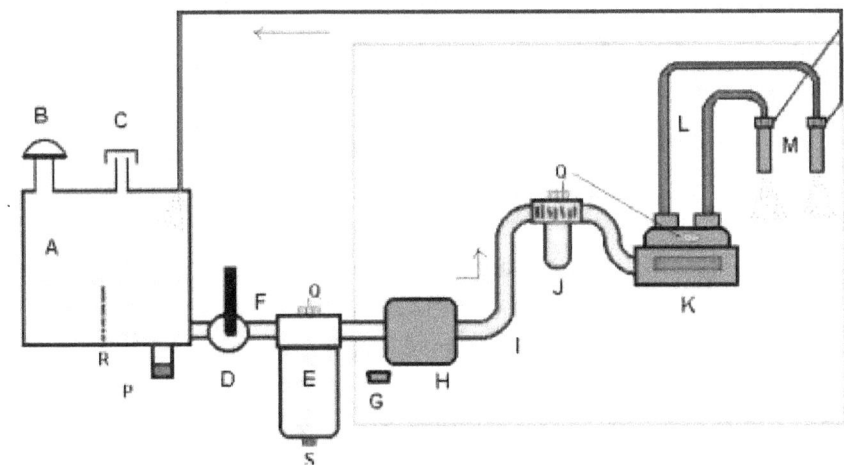

Figure 5-41: Fuel injection system schematic

[A. Diesel tank B. Tank filters C. Tank filler. D. Tank shutoff valve E. Primary fuel filter/water trap F. Suction side. G. Manual fuel pump lever H. Fuel-lift pump I. Low pressure side. J. Secondary fuel filter K. Fuel injections pump L. Injection pipes M. Injector. N. Fuel return line P. Fuel tank drain Q. air bleed screws R. Tank baffle S. Filter water drain.]

The sensors monitor the engine, transmission, air-intake, exhaust, vehicle speed, fuel flow, and many other features that affect engine performance. The ECM uses this information to control fuel supply, the timing of fuel injections, and the mix of fuel and air in the engine's combustion chambers. Modern ECMs also control the engine's ignition system, which ignites fuel in the combustion chambers at precisely timed intervals to operate the engine efficiently.

Types of fuel injection

Throttle-body injection (TBI): This type of fuel injection sprays a fuel mist through one or more injectors located near the start of the intake manifold, which carries air to the individual cylinders. The pistons in the cylinders create a partial vacuum that pulls fuel and air through the manifold into the combustion chambers.

Note: The injectors in a TBI system are located in the intake manifold close to where a carburetor was typically housed.

Port-type fuel-injection systems, sometimes called multi-port systems, have injectors that spray fuel directly into intake port (an opening through the engine into the combustion chamber) of each cylinder.

Note: A port-type injection is better and able to deliver equal amounts of fuel to each cylinder. Moving the injectors to the ports also permits alterations to the intake manifold that improve engine performance.

Direct fuel injection: In this design, the nozzles open directly into the combustion chamber. Atomized spray is injected directly into the open combustion chamber.

c. *Valve system*

The valve system functions as the gateway controlling the intake of air/fuel mixture into the combustion chamber and the exhaust of waste gases from the combustion chamber.

Owing to their function, they operate under very high temperatures and are susceptible to wear and pitting when not well maintained. They need constant lubrication to operate without making noise. Tapping noise in engine is largely due to non-feeding valve system.

The valve system includes the valves, timing gears, crankshafts, rocker arm assembly etc. Valves are used for opening and closing ports leading into or out of the combustion chambers.

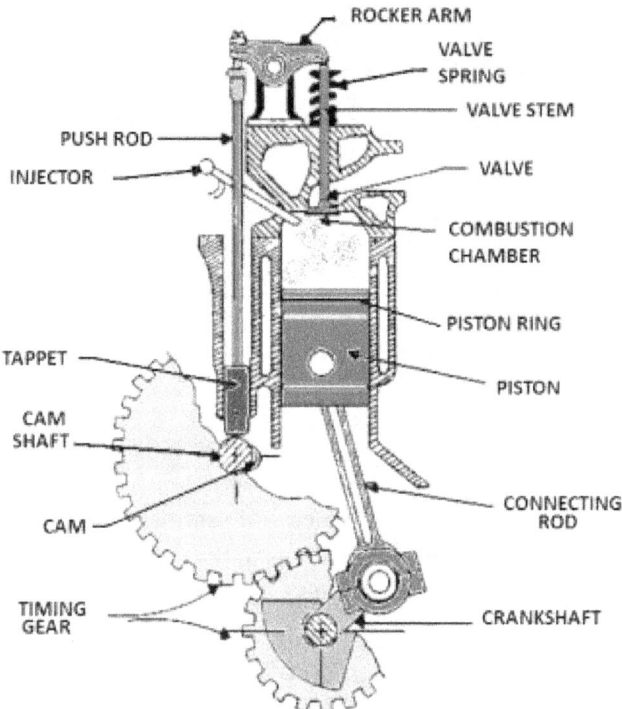

Figure 5-42: The valve system schematics

There are two sets of valves attached to one piston in conventional straight engines; the intake valve and the exhaust valve. One form of identification is through the size of each valve head. The exhaust valve head is smaller than the intake valve head because the intake needs wider surface area for air intake into the combustion chamber.

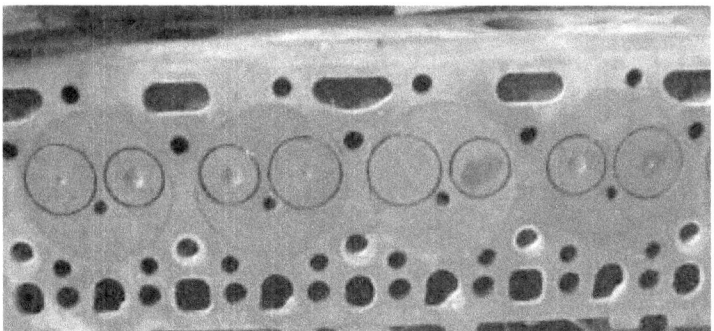

Figure 5-43: Intake and exhaust valve arrangement

Component description

Poppet valve consists of a stem and a mushroom shaped head that when closed fits tightly on to the circular seat.

Valve head is made of special alloy steel chosen to withstand hammering action and the high gas temperatures to which it is exposed.

Valve stem is a round steel rod to which the valve head is attached.

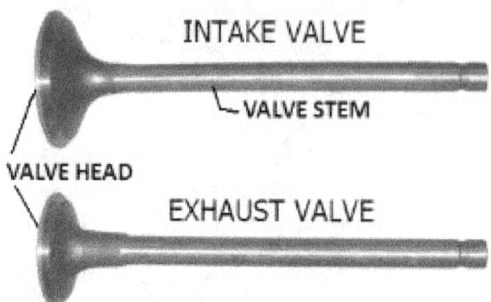

Figure 5-44: Intake and exhaust valves

Valve seats may be formed in the cylinder head or block or may be of removable inserts.

Valve-stem guides fits tightly in the cylinder block and serve as guide to the poppet valve in its motion. The valve lifters or tappets raise the valves in the L-head and T-head types of engines. They derive their motion from the cams mounted on a camshaft or cam gear.

Valve-lifter guide is made of cast iron and serves as guide to the lifter or tappet in its motion.

Rocker arm is made of forged steel or cast iron and is pivoted at about its center. One end contacts the end of the valve stem and the other contacts the upper end of the tappet rod.

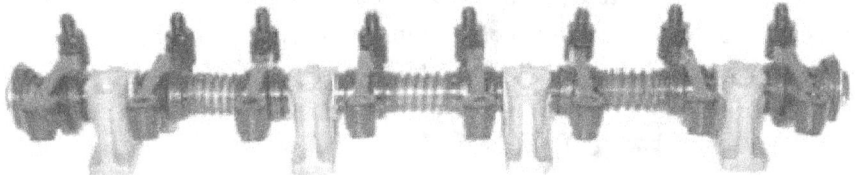

Figure 5-45: The rocker arm assembly

Cam is a wheel with a lobe or projection on its surface. It changes the rotational motion of the camshaft to translation motion in the tappet rod.

Camshaft is the shaft to which cams are attached. They are supported in cast iron, Babbitt or bronze bushings or ball bearings and it is mounted on the crankcase parallel to the crankshaft.

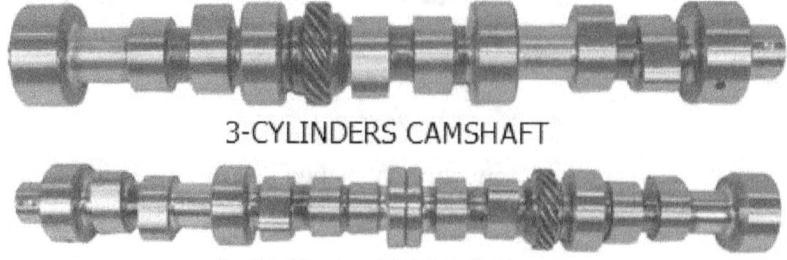

Figure 5-46: Ford/Newholland overhead camshafts

The timing gears are either helical or spur gear type are made of cast iron, cast steel or forged steel or fiber.

Note: The mechanically operated valve is closed and held by a strong spring and is opened by means of a cam. An automatic inlet valve is held closed by a weak spring and opened by atmospheric pressure.

Camshaft arrangements

The type of valve most commonly found in present-day IC engines is the poppet valve. The common valve arrangement found in tractor engines is the overhead arrangement. Two stroke engines usually do not conventional poppet valves, instead the intake and exhaust processes are through ports in the cylinder walls.

- *Single overhead camshaft*: The camshaft is located in the cylinder head (Figure 5-47). The intake and the exhaust valves are both operated from a common camshaft.

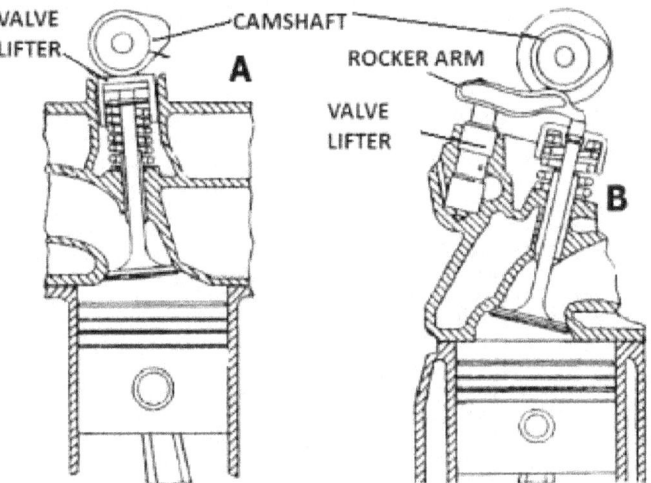

Figure 5-47: Single overhead camshaft

The valve train may be arranged to operate directly through the lifters, as shown in view A, or by rocker arms, as shown in view B. This configuration is becoming popular for passenger car petrol engines.

- *Double overhead camshaft*: When the double overhead camshaft is used, the intake and the exhaust valves each operate from separate camshafts directly through the lifters (Figure 5-48). It provides excellent engine performance and is used in more expensive automotive applications.

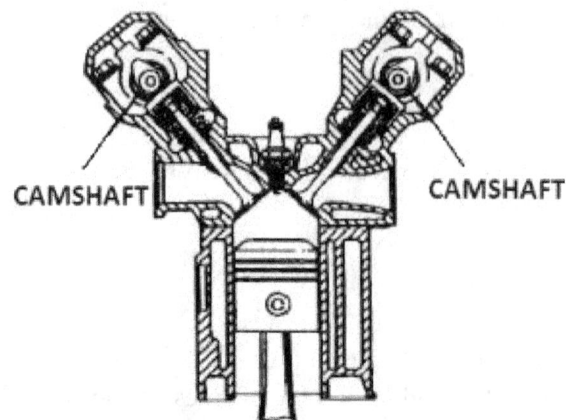

Figure 5-48: Double overhead camshaft configuration

Valve head arrangements

The following are types of valve arrangements are classified according to valve head arrangement and camshaft arrangements. This arrangement equally defines the various types of combustion chamber arrangements in use in engines.

- *L-Head*: The intake and the exhaust valves are both located on the same side of the piston and cylinder. The valve operating mechanism is located directly below the valves, and one camshaft actuates both the intake and the exhaust valves (Figure 5-49).

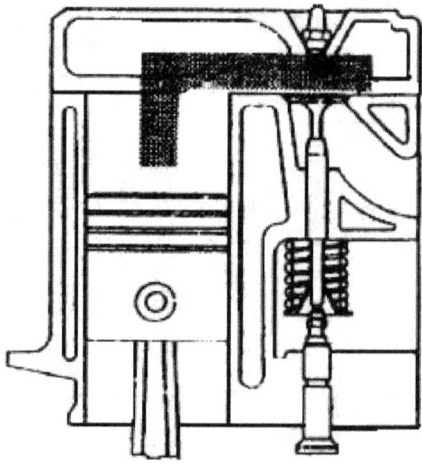

Figure 5-49: L-head engine

- *I-Head*: The intake and the exhaust valves are both mounted in a cylinder head directly above the cylinder (Figure 5-50). This arrangement requires a tappet, a pushrod, and a rocker arm above the cylinder to reverse the direction of valve movement. Although this configuration is the most popular for current gasoline and diesel engines, it is rapidly being superseded by the overhead camshaft.

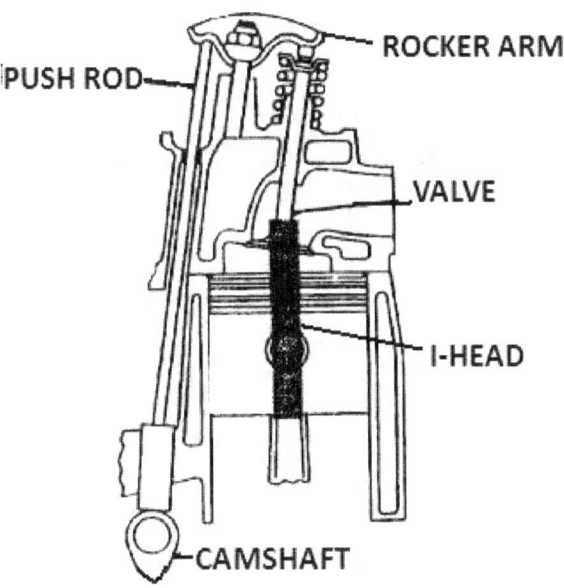

Figure 5-50: I-head engine

- *F-Head*: The intake valves are normally located in the head, while the exhaust valves are located in the engine block (Figure 5-51). The intake valves in the head are actuated from the camshaft through tappets, pushrods, and rocker arms. The exhaust valves are actuated directly by tappets on the camshaft.

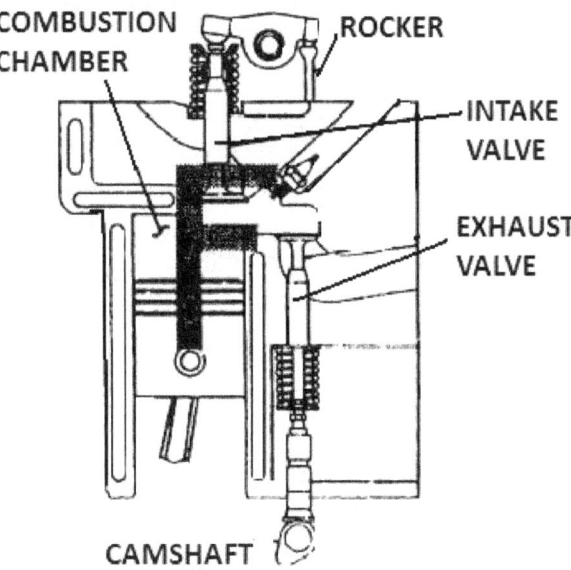

Figure 5-51: F-head engine

- *T-Head*: Intake and the exhaust valves are located on opposite sides of the cylinder in the engine block; each requires their own camshaft (Figure 5-52).

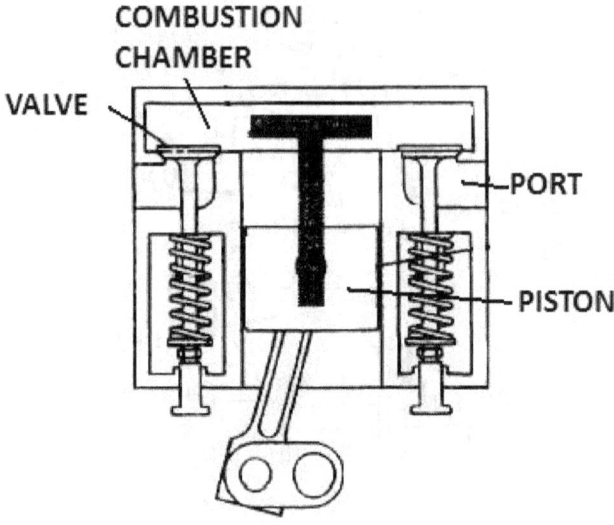

Figure 5-52: T-head

d. *Engine governing system*

The engine control device is usually called a governor. These parts function as a timer for the operating system, programmes the sequence of operation and enhance engine performance. The governor can be mechanical or electronically controlled. In general, governors for internal combustion engines are of the centrifugal-force, spring-loaded type.

Principles of centrifugal action

All governing systems operate on the same principle as the original governor designed by Watt for steam engines. The regulation of a centrifugal governor results from a change in centrifugal force when the speed of rotation changes. For equilibrium to be attained, the masses must assume a new position in response to any change in speed, and this movement controls the device that varies the flow of fuel to the engine, thus restoring its speed to the normal value.

The performance of a governor depends on the design, and certain definitions relating to design will now be discussed.

Stability: A governor is said to be stable when it occupies a definite position of equilibrium and does not oscillate for each speed within its working limits.

Load regulation: Speed of an engine at no load will be higher than the speed at full load. This effect drops off with increase in load (often referred to as speed droop).

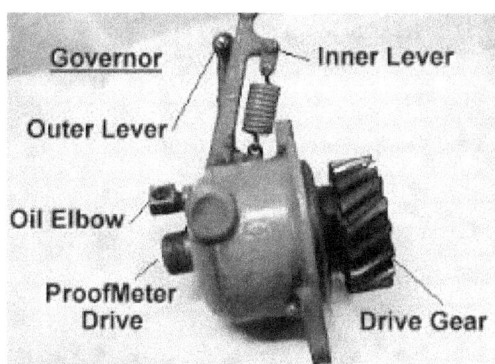

Figure 5-53: Tractor governor system

Speed drop is not an entirely undesirable quality in a governor because it helps prevent overshooting or shunting of the governor when fuel quantity is being corrected during the load changes.

Governor strength: For a simple mechanical governor, the flyballs (masses) must do all the work. The flyballs must not only move to bring about a speed change but also supply the force to move the connecting links to the fuel-metering shaft in the engine.

4. Auxiliary engine systems

Engine auxiliary systems are those systems that enhance engine performance and include the followings:

a. Exhaust system and components

The exhaust system channels the exhaust gases from the engine's combustion chamber to the atmosphere. Exhaust gases leaves the engine in a pipe, traveling through a catalytic converter and a muffler before exiting through the tail pipe.

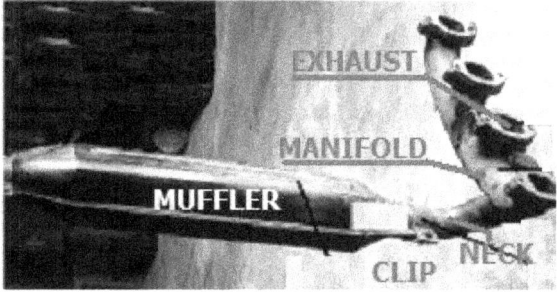

Figure 5-54: Exhaust assembly

The conventional muffler is an enclosed metal tube packed with sound-deadening material to reduce, or muffle engine noise. Most conventional mufflers are round or oval-shaped with an inlet and outlet pipe at both ends. Some contain partitions to help reduce engine noise.

The exhaust system consists of the followings;

- *Exhaust valves*-these allows exhaust gases to pass out of the cylinder.
- *Exhaust manifold* collects exhaust gas from the cylinder.

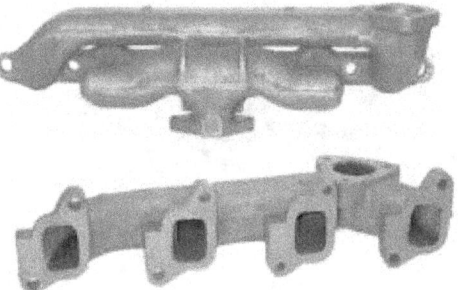

Figure 5-55: Exhaust manifolds

- *Exhaust pipe and muffler:* The pipe conducts hot poisonous exhaust gas coming from the cylinder into the muffler. The muffler deadened the loud noise of the engine exhaust.

Figure 5-56: The muffler and exhaust pipe

- *Exhaust gasket*: Seals the mating surface against pressure leakage

Figure 5-57: The exhaust gaskets

- *Fire arrester*: Present in some automobiles. It puts off fire coming out of the exhaust manifold before sending hot gasses to the muffler.

b. Turbocharging and intercooling systems

Because combustion depends upon both air and fuel, the power of an engine is limited by the amount of air reaching the cylinders. Hence, to increase power, an engine may be supercharged or turbocharged.

A *supercharger* is an engine-driven pump that forces extra air into the cylinders to increase combustion and thus the engine power. A *turbocharger* on the other hand, is an exhaust-driven pump. Turbocharger increases air charge in the intake manifold or an air and fuel mixture to the cylinder for more efficient burning.

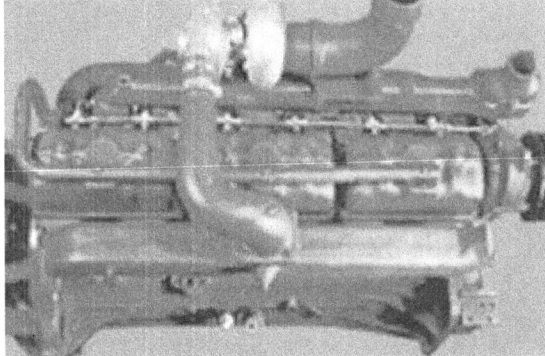

Figure 5-58: A turbocharger and intercooler unit

Exhaust gas is forced through turbine housing to drive a turbine wheel on its way out of the exhaust. The turbine wheel is attached to a shaft that rides on floating bearings and drives a compressor wheel. The compressor wheel pulls air through the air filtration system and channels it through a compressor housing, where it is compressed and directed into the engine intake manifold.

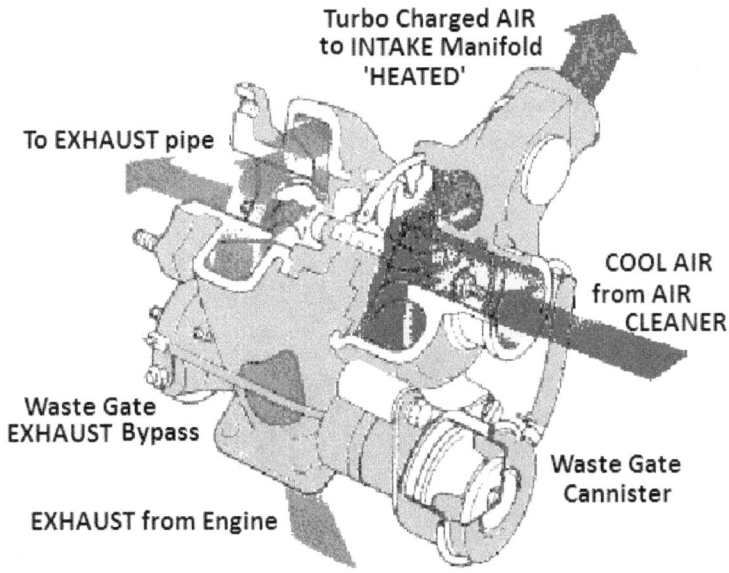

Figure 5-59: Schematic diagram of turbocharger

The *intercooler* cools the combustion chamber as hot air is being returned into the engine from the turbocharger. Aftercooler or intercooler reduces the temperature of air entering the combustion chamber, thereby increasing the density of air and in turn increasing the power output. In most mobile diesel engine installations, the aftercooler also uses the engine water jacket coolant to cool the air.

Using air to cool the air from turbocharger is more effective in lowering the temperature; however, the size of an air-cooled aftercooler may prohibit its use on some mobile equipment. Turbocharger and intercoolers are optional systems in an engine and are commonly found in heavy duty diesel engines.

c. Ignition systems

Unlike steam engines and turbines, internal-combustion engines develops no torque when ignition is turned on, and therefore provision must be made for turning the crankshaft so that the cycle of operation can begin. Small engines are sometimes started manually by turning the crankshaft with a crank or by pulling a rope wound several times around the flywheel.

Types of ignition systems

Breaker point ignition systems: Breaker Point Ignition Systems were, until the advent of electronic ignition systems, used on millions of engines. From the engines powering rum runners of the 1930s to all those Jeeps in World War II, all of them had breaker point systems. Simple to troubleshoot and repair, they are, like anything else, infinitely complex if you don't understand the basics of how they work.

Methods of starting engines (kick mechanism)

Methods of starting large engines include the *inertia starter* and the *explosive starter*. The *inertia starter* consists of a flywheel that is set in motion by hand or by means of an electric motor (Kick) until its kinetic energy is sufficient to turn the crankshaft, while the *explosive starter*, employs the explosion of a blank cartridge to drive a turbine wheel that is coupled to the engine.

Figure 5-60: Starter kick

FARM TRACTOR SYSTEMS

d. *Electrical system*

The functions of the electrical system are to start (switch on) and operate the engine, monitor and control some aspects of the vehicle's operation, and also to power electrical systems and components such as headlights, air conditioner and radio. The automobile depends on electrical system for fuel ignition, headlights, vehicle traffic control signals, horn, radio, windshield wipers, and other accessories.

Battery and the alternator

Battery and the alternator supply electricity for engine use. The battery stores electrical energy for starting the car. The alternator generates electric current while the engine is running, recharging the battery and powering the rest of the vehicle's electrical needs.

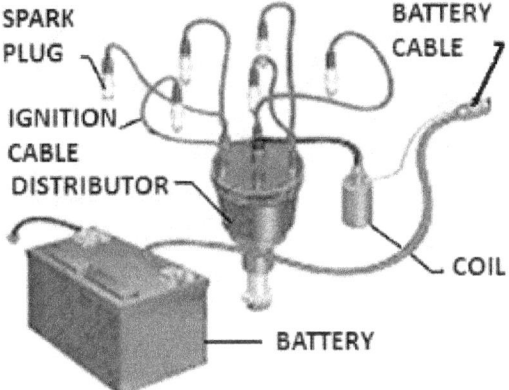

Figure 5-61: Ingition system

When the ignition switch is turned on, the battery sends current to a starter motor that turns the engine's crankshaft. Turning the crankshaft moves pistons up the cylinders, compressing fuel vapour for combustion. While the engine is running, an alternator (electric generator) recharges the battery and supplies power to other electrical components.

Battery maintenance includes cleaning and greasing of terminals. Keep batteries topped up with distilled water. Occasional running of engines keeps the battery charged. If battery becomes low, re-charge using workshop re-charger. Disconnect terminals to avoid alternator damage during re-charging.

e. *Cooling system*

Combustion inside an engine produces temperatures high enough to melt cast iron. Therefore, a cooling mechanism is required to conduct this heat away from the engine's cylinders and radiate it into the air. Cooling systems are frequently overlooked in maintenance checks. 40% of engine problems emanated from inadequate cooling system maintenance. Today's high tech engines demand glycol containing coolant with the additive package and deionized water already premixed to avoid mixing mistakes. Because of the heat of combustion, all engines must be equipped with some type of cooling system either air or water cooling systems.

Air-cooling system

In air cooling system, the outside surfaces of the cylinder are shaped in a series of radiating fins with a large area of metal to conduct heat from the cylinder.

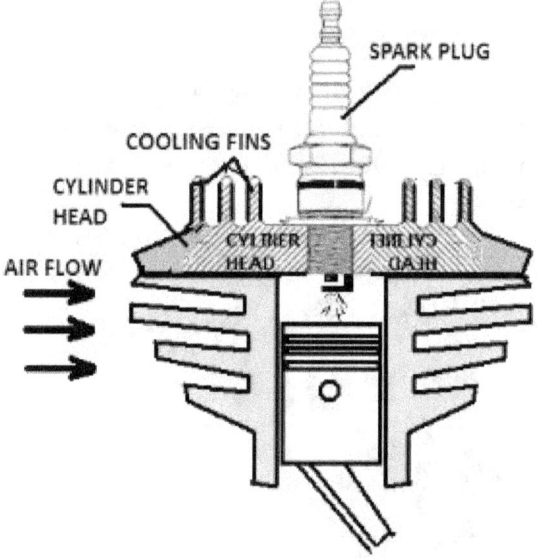

Figure 5-62: An air cooled single-cylinder engine

Air-cooling system is used on some power units on balers and other field machinery and on one cylinder garden tractor and lawn mower engines. In an air-cooled engine, shrouds to fins surrounding the cylinders direct a string blast of air from a fan. The fan is usually incorporated in the flywheel.

Figure 5-63: Air cooled single-cylinder engine block

Water-cooling system

Other engines are water-cooled and have their cylinders enclosed in an external water jacket. In automobiles, water is circulated through the jacket by means of a water pump and cooled by passing through the finned coils of a radiator.

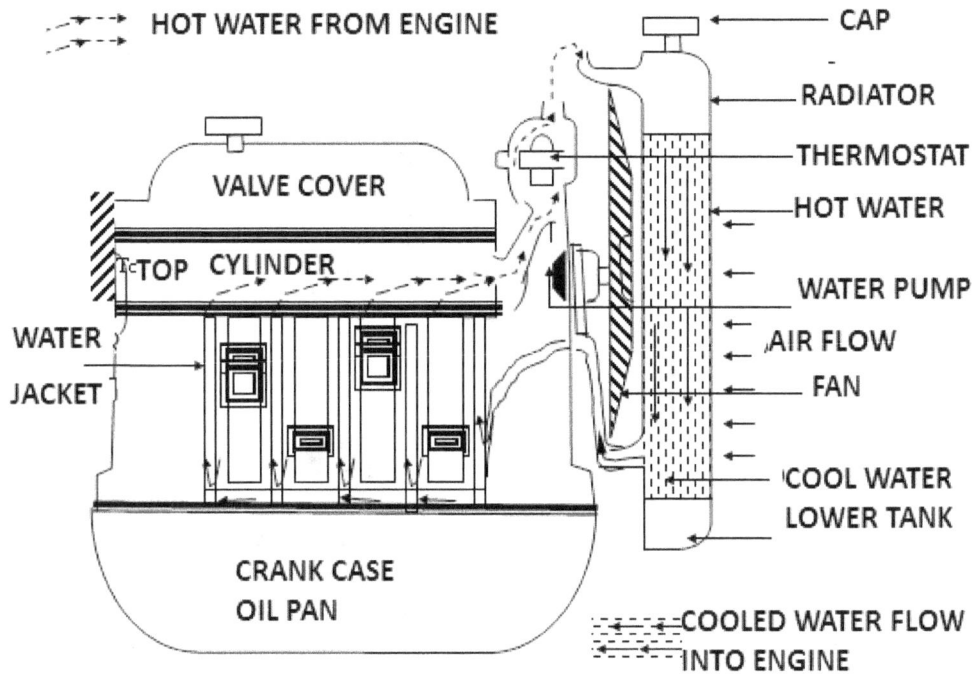

Figure 5-64: Water cooling system

Components and function of water cooling system

Radiator: Radiator is the part of the cooling system in a liquid-cooled engine that removes excess heat from the engine. The radiator is important in internal combustion engines because the engine cannot run properly when it is overheated. At high enough temperatures, oil lubricating the engine's moving parts breaks down and burns away. Eventually, some of these parts jam or melt and the engine stops running.

Fluid flow in radiator: Coolant movement from the radiator top tank through the radiator water jackets to the lower tank happens in two phenomena: gravity flow and thermo-siphon flow.

1. *Gravity flow*: When coolant is poured into the radiator, it flows under its own weight (*gravity flow*) to occupy empty spaces in the lower radiator tank and the engine hoses. When the spaces are filled up, and the engine is running, the water pump pumps the water through the engine and return hot water to the top tank of the radiator.
2. *Thermo-siphon flow:* The force-effect of the pump is not enough to drive the pressure built up in the top tank but a process of *thermo-siphon flow* is enforced as follows: The force (thermo-flow) of the hot water (with light weight) returning from the engine to the

radiator and the suction effect (siphon) of the water pump forces the water to flow in a process known as thermo-siphon flow.

Water jackets: In a liquid-cooled engine, a coolant—usually a mixture of water and chemicals—circulates through hollow chambers (water jackets) that surround the engine's cylinders. Heat produced by burning fuel is transferred from the metal into the coolant.

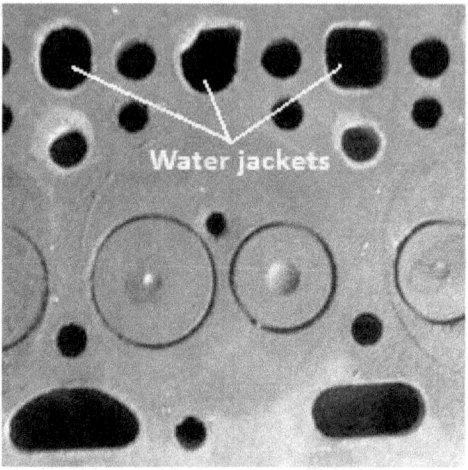

Figure 5-65: Water jackets on the cylinder head

Water pump: A water pump circulates coolant through the engine to the radiator. Hot coolant arrives at the radiator, which exposes the coolant to cooler metal and lowers its temperature. Afterwards, the coolant is pumped through the engine to repeat the cycle.

Figure 5-66: Water pump

Coolant: Coolants normally consist of a mix of water, antifreeze, and a conditioner or inhibitor. A coolant must be able to:

- Effectively remove heat
- Prevent freezing and resultant block damage
- Prevent deposits of scale and sludge on interior passages
- Inhibit corrosion

- Prevent cavitation erosion. Cavitation occurs when the amount of fluid flowing into the pump is restricted or blocked. This is the result of vapour bubbles imploding.
- Lubricate components such as water pumps
- Be compatible with hoses and seals

Operators should know that a 50-50 mixture of ethylene glycol antifreeze and water provides good heat transfer and freeze protection. (It's usually yellow or green. Propylene glycol (a brand of antifreeze agent, pink in colour) is nontoxic for drinking water systems but transfers heat poorly and should not be used as a coolant. Unfortunately, the colour does not always indicate the contents, so read the label.)

Engine operating conditions

Correct operating temperature: An engine cannot last long or operate efficiently when it is not operated at the correct temperature. The engine should operate at a temperature high enough so that water will not quickly boil away in the top of the radiator. At this temperature, wear is reduced to a minimum, fuel consumption is reduced and power is increased. Most tractor operators have observed the lack of power with a cold engine when starting ploughing operation or doing some similar heavy work.

Overheating: This is excessive increase in radiator water during engine operation. Water in the radiator could rise above the boiling point up to superheated temperatures in extreme cases. This is an indication of rapid boiling away of the water, with resultant effect of increase in engine operating temperature which could lead to knocking sound and loss of power. There are cases of overheating causing rapid damage to the engine.

If an engine over heats, it should be shut down immediately or serious damage may result. Idle the engine a few minutes to permit exhaust valve and piston to cool so that expanded piston will not seize and valves will not warp or crack because of overheating.

Causes of overheating: Engine overheating has been ascribed to one or multiple of these causes;

1. Low water level in the radiator
2. Using a fuel that knocks
3. A slipping fan belt
4. Overloading of the tractor
5. Clogged radiator fins
6. Collapse radiator losses
7. Dirt accumulation in the cooling system
8. Ignition out of time
9. Carburetor out of adjustment (too lean a mixture)
10. Faulty water pump
11. Thermostat stuck closed.

FARM TRACTOR SYSTEMS

e. Lubrication system

Heat produced when an engine runs must be removed or the engine will overheat and eventually seize. One way to do this is by removing friction between two contacting surfaces and quick exhaust waste gases discharge from the combustion chamber. Engine lubrication is the greatest single factor contributing to the life of any tractor/automobile and also to its satisfactory operation. The properties of engine oil and the design of modern engines allow the lubrication system to accomplish lubrication functions.

Process of oil lubrication

Oil lubrication of a shaft rotating on a bearing in three processes as follows:

Step 1: Oil begins to flow into and around a shaft resting on a bearing,

Step 2: Oil entering rotating bearing shaft,

Step 3: Oil has wedged shaft and bearing.

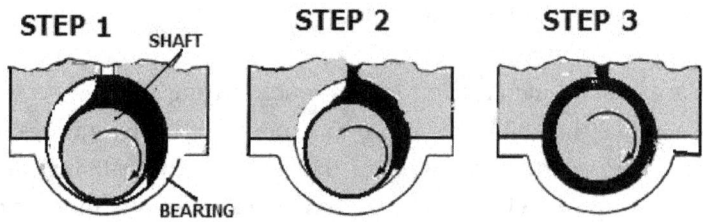

Figure 5-67: Oil lubrication process

Types of lubrication

Lubrication system differs from engine to engine depending on the manufacturer and engine capacity. Generally, lubrication can be achieved in engines in the following four ways:

a. Splash circulation

In the splash lubricating system (Figure 5-69), oil is splashed up from the oil pan or oil trays in the lower part of the crankcase. The oil is thrown upward as droplets or fine mist and provides adequate lubrication to valve mechanisms, piston pins, cylinder walls, and piston rings.

The splash system is no longer used in automotive engines. It is widely used in small four-cycle engines for lawn mowers, outboard marine operation, and so on.

Figure 5-69: Splash-type lubrication system

b. *Combination splash and force feed*

In a combination splash and force feed system (Figure 5-70), oil is delivered to some parts by means of splashing and other parts through oil passages under pressure from the oil pump. The oil from the pump enters the oil galleries. From the oil galleries, it flows to the main bearings and camshaft bearings.

The main bearings have oil-feedholes or grooves that feed oil into drilled passages in the crankshaft. The oil flows through these passages to the connecting rod bearings. From there, on some engines, it flows through holes drilled in the connecting rods to the piston-pin bearings. Cylinder walls are lubricated by splashing oil thrown off from the connecting-rod bearings.

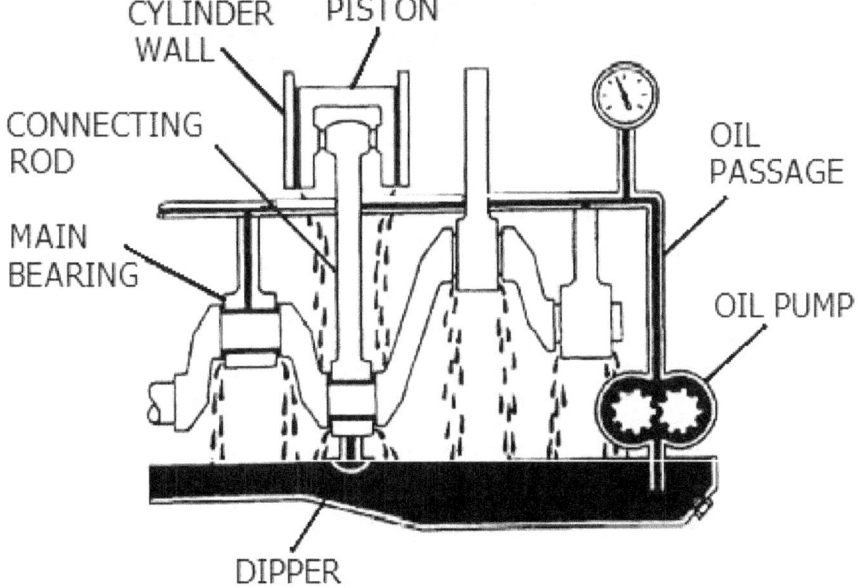

Figure 5-70: Splash-force feed-type lubrication system

Some engines use small troughs under each connecting rod that are kept full by small nozzles which deliver oil under pressure from the oil pump. These oil nozzles deliver an increasingly heavy stream of oil as speed increases. At very high speeds these oil streams are powerful enough to strike the dippers directly. This causes a much heavier splash so that adequate lubrication of the pistons and the connecting-rod bearings is provided at higher speeds. If a combination system is used on an overhead valve engine, the upper valve train is lubricated by pressure from the pump.

c. *Force feed*

A more complete pressurization of lubrication is achieved in the force-feed lubrication system (Figure 5-71). Oil is forced by the oil pump from the crankcase to the main bearings and the camshaft bearings.

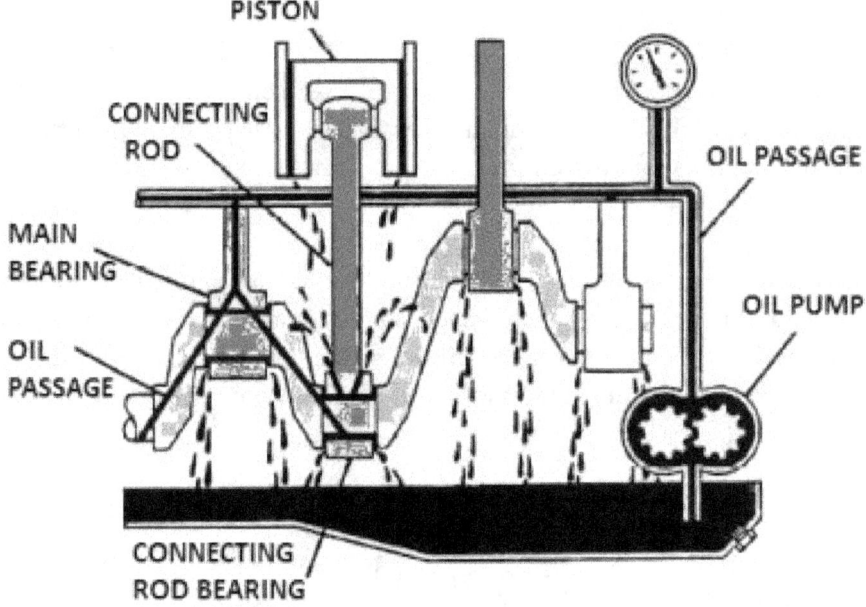

Figure 5-71: Force-feed lubrication system

Unlike the combination system the connecting-rod bearings are also fed oil under pressure from the pump. Oil passages are drilled in the crankshaft to lead oil to the connecting-rod bearings. The passages deliver oil from the main bearing journals to the rod bearing journals. In some engines, these openings are holes that line up once for every crankshaft revolution. In other engines, there are annular grooves in the main bearings through which oil can feed constantly into the hole in the crankshaft.

The pressurized oil that lubricates the connecting-rod bearings goes on to lubricate the pistons and walls by squirting out through strategically drilled holes. This lubrication system is used in virtually all engines that are equipped with semi floating piston pins.

d. Full force-feed

In the full force-feed lubrication system (Figure 5-72), the main bearings, rod bearings, camshaft bearings, and the complete valve mechanism are lubricated by oil under pressure.

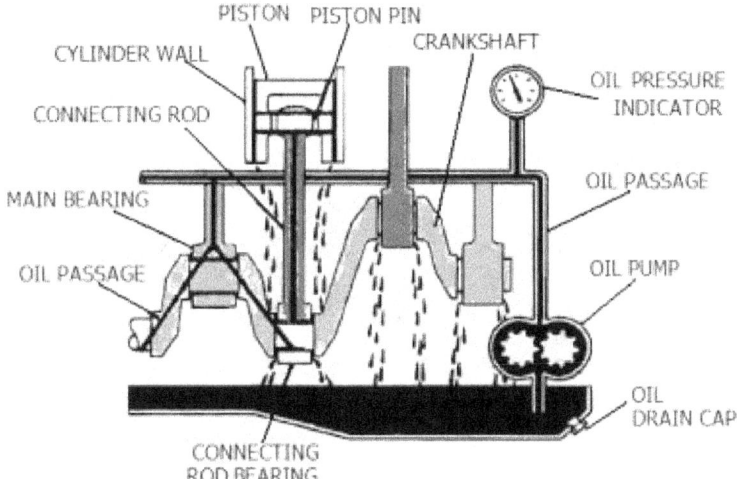

Figure 5-72: Full force-feed lubrication system

In addition, the full force-feed lubrication system provides lubrication under pressure to the pistons and the piston pins. This is accomplished by holes drilled the length of the connecting rod, creating an oil passage from the connecting rod bearing to the piston pin bearing. This passage not only feeds the piston pin bearings but also provides lubrication for the pistons and cylinder walls. This system is used in virtually all engines that are equipped with full-floating piston pins.

e. Oil mist application

Oil mist lubrication is common in 2-stroke engines. The oil is mixed with fuel in required mix ratio of 50ml to 1 liter of fuel. The oil mixture is taking in through the intake port into the crankcase where it is compressed into the combustion chamber. The mixture lubricates the crankshaft and the connecting rod and other moving parts as fuel is supplied into the engine.

Components of the lubrication system

The lubrication system comprises of the following parts:

a. Oil filter

The oil filters removes dirt, wear particles and also conducts heat away from the engine. Oil filters that are used in automobiles are of the disposable type (Spin-on type) while most tractors filters are of the replaceable cartridge type.

Spin-on filter (Figure 5-73) is completely self-contained, consisting of an integral metal container and filter element. This type of filter is screwed onto its base and is removed by spinning it off.

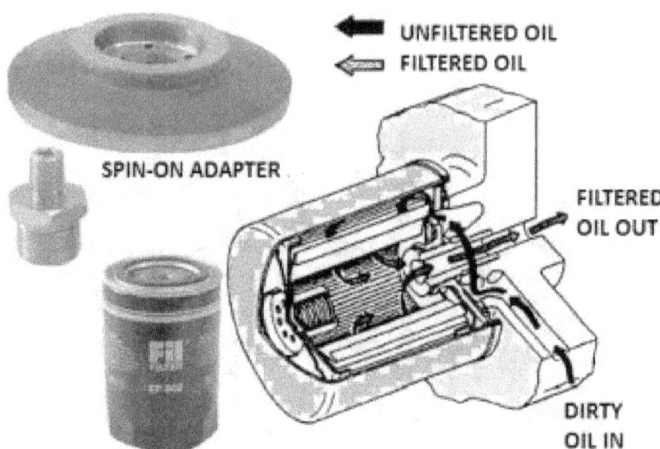

Figure 5-73: spin-on filter

Cartridge-type element (Figure 5-74) fits into a permanent metal casing. Oil is pumped under pressure into the container where it passes from the outside of the filter element to the center. From here, the oil exits the container. The element is changed easily by removing the cover from the container.

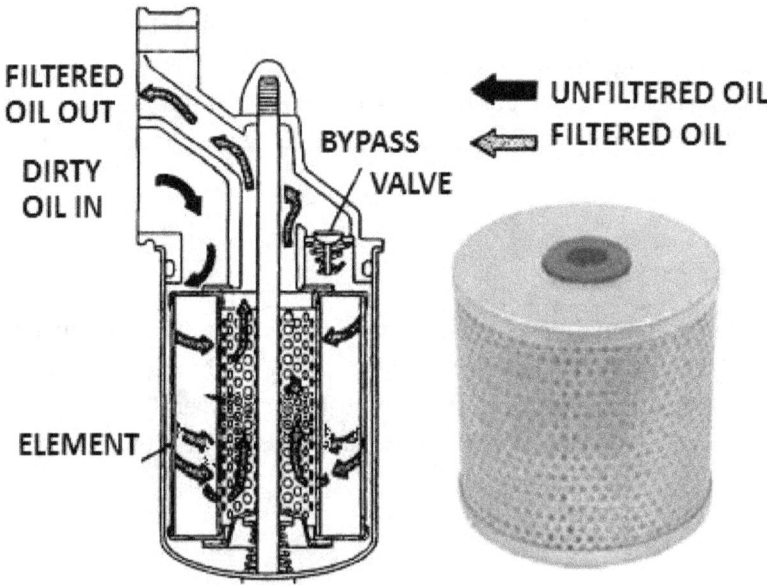

Figure 5-74: Cartridge-type filters

The elements may be paper folded or it may be of some kind of cloth or waste with a porous case around it. The paper type is most common. The filter element must fit the filter case to function properly. Oil filters are installed in two configurations either as bye-pass flow or full-flow installation.

Oil filter configurations

There are two types of filter configurations used in engine lubrication systems. These are the full-flow system and the bypass system. Operations of each system are as follows:

Bypass flow system (Figure 5-75) diverts only small quantity of oil each time it is circulated and returns it directly to the oil pan after it is filtered. This type of system does not filter the oil before it is sent to the engine.

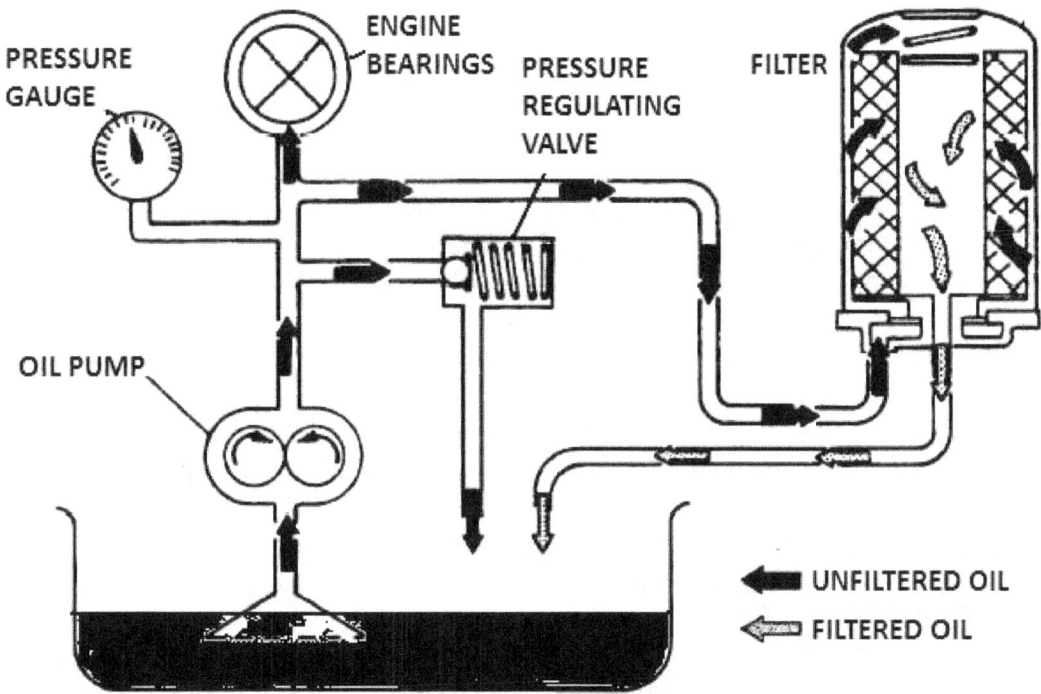

Figure 5-75: Bypass flow system configuration

The oil from the main oil gallery enters the filter and flows out through the filter element into the collector in the center of the filter. The filtered oil then flows out through a restricted outlet preventing the loss of pressure. The oil then returns directly to the oil pan.

Full-flow system (Figure 5-76) is the more common of the two configurations. All oil in a full-flow system is circulated through the filter before it reaches the engine. When a full-flow system is used, it is necessary to incorporate a bypass valve in the oil filter to allow the oil to circulate through the system without passing through the element in the event that it becomes clogged. This prevents oil supply from being cut off to the engine.

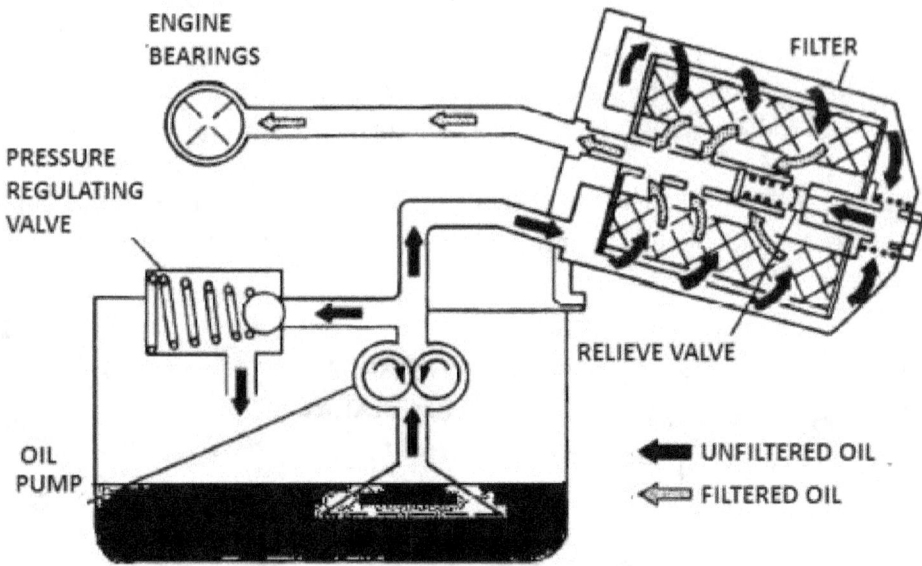

Figure 5-76: Full flow system configuration

b. *Oil pump*

The oil pump sucks in oil from the oil pan through a sieve and then circulates it to other parts and components of the engine through the oil ways and galleries. There are two basic types of oil pumps—rotary and gear.

Figure 5-77: Oil pump

Rotary pump (Figure 5-78) has an inner rotor with lobes that match similar shaped depressions in the outer rotor. As the oil pump shaft turns, the inner rotor causes the outer rotor to spin. This action squeezes the oil and makes it spurt out under pressure. As the pump spins, this action is repeated over and over to produce a relatively smooth flow of oil.

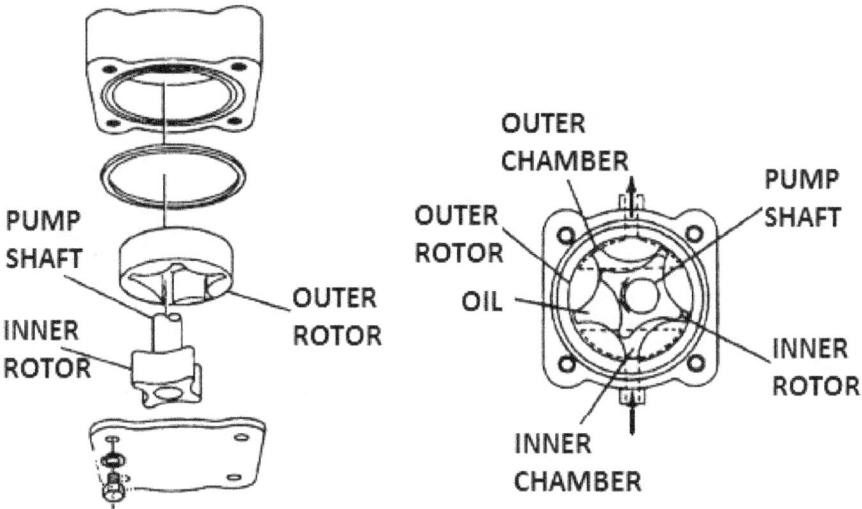

Figure 5-78: Rotor-type oil pump

Gear pump (Figure 5-79) consists of two pump gears mounted within a close-fitting housing. A shaft, usually turned by the distributor, crankshaft, or accessory shaft, rotates one of the pump gears. The gear turns the other pump gear that is supported on a short shaft inside the pump housing.

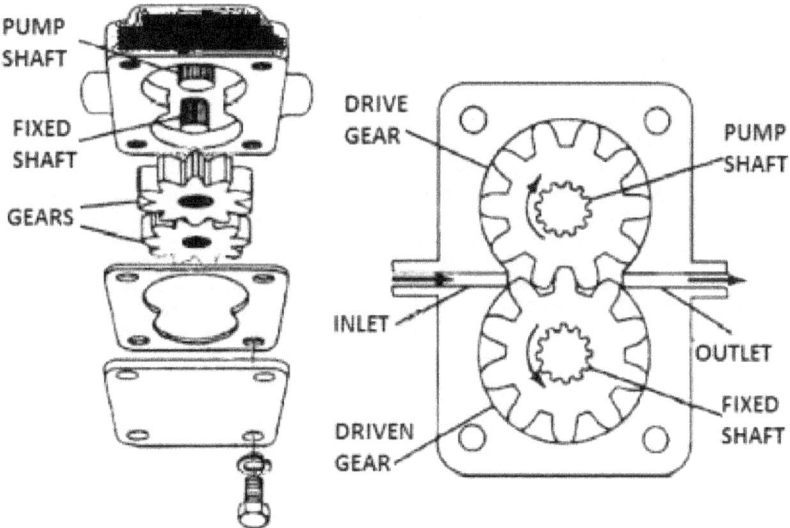

Figure 5-79: gear-type oil pump

c. *Oil level gauge/dip stick*

The dip stick is used to measure the quantity of the oil in the oil pan. The tip of the dip stick has two graduations marked on it indicating the minimum oil level, and the maximum oil level in the oil pan.

FARM TRACTOR SYSTEMS

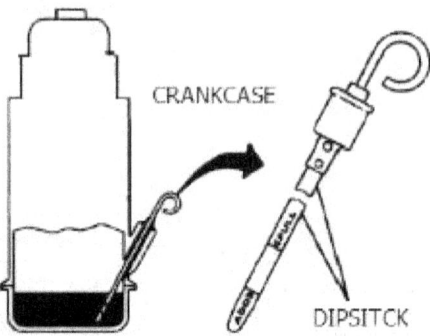

Figure 5-80: oil level gauge

When the oil level is above the maximum oil mark, there is likelihood of pressure build-up and over pumping of oil which could lead to oil waste. When the oil level is below the minimum oil level on the dip stick, the pump will be operating at low pressure thus not pumping oil to the essential parts of the engine. This could cause excessive engine heating, valve tapping, and knocking sound.

d. *Oil pan*

This is the reservoir or storage area for engine oil. The oil pan houses the oil with the lower part of the crankshaft completely immersed in the oil. It has a draining plug trough which used oil is drained without removing the oil pan at each service. The dip stick hangs freely in the oil pan and not touching the pan or any other components.

Figure 5-81: Lip seal style oil pan

Oil pan (bottom plate) gasket

The surface between the oil pan and the crankcase is sealed by fiber or gasket reinforced with metal to prevent leakage of oil.

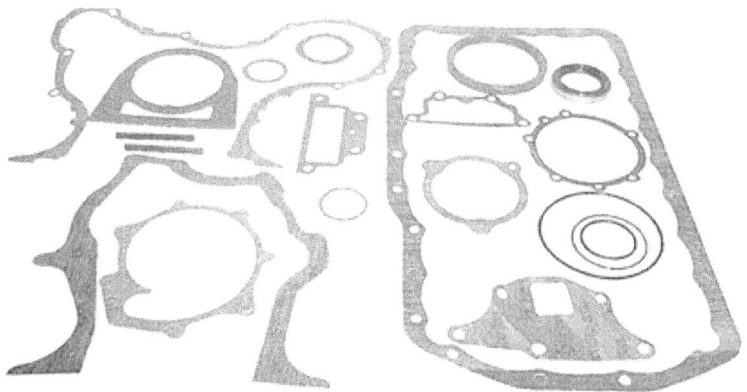

Figure 5-82: 4-cylinder bottom gasket set and seals

e. *Oil Pickup and strainers*

The oil pickup is a tube that extends from the oil pump to the bottom of the oil pan. One end of the pickup tube screws into the oil pump or to the engine block. The other end holds the strainer.

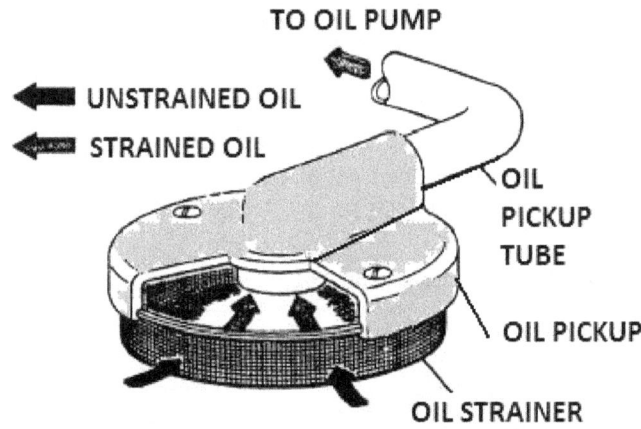

Figure 5-83: Oil pickup and strainer

The oil pickup provides the channel through which oil is supplied to the pump while the strainer removes large particles. The strainer has a mesh screen suitable for straining large particles from the oil and yet passes a sufficient quantity of oil to the inlet side of the oil pump. The strainer is located so that oil entering the pump from the oil pan flows through it. Strainer may be either the floating or the fixed type.

Floating strainer has a sealed air chamber that is hinged to the oil pump inlet, and floats just below the top of the oil. As the oil level changes, the floating intake will rise or fall accordingly. This action allows oil taken into the pump to come from the surface. This design prevents the pump from drawing oil from the bottom of the oil pan where dirt, water, and sludge are likely to have settled. The strainer screen is held to the float by a holding clip.

Fixed strainer (Figure 5-84) is simply an inverted funnel-like device, placed about 1/2 inch to 1 inch from the bottom of the oil pan. This device prevents any sludge or dirt that has accumulated from entering and circulating through the system. The assembly is attached solidly to the oil pump in a fixed position.

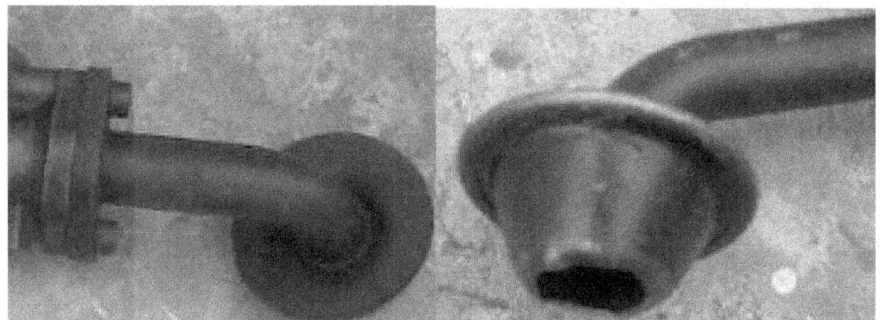

Figure 5-84: A fixed oil strainer

f. *Oil galleries*

Oil galleries are small passages within the cylinder block corresponding with those within the cylinder head for lubricating oil circulation. They are cast or machined passages that allow oil to flow to the engine bearing and other moving parts. The main oil galleries are large passages through the center of the block. They feed oil to the crankshaft bearings, camshaft bearings, and lifters. The main oil galleries also feed oil to smaller passage running up to the cylinder heads.

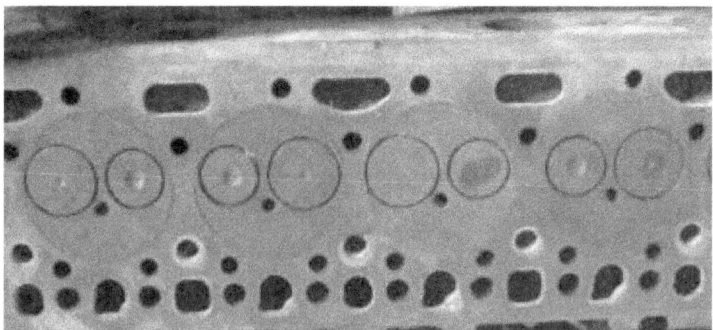

Figure 5-85: Intake and exhaust valve arrangement

g. *Other oil lubrication accessories*

Oil pressure gauge registers actual oil pressure in the engine. The gauge indicates how regularly and evenly the oil is being delivered to all vital parts of the engine and warns of any stoppages in this delivery. Pressure gauges may be electrical or mechanical.

Oil temperature regulator controls engine oil temperature on diesel engines. The oil temperature regulator (Figure 5-86) in diesel engine lubricating systems prevents oil temperature from rising too high in hot weather, and assists in raising the temperature during cold starts in wet weather. It provides a more positive means of controlling oil temperature than does cooling by radiation of heat from the oil pan wells.

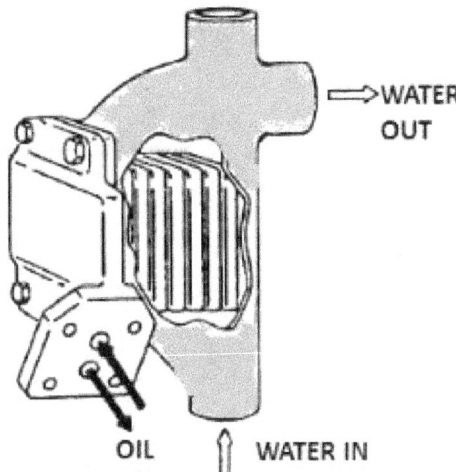

Figure 5-86: Oil temperature switch

Oil pressure indicator/switch warns the operator of low oil pressure. Oil pressure switch sound a continuous alarm when the pressure is low. Because the engine can fail or be damaged in less than a minute of operation without oil pressure, the warning light is used as a backup for a gauge to attract instant attention to a malfunction.

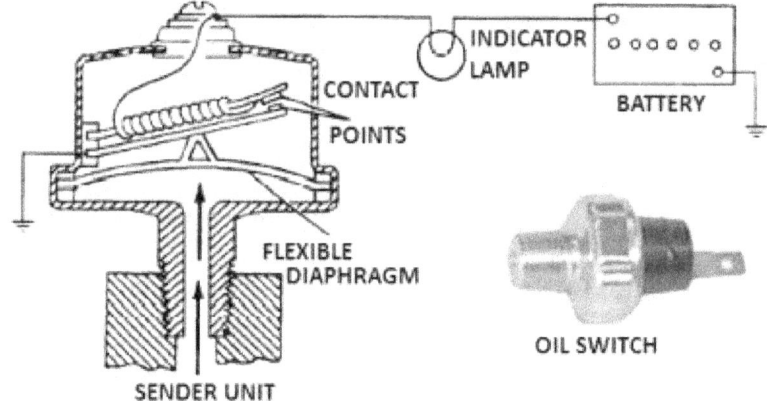

Figure 5-87: Oil pressure light and switch

Lubricating oil and oil classification

a. *Function of lubrication oil*

The engine lubricating oil performs the following main functions in an engine system:

- They help extend the life of the engine.
- It reduces friction between moving parts in the engine and provides a shield against all forms of wear and reduces wears much as possible.
- It lubricates the cylinder walls and piston rings and forms a seal between them.
- It absorbs heat and help conducts it to the cooling system.
- It acts as a cushion to absorb chock loads and helps deadened engine noise.

Other lubricant benefits include:

- Helps maintain fuel economy,
- Helps reduce oil consumption,
- Outstanding soot control, and
- Consistent oil pressure for thorough engine lubrication.

b. Oil Classification

Oils are classified according to their viscosity (thickness). Oil viscosity, also called oil weight, is the thickness or fluidity (flow ability) of the oil. Classification according to viscosity is the most prevalent method of describing oils, and the most common classification systems are those of the SAE (Society of Automotive Engineers), API (America Petroleum Institute), and ISO (International Standard Organization). Each organization uses a different kinematic viscosity range numbering systems. In the SAE classification, the center of the donut shaped logo indicates the oil viscosity. The higher numbers indicates thicker oil. Thin oil has a low viscosity number and thick oil has a high viscosity number. The viscosity range includes:

SAE 20 and SAE 50   -   Engine oil

SAE 40 and SAE 50   -   Transmission (gear) oil

SAE 70- SAE 140   -   very thick oil, sometimes used for transmission

An indication of oil classified with a 10W is described as follows

- 10 indicates the viscosity or number rating and,
- "W," indicates the oil viscosity when the engine is cold during **W**inter (dry season) conditions. A number rating without a "W" such as SAE- 40 describes the oil viscosity during normal engine temperatures in non-winter (wet) conditions.

Single viscosity oil

Single viscosity oil is the oil that operates at a given temperature under a given operating condition. Example includes 5W or SAE 20W. Viscosity of each oil increases in this order: 5W,

10W, 20W etc. small engines used in lawn mowers, generators and even some larger diesel engines or racing engines "require" either a SAE 30 or SAE 40 oil. This is because they operate under extremely high speed, load and heat.

Multi-viscosity oils

There are other oils with dual viscosity rating. They are called multi-grade oils and meet the requirements of one viscosity during warm weather temperatures and a lower viscosity in the cold weather. The oil becomes thicker as its temperature falls.

Multi-viscosity petroleum oils are full of viscosity improvers (VI's) that create weak links among the oil molecules. As these agents are subjected to heat, load and high speed, the oil cannot reach its intended high temperature viscosity, which could results in shearing and thus results in loss of oil film strength, increased wear rates, temperatures and oil consumption issues.

Multi-viscosity oil, such as 10W-30, means that the oil will flow and pour like a 10W-weight oil at very low temperatures yet maintain viscosity similar to 30-weight oil at operating temperatures because of the oil's naturally high viscosity or it is additive improved. Multi-viscosity oils are often used in automobiles operating under low temperatures.

Oil operating temperatures

When 5W-30 oil is under high shear loads at temperatures above $225°F$, it can constitute problem. It becomes a potentially damaging problem when oil temperature approaches $300°F$. If the oil temperature in the pan is $250°F$ or above, the corresponding oil temperature in bearings will be approaching $300°F$.

Lubricants with narrow viscosity range could have problem of relatively high evaporation rate and high temperatures while working under severe service conditions. A higher evaporation rate contributes to increased oil consumption and substandard lubrication in areas of high temperature. Actually a great deal of personal judgment must be used when selecting the seasonal grade of oil to put into the engines.

Oil treatment

Oil treatment is specially formulated to keep old engine running like a new one with oil viscosity improvers, anti-friction and oxidation additives. Oil treatment help to stop oil-burning, engine wear and oil foaming. It serves the following functions in an engine:

- It seals worn rings to help improve engine compression.
- It fights engine oil breakdown and thinning at high temperatures.
- It helps worn engine run smother and quieter.

- It added engine protection to new and old vehicles.
- It blends with all petroleum based oils.

Adding oil treatment during servicing

In 2-cycle engines: Add 10% oil treatment to oil and mix with fuel in the normal ratio as instructed.

During routine oil change oil: Pour one full can (443ml) of oil treatment into the crankcase when your engine is idling and has warmed up to the operating temperature.

For engines with excessive wear, use 2 cans of 15 oz (443ml) fluid cans of oil treatment.

In rebuilt (overhauled) engine: Mix 50% oil treatment with 50% engine oil to coat parts or components before assembling.

Oil contamination

Oil contamination is the degrading of oil by the presence or infiltration of foreign particles into the oil. These foreign particles are referred to as contaminants. The following sources of contamination have been identified

Sources of oil contamination

1. *Blow-by*: The term *blow-by* connotes the escape of compression and combustion gases past the pistons and piston rings into the crankcase (Figure 5-88). All engines have some degree of blow-by. Excessive blow-by is usually detected by higher oil consumption.

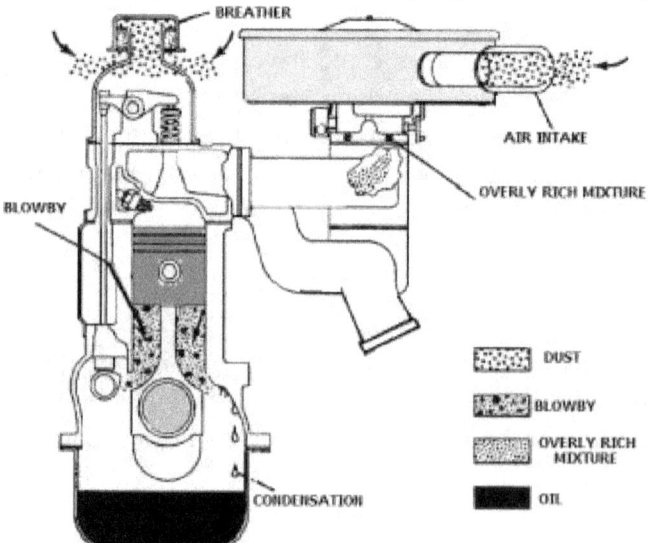

Figure 5-88: Sources of contamination in engine

2. *Particles of natural or aspirated air*: Air intake and crankcase breather are regular channels through which oil can be contaminated. Some small particles of natural or aspirated air escaping from the air filter easily mixed with oil and could cause contamination.
3. *Condensation*: Moisture condensation in the crankcase is another source of oil contamination. Excessive oil burn could cause condensation within the crankcase when the engine cools down. *Note:* The practice of a quick few moment idle run of the engine without full warm-up may cause more harm than good, as metal temperatures at such short time are insufficient to drive off water vapour condensates on the metal surfaces.

Complete warming-up of engine systems before loading the system helps to keep bare metal surfaces covered with protective oil film and drive off condensed water vapour in the oil. Thus engines and other system should be thoroughly warmed-up periodically during periods of no usage.

Effect of oil contamination

Oil contamination has significant effect on oil as well as engine performance. Such effects include

1. Reduced oil viscosity leading to poor lubrication
2. Excessive engine component wear
3. Noisy engine operation
4. Increased engine operating temperature
5. Loss of power/compression

Sludge formation

Sludge is primarily formed when the engine oil is contaminated as a result of a very complex interaction of components which include mechanical and thermal stress and multitude of chemical reactions.

Types of sludge

There are different "types" of sludge with different appearance ranging from light brown to opaque black; they range from semi-liquid to solid. "*Black sludge*" is typically found in rocker cover, cylinder head, timing chain cover, oil sump, oil pump screen, and oil rings in variable quantities.

Sludge in gasoline engines is usually black emulsion of water and other combustion by-products, and oil formed primarily during low-temperature engine operation.

Sludge in diesel engines is soot combined with other *combustion* by-products which can thicken the oil to gel like sludge. This sludge is typically soft, but can also polymerize to very hard substance.

Causes of sludge

Sludge formation is as a result of one or more of these factors:

1. *Severe service driving*: The term "severe service" refers to:

    a. *Making short trips*
    b. *Stop and go driving*: Slow driving speeds and long idling periods lead to high under hood temperatures due to limited air flow
    c. *Extended idling*: sitting in traffic, delivery truck operation
    d. *High temperature operation* such as driving at high ambient temperatures, towing and driving at maximum engine power output (high speeds or uphill)
    e. *Extreme cold operation* such as starting engine below $0°F$ and engine coolant and engine oil never reach "normal" operating
    f. *Heavy loads* such as operating in hilly regions or trailer towing

Severe conditions are not all that uncommon. It is estimated that we operate our tractors 80% of the time under one of above "severe service" conditions!

2. *Mechanical malfunctions:* Mechanical malfunctions such as:

    a. Leakage of coolant from cooling system into oil in the crankcase can create harmful deposits.
    b. If thermostat sticks and does not allow coolant flow when needed, the engine will run either too cold or too hot.
    c. Constant elevated temperatures promote oil thickening after thousands of miles and can create sludge.

3. *Oil factor*: As complex as the issue of sludge formation is, the most common cause of the sludge problem is the oil itself. The followings are the enemies of your engine oil and contributors to sludge formation:
    i. Soot,
    ii. Heat,
    iii. Fuel,
    iv. Water,
    v. Acid,
    vi. Dirt,
    vii. Blow-by and

viii. Engine coolant.

Each of these factors deserves attention.

Effect of sludge in engine and control

Effect of sludge in engine include

1. Plugging of oil filters, oil lines and filter screens, and
2. Accelerated wearing of engine parts.

Sludge control in engine

Sludge deposits can be controlled with a dispersant additive that keeps the sludge constituents finely suspended in the oil.

Practical exercise

1. In the workshop, identify various parts on components on engine display table; make a full sketch of a particular tractor engine component and label appropriately. Observe their constructional features.
2. Identify the various components of any selected operating systems in tractor engine, give full-scale description and sketches. Report your work in the work book.
3. Perform lubrication system servicing and service the air cleaner systems. Identify the parts serviced

CHAPTER 6

Tractor Power Transmission Systems

Introduction

The descriptions of tractor engine systems, component and functions have been discussed in previous chapter. This chapter focused on the description of various power transmission elements and mechanisms found in tractor. The tractor still remains the most important farm machine because of its transmission system which is a principal component accounting for about 25-30% of the total initial cost of tractor (Renius and Rainer, 2005). Tractor power transmission has developed from the stepped transmissions used since the beginning of tractor revolution to infinitely variable drives with automatic controls introduced in Europe in 1996 for standard tractors thereby opening a new era of power train designs.

Tractor power drive systems

Power is transmitted in tractors and agricultural machinery through three basic ways:

1. Mechanically through friction drives including wheel/pulley and belt, chain and sprocket, shaft and gear etc.
2. Through hydraulic power drives e.g. hydrostatic and hydrodynamic systems and
3. Power shafts e.g. PTO shaft, transmission shafts etc.

Mechanical transmission (Friction drives)

When two surfaces meant to transmit power are joined together without slippage, motion is generated in the same direction.

Friction drives are mechanical devices that generate and transmit power when two surfaces come in contact either directly or indirectly. For instance, a friction drive connecting two shafts by a belt that is drawn tight enough to grip each shaft will turn them. If each shaft has a pulley attached to them, the belt is pulled tight to create friction with the pulleys. As the drive pulley turns, the belt moves the connected driven pulley.

Examples of friction drives include: belt and pulley, chain and sprocket, gears and wheels.

Belts and pulley drives

Belt drive is a method of transferring rotary motion between two shafts. A belt drive includes one pulley on independent shafts and one or more continuous belts over the two pulleys. The motion of the driving pulley is, generally, transferred to the driven pulley via the friction between the belt and the pulley. The success of the friction drives depends on maintaining frictional contact between belt and pulleys.

Figure 6-1: Belt and pulleys

Belt drive friction is dependent on several factors. These factors include:

- Tight side and loose side belt tensions.
- The coefficient of friction between the belt material and the pulley material.
- The total angle of contact around the pulley, which depends on the pulley diameters and their distance apart.
- The centrifugal force lifting the belt off the pulley produced by the rotation of the pulley.
- The angle of the Vee in the pulley, which acts to wedge the belt in place. It is usually between $17°$ and $19°$.

Types of belt

Basically, there are four types of belts in use in power transmission; flat belt, vee-belt, round belt and cog belts.

Flat belt

In the past, flat belts have been widely used to drive machines in factories. They are convenient to install and operate and are reliable. However, in the modern times, machines are being driven individually generally using electric or hydraulic drives.

Figure 6-2: Flat belt

Flat belt drives are now mostly used in low power high speed applications in specialized industries including the textile plants, paper making processes, and in office machinery. Flat belts are also used in conveyor applications, they are susceptible to slip. Flat belt transfers torque by belt friction over pulley. Flat belts are made from leather, woven cotton, rubber, and balata.

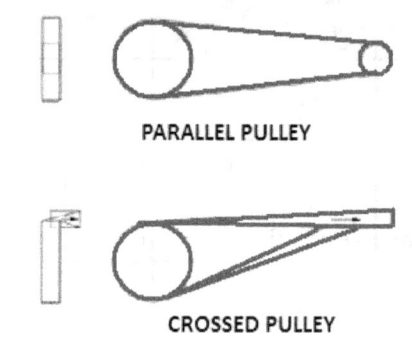
Figure 6-3: Types of flat belt arrangements

Timing/synchronous belt

Timing/synchronous belt are toothed on the inside driving via grooved pulleys. This enables positive drive. Limited power capacity compared to chain and Vee-belt derivatives. Does not require lubrication and are extensively used in low power applications.

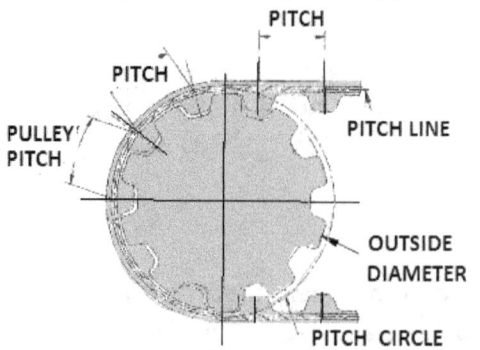
Figure 6-4: Features of timing/synchronous belt

Synchronous/timing belts are basically endless flat belts which pass over pulleys with the belts having grooves which mate with teeth on the pulleys. These types of belt drives, unlike flat and vee belt drives are positive. Any slip of the belt relative to the pulleys is minor in degree and is due to belt stretch, or erosion of the grooves. These belts are used for power transfer and for synchronized drives to ensure that the driven pulley is always rotating at a fixed speed ratio to the driving pulley.

Vee-belt

Vee belt drives replaced flat belt drives for many applications because higher power drives could be transmitted with more compact drive arrangements. With a flat belt drive only one belt is used. With a Vee belt drive a number of belts are used. Flat belts and Vee belts may, and do, slip as the loading increases. For belt drives which drive without slip timing belts should be used. Vee belts on higher power duties generally have to be matched to ensure the drive power is shared.

Vee Belt has better torque transfer compared to flat belt. Generally they are smooth and reliable and are made from hi-text woven textiles, polyurethane, etc. Vee Link Belts can be used in place of Vee belts. They have the advantage that the length can be adjusted and the belt can be easily installed with removing pulleys. They are expensive and have limited load capacity. Figure 6-5 shows two views (elevation and plan) of a twin belt drive arrangement.

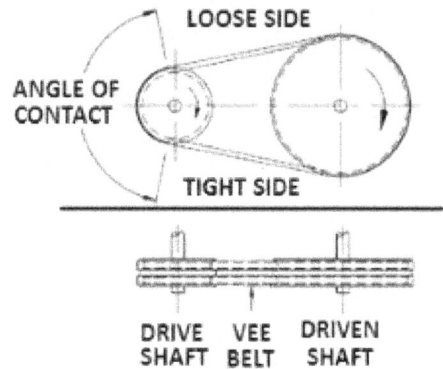

Figure 6-5: Twin Vee-belt drive arrangement

Conventional Vee-belts are made of rubber reinforced with imbedded plastic, fiberglass or steel cords. The Vee grooves are machined into the solid billet around the circumference of the pulley. The cross-sectional area of the belt depends on the power to be transmitted through the belt. Multiple belts are used in combination to transmit large amounts of power.

Advantages of Vee-belts

Vee-belts offer a number of advantages over other types of belts in their use.

- Easy, flexible equipment design, as tolerances are not critical.

- ☞ Reduce shock and vibration transmission.
- ☞ Changing pulley sizes changes the driven shaft speed.
- ☞ They require no lubrication.
- ☞ Maintenance is easy provided unrestricted access is available to the drive arrangement.
- ☞ The pulley alignment is quickly done with a straight edge or a string line spanning across both pulley faces.

Gear drives

Gears (in mesh): These are the most common ways through which power is transmitted in modern machines. When two gears are in mesh all slippage is gone.

Figure 6-6: Meshing gears

Gear oil

Gear oils in enclosed gearboxes lubricate mechanical transmissions, differentials, and steering gears. Society of Automotive Engineers (SAE) classification is based on viscosity alone and is not indications of quality or service. SAE numbers of gear oils are higher than SAE numbers of engine oils; gear oils are not necessarily higher in viscosity. For example, SAE 80 gear oil actually has about the same viscosity as SAE 20 engine oil.

Figure 6-7: Transmission oil

Multi-grade gear oils are presently available from some suppliers in grades of SAE 75/80, SAE 85/90 etc. Some manufacturers recommended engine oil for use in standard transmission service, while some transmissions may use SAE 50 engine as an alternative for SAE 90 gear oil.

Performance, protection, and prevention are the challenges faced by today's industrial gear oils. Indeed it is not even enough that gear oils lubricate effectively, it should also reduce friction and wear, including micro-pitting fatigue in case-hardened steels, protect against rust and corrosion, dissipate heat, help prevent scouring and welding whilst at the same time keeping gear system clean.

Chain and sprocket drives

Chain drive: This is another variation from the gear drive. The gears (sprockets) are not in mesh but are connected by a linked chain, which also eliminates slippage and allow power transmission between two shafts over a distance.

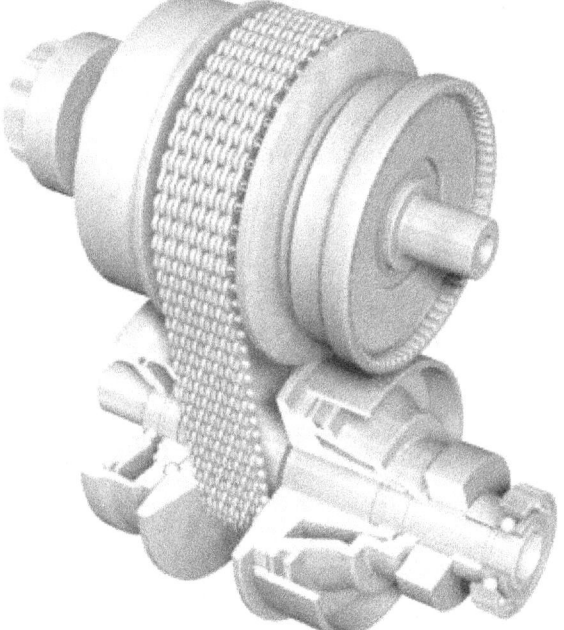

Figure 6-8: Pull type chain CTV concept

Hydraulic power drives

Most farm machines have hydraulic systems to make steering and braking easier, and also provide the power for implements, loaders, trailers and other implements. Basic hydraulic system has three parts: the reservoir, the pump and cylinder.

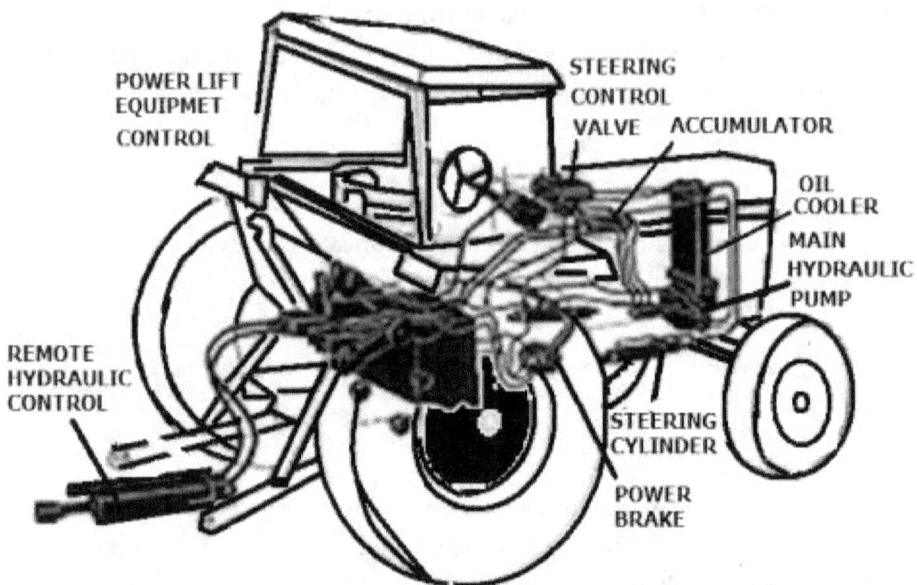

Figure 6-9: A phantom tractor hydraulic system

Hydraulic pump: The hydraulic pump generates power to force fluid through valves and actuators. The pump converts mechanical force to hydraulic power, while the cylinder converts the hydraulic power back to mechanical force to do work.

Figure 6-10: Tractor main hydraulic pumps

Figure 6-11: Tractor steering hydraulic pumps

The reservoir stores the oil.

Figure 6-12: A hydraulic reservoir mounted pump

For continuous and synchronized operation, the following systems were added.

Check/safety valve: This holds the oil in the cylinders between strokes and prevents oil from returning to the reservoir during pressure stroke.

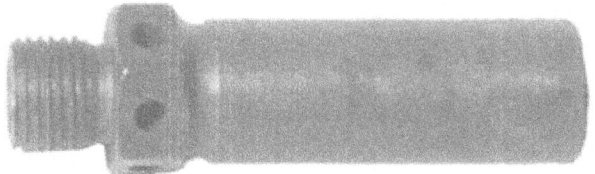

Figure 6-13: Fiat check (safety) valve

The control valve: This allows the operator to control the constant supply of oil from the pump to and from the hydraulic cylinder. In mental position the flow of oil from the pump goes directly through the valve to line carrying the oil back to the reservoir.

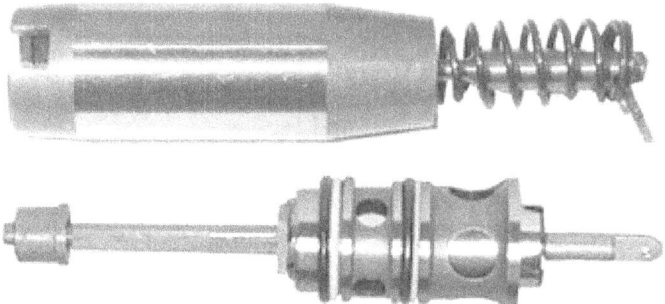

Figure 6-14: Tractor hydraulic control valve assembly

Pressure relief valve - Protects the system from high pressure. If the pressure required to lift a load is too high, this valve opens and relives the pressure by dumping the oil back to the reservoir. It is also required when the piston reaches the end of the stroke. At this time, there is no other path for the oil and it must be returned to the reservoir through the relief valve.

Figure 6-15: Tractor relief valve (Ford/New Holland)

Hydraulic pump repair kit

The following kits are necessary for hydraulic pump repairs and overhauling

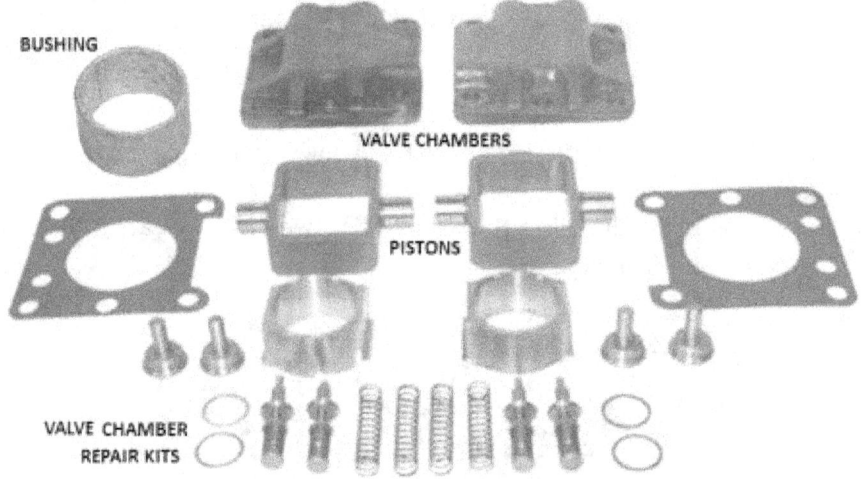

Figure 6-16: Pump repair kits

Tractor drive shafts

Drive shafts in tractor include the transmission drive shafts and the PTO drive shafts. The transmission drive shafts are used to transmit gear loads to the bearings and to transmit power from one component to another. Tractor transmission drive comprises of drive shafts, and various types of transmission gears which are arranged in the transmission housing in combinations to produce the forward speeds and reverse.

These shafts are used in clutch engagements, power shifting etc. These shafts require an extremely hard surface for wear resistance, which is obtained by carburizing a lower carbon steel though not as strong as the higher carbon induction-hardened shafts.

Tractor power takeoff system (PTO)

The Power-Take-Off (PTO) is an attachment in the power train (transmission) of the tractor designed to transmit power from the tractor transmission to power coupled implements such as slasher, bailers, machinery etc. There are three standards of Power Take Off arrangements:

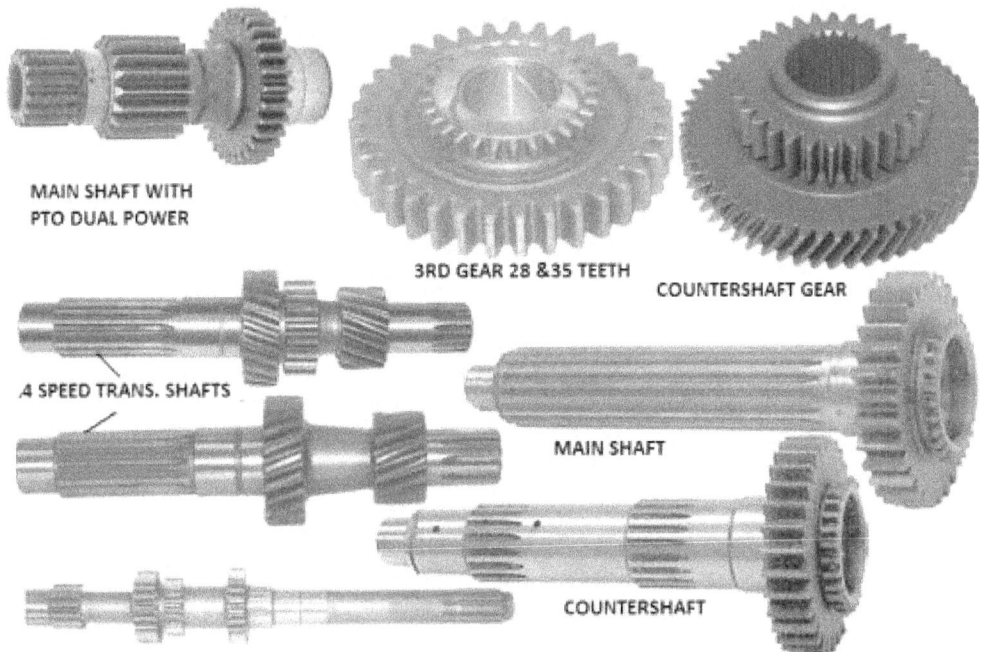

Figure 6-17: Transmission drive shafts and gears

540 rpm with a ($1^3/_8$) inch (35m diameter shaft, 6 splines)

1000 rpm with a ($1^3/_8$) (35mm diameter shaft, 21 splines) and

1000 rpm with a ($1^3/_4$) inch (45cm) shaft.

PTO conversion/ standardization

The tractor PTO drive can be converted from one rpm to another to adapt tractors with different rpm to implements with different rpm. This process is called *PTO standardization* and can be done by replacing some gears and shafts and the draw bar. This is called permanent conversion. In some cases, kits are available to convert 540 rpm implements to 1000 rpm operation for use with a new tractor. Also an external gearbox is made available in some tractors for such conversions.

Figure 6-18: A 6-spline PTO conversion shaft

The following process is carried out to convert a live PTO to independent PTO:

FARM TRACTOR SYSTEMS

Park the tractor such that its front is lower than its rear to avoid any oil leakage through the shaft housing.

Remove the shaft protective cover (if any)

Using gripping pliers, compress the elastic ring (Figure 6-19) by its ends, releasing it from the housing

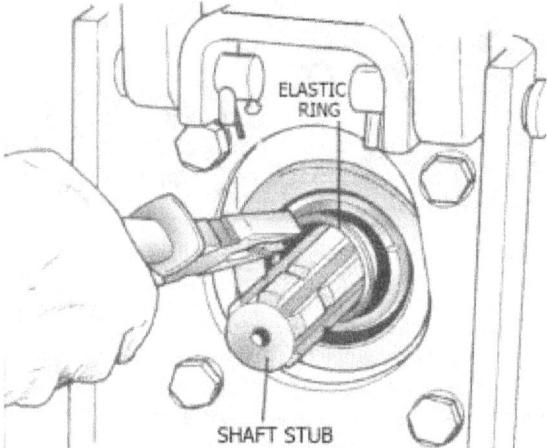

Figure 6-19: Removing PTO shaft for exchange

Pull the shaft stub by hand out of its casing.

Then introduce the outer shaft carefully into the internal gear grooves.

Re-install the locking ring into its groove. The PTO is now ready to operate at the speed related to the installed shaft.

PTO drive control

Tractors are manufactured with some arrangements for a 'live' PTO; that is the ability to stop forward motion of the tractor without stopping the rotation of the PTO.

The different types of PTO arrangements include:

a. *Transmission driven PTO* operates only when the engine clutch is engaged. To engage live PTO, depress the clutch pedal to disconnect the drive from the engine to the transmission without interrupting the operation of the PTO. Move the engagement lever forward or rearward according to tractor design and release the clutch pedal slowly. Set the engine speed to obtain 540 or 1000 rpm PTO shaft speed.

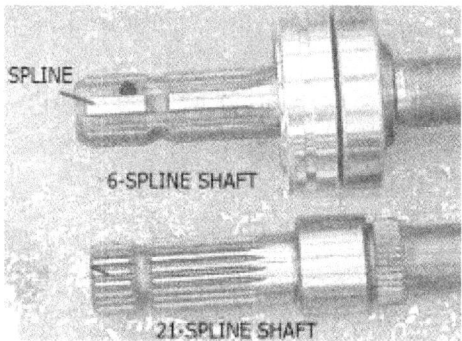
Figure 6-20: PTO transmission shafts

b. *Independent PTO*: The forward motion clutch and PTO clutch are stopped independently. An independent PTO (IPTO) lets you engage or disengage implements on the motion. The IPTO features an independently controlled hydraulic clutch that works with just the touch of a lever.
c. *Continuous running PTO*: An auxiliary clutch is provided to operate the forward motion of the tractor. The PTO clutch cannot be engaged or disengaged without stopping forward motion of the tractor.
d. *Relative ground speed PTO*: The speed of the PTO is controlled by the forward speed of the tractor, regardless of the transmission gear being used.

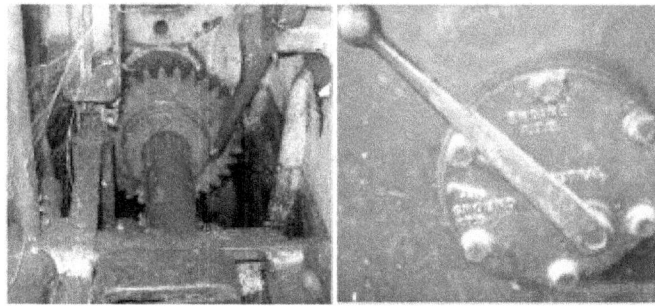
Figure 6-21: PTO gear and engagement lever

PTO components

Major components of a PTO system found on the tractor include PTO drive that consists of a short or a long shaft (Figure 6-22), three universal joints, slip clutch, safety shields, etc.

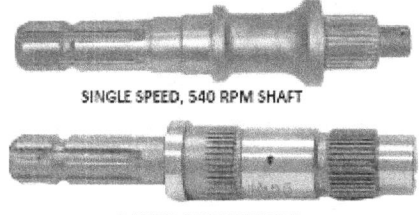
Figure 6-22: Short PTO shafts

- *PTO shaft*: Most pull-type machines have short and long shafts ending in stub for coupling to implement.
- *Drive shaft* has a telescoping section, an adjustable support for the front end of the longer shaft, and a coupler for attaching the shaft to the tractor (Figure 6-23).

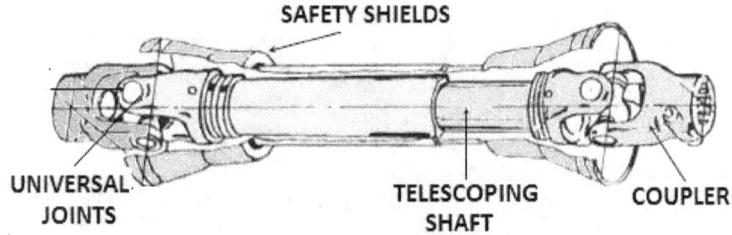

Figure 6-24: Drive shaft with telescoping section

Most PTO shaft have the ability to 'telescope' (Figure 6-24) in and out for use in turns or over uneven terrain. They are constructed with a square or triangular shaft which is inserted into housing. At least $5^1/_2$ inches (140mm) of the sliding shaft remain within the housing when the power shaft is connected to the tractor to reduce the possibility of the shafts separating while the tractor is in motion.

- *PTO stub*: This transfers power from the tractor to the machine and Rotates at 540 rpm (9 times /sec.) or at 1,000 rpm (16.6 times /sec.) *Drive gears*:

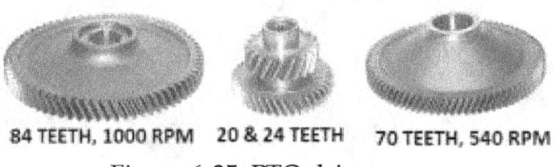

Figure 6-25: PTO drive gears

Figure 6-23: PTO shaft stub and guard

- *Master shield*: Protects the operator from the PTO stub. It is often damaged or removed and never replaced.
- *Drive plate, trust washer and piston sealing rings*

FARM TRACTOR SYSTEMS

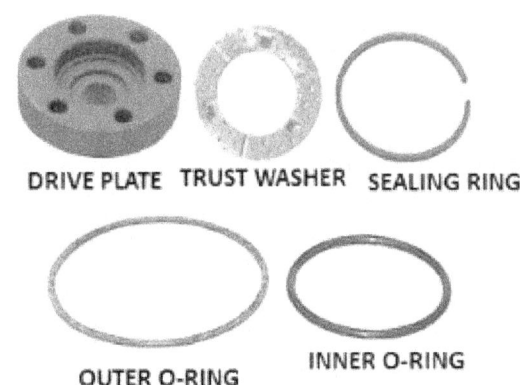

DRIVE PLATE   TRUST WASHER   SEALING RING

OUTER O-RING   INNER O-RING

Figure 6-27: PTO shaft accessories

- *PTO coupler*

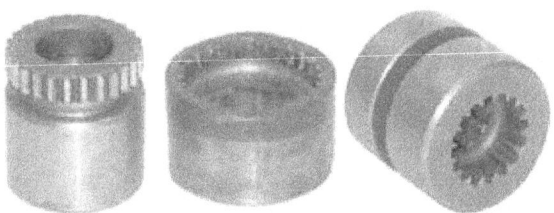

Figure 6-28: PTO shaft coupler

- *PTO shifter accessories*

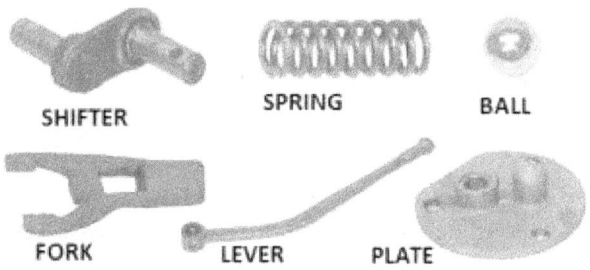

SHIFTER   SPRING   BALL

FORK   LEVER   PLATE

Figure 6-29: PTO shifter accessories

- *PTO Clutch Pack*

Figure 6-30: PTO clutch pack

PTO shaft connections

When hooked up to a piece of equipment, the complete PTO system or implement input driveline may connect either directly to the tractor spline or stub, or indirectly through a pedestal connection as shown in Figure 7-35. Coupling devices are used to attach the shaft to the tractor and to the implement.

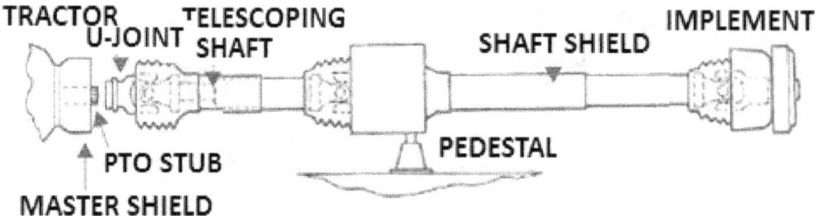

Figure 6-31: Pedestal connection of PTO shaft

A spline collar slides over the tractor's spline, which extends from the tractor differential. The collar is held in place by a spring-loaded pin that latches into a recess on the spline. The telescoping feature of the shaft allows the collar to slide easily onto the spline. Together with the universal joints, the telescoping shaft allows the PTO system to flex and adjust when the tractor turns or travels over uneven terrain.

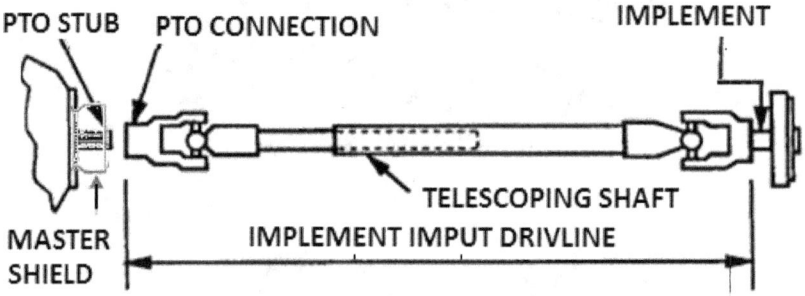

Figure 6-32: The major components of a PTO system

The implement's input driveline connects directly to the tractor PTO stub. This arrangement is most common with tractors and implements in use around here. Examples of this type of connection include three-point hitch-mounted equipment, such as post-hole diggers, small rotary mowers, fertilizer spreaders, and augers.

Connections from the tractor to the implement are made through the flexible universal joints. The "U-joints" are connected by a square rigid shaft, which turns inside another shaft. The combination of universal joints and turning shafts provides the remote power source to a farm implement. Without proper guarding, a serious threat to the operator's safety is created.

PTO safety guards

The major guards of a PTO system include (Figure 6-33):

1. *Implement input connection (IIC) shield* (Figure 6-33-1). This protects the operator from the IIC, including the implement input stub and the connection to the Implement Input Drive (IID)

Figure 6-33: The major guards of a PTO system

2. *Safety chain* (Figure 6-33-2): This keeps the integral journal shield from spinning
3. *Master shield*(Figure 6-33-3): Protects the operator from the PTO stub and the connection of the IID to the PTO stub
4. *Integral journal shield* (Figure 6-33-4): This completely encloses the IID. It may be made of plastic or metal mounted on bearings to allow it to spin freely from the IID

Types of PTO safety guards

Two types of guards are in general use in guarding shafts:

1. Cone and tube type guard

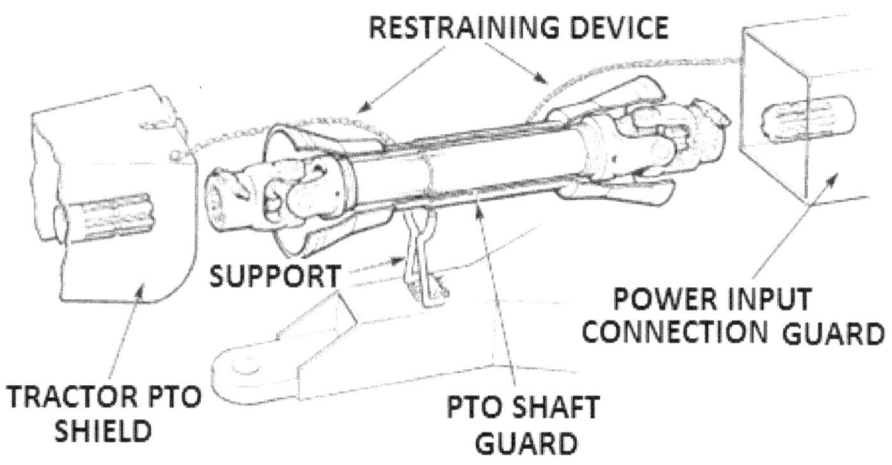

Figure 6-34: Cone and tube guards

2. Bellows type guard

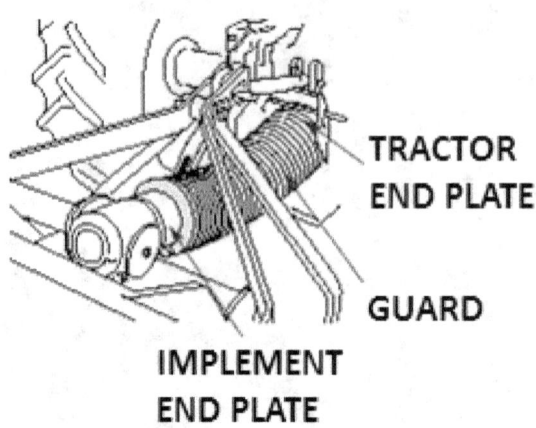

Figure 6-35: Bellows guard

Tractor transmission systems

The major components of the power drive train in mechanical drive systems include the transmission, one or more drive shafts, differential gears, and axles. In tractors, the transmission is usually located behind the engine, although some automobiles were designed with the transmission mounted on the rear axle. The complete tractor transmission (also called transaxle) is usually defined as demonstrated as shown in Figure 6-36. A combination of the vehicle speed change gearbox, the rear axle with brakes, the power take off (PTO) and - if required - arrangements for the front axle drive and for the drive of auxiliary units (mainly hydraulic pumps).

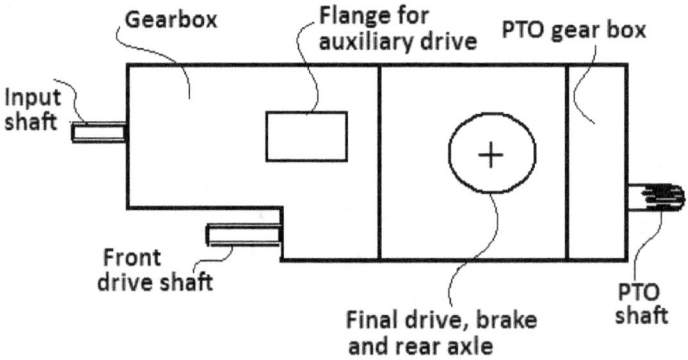

Figure 6-36: Tractor transmission system (Renius, 1999)

There are three basic tractor transmission drive train deigns in use which include: the manual, automatic, continuously variable transmissions and transaxle or tractor transmission.

3. A manual *transmission (MT)* has a gearbox from which the operator selects specific gears depending on road speed and engine load.

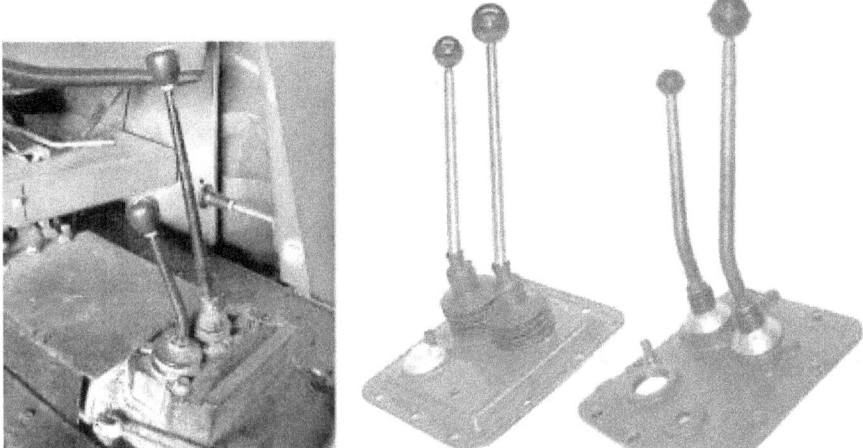

Figure 6-37: Gear lever or selectors

Gears are selected with a shift lever located on the floor next to the operator or on the steering column. The operator presses on the clutch to disengage the transmission from the engine to permit a change of gears. The clutch disk is attached to the transmission's input shaft. It presses against a circular plate (pressure plate) attached to the engine's flywheel. When the operator presses down on the clutch pedal to shift gears, a mechanical lever called a clutch fork and a device called a throw-out bearing separates the two disks. Releasing the clutch pedal presses the two disks together, transferring torque from the engine to the transmission.

4. An *automatic transmission (AT)* selects gears itself according to road conditions and the amount of load on the engine. Instead of a manual clutch, an automatic transmission uses a hydraulic torque converter to transfer engine power to the transmission. Instead of making distinct changes from one gear to the next, a continuously variable transmission uses belts and pulleys to smoothly slide the gear ratio up or down.

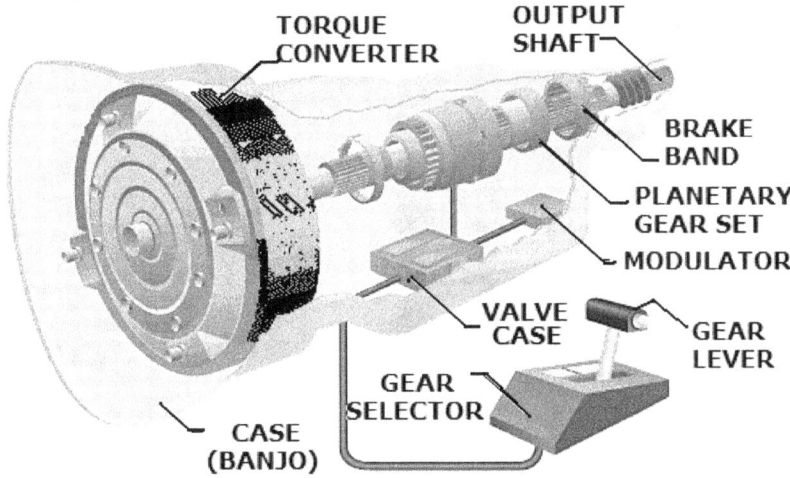

Figure 6-38: Automatic power transmission system

5. *Continuously variable transmissions (CVT):* CVT transmission has the ability to transmit power through continuously varying ratios. It does this by splitting the drive in a planetary gear system between a hydrostatic system, which is driven through a ring gear, and a mechanical gear system that is driven through a sun gear. The amount of force required to drive the hydraulic pump determines how the power from the engine will be split.

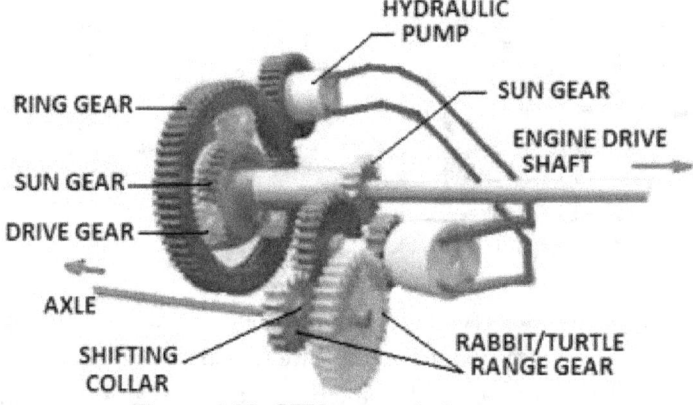

Figure 6-39: CTV transmission system

CVT operation

Power from the engine is carried through the *carrier gear (drive gear)*. When the transmission is in neutral, the *ring gear* spins freely, since the hydraulic pump is at idle and provides no resistance. As the transmission is engaged, the angle of the hydraulic pump is altered to begin driving the hydraulic motor. This motor, in turn, drives the axle. By changing the angles of the hydraulic pump and motor, via the control, more hydraulic drive is provided to the tractor.

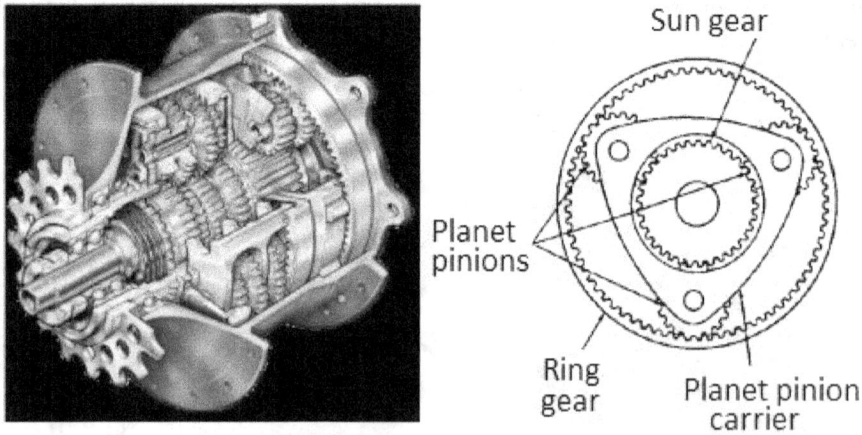

Figure 6-40: Planetary gear

6. *Transaxle or tractor transmission:* The tractor transmission (also called *transaxle*) is usually defined and include; a combination of the vehicle speed change gear box, the rear axle

with brakes, the power take off (PTO) and – if required – arrangements for the front axle drive and for the drive of auxiliary units (mainly hydraulic pumps).

Figure 6-41: Section of a tractor transmission drive

Tractor steering systems

The three most widely used steering concepts for two-axle tractors are demonstrated by Figures below with the center of motion defined for static conditions.

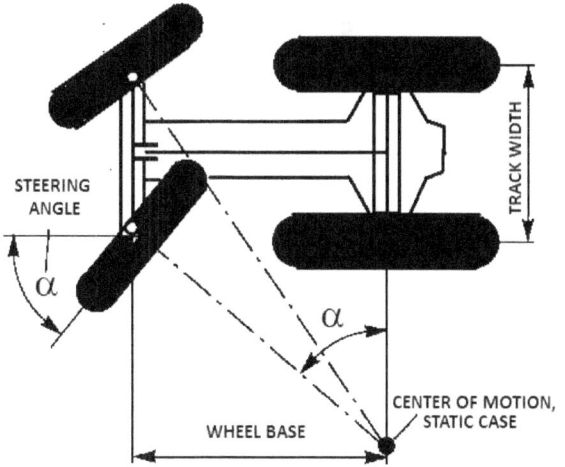

Figure 6-42: Ackerman (front-axle) steering

Concept I: front-axle steering ("Ackerman steering"), is the most widely used steering concept, and typical for standard tractors. Steering linkages of front-axle steering are mostly not able to offer the exact geometry of *steering angle a,* always refers to the curveinner wheel, as shown in Figure6-42. Therefore a steering angle error is defined to be the angle error of the curve-outer wheel.

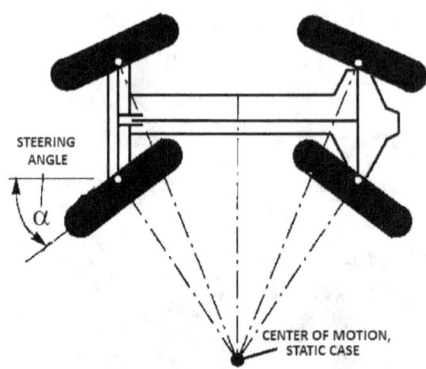

Figure 6-43: Front and rear-axle steering

Concept II: *front and rear-axle steering,* is used for special vehicles with four equal tyres such as hillside multipurpose tractors. It typically offers three steering modes: front steering (on the road), front and rear opposite and front and rear parallel (for example to compensate hillside drifts). The full use of the steering angle is often not possible due to the required clearance for the lower links of the rear three-point hitch. An advantage can be seen in the *track-in-track driving* (multi-pass, no transmission wind-ups).

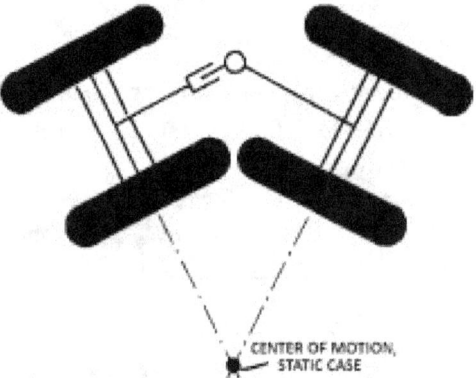

Figure 6-44: Steering by articulated frame

Concept III: *steering by articulated frame* is typically for big four-wheel-drive tractors, offering again track-in-track driving. Main disadvantages relates to the high level of required hydrostatic power (losses) and the poor ability for high speeds on the road. Concept III is sometimes also applied for vineyard and orchard tractors.

Steering systems

Three major types of steering systems are in use in today's machines:

1. *Manual steering*

Manual steering system incorporate a wheel linked directly to the turning wheels and operator does all the steering with the help of mechanical drag link.

2. Power steering

With full power steering, the only force required from the operator is enough steering wheel force to open valves. Power is supplied by pump, which gives all steering force up to the capability of the system. Power steering increases not only the comfort for the driver but also offers a much higher degree of design flexibility (pipes and hoses instead of linkages) also favouring cab noise reduction.

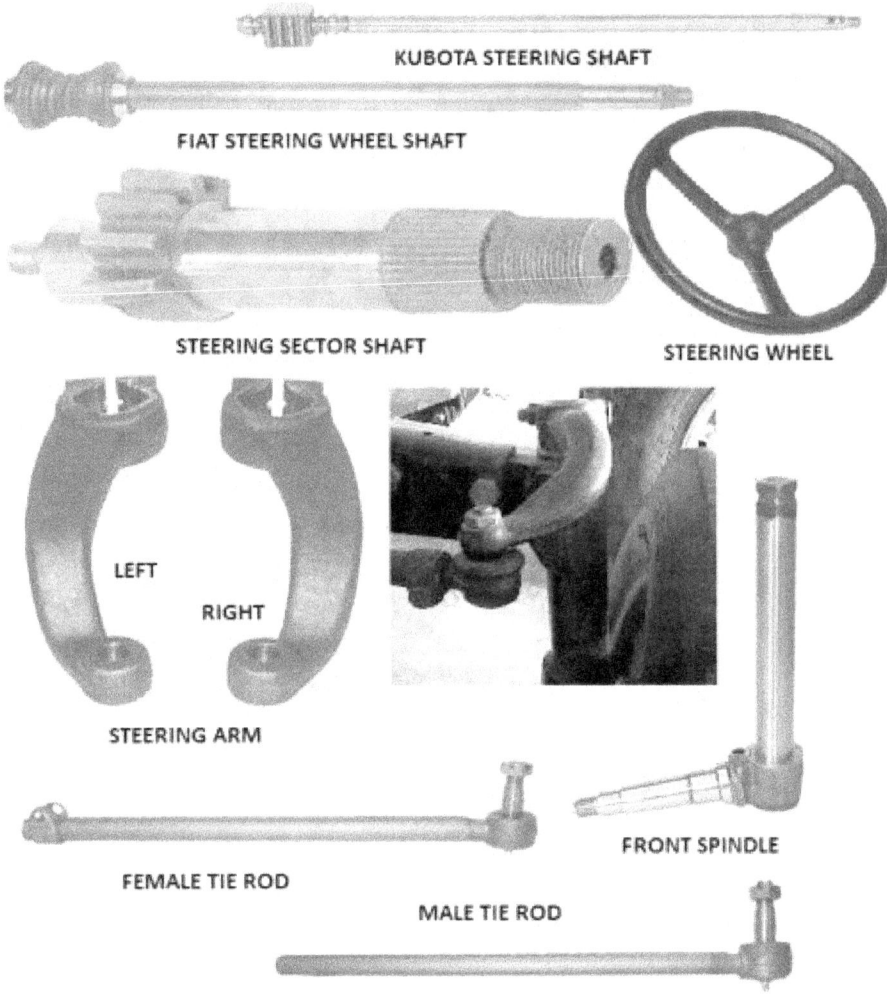

Figure 6-45: MF steering components

3. Hydraulic steering with mechanical drag link

When the operator turns the steering wheel to the right, shaft is forced up out of the worm nut because of the turning resistance of the steering wheel. The spool valve (control valve) and steering shaft is lifted forward, which directs oil to the cylinder, rotates the rack and pinion device, which turns the front wheel. Oil from the cylinder is returned to the reservoir through the spool.

As soon as the steering wheel motion is stopped, the hydraulic pressure will turn the wheels slightly further to the right, moving the steering linkage forward and pulling the valve back to the neutral position.

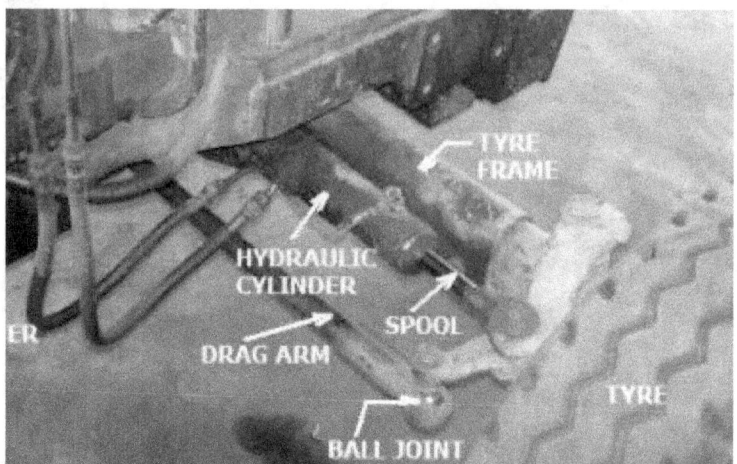

Figure 6-36: Hydraulic steering system

Engines and hydraulic systems should be thoroughly warmed-up periodically during periods of no usage. Complete warming-up of both systems helps to keep bare metal surfaces covered with protective oil film and drive off condensed water vapour in the oil.

Figure 6-37: Hydraulic steering cylinders

Note: The practice of a quick five minute run without full warm-up may cause more harm than good, as metal temperatures reached are insufficient to drive off water vapour.

In operation, the farmer requires that the power steering system is just able to turn the maximum steering angle on dry concrete with maximum axle load even if the tractor is not moving (important for front end loading).

In the case of four-wheel drive, the steering behaviour is substantially influenced by the power train concept. If both axles are connected by a fixed-ratio gearing (which is the most popular concept for standard tractors), the front axle input speed is in sharp turns too low and therefore creates braking effects instead of traction.

Tractor track width

The track width is the distance at ground level between the middle planes of the wheels on the same axle, with the tractor stationary and with the wheels in position for traveling in a straight line. Track width is important mainly for operating the tractor in row crops and to enable *multi-pass effect* for off-road tractor-trailer operations.

Track wheel adjustment

The tractor track should be adjusted according to field operational requirements, bearing in mind the following factors; type or operation, type of cultivation and type of soil or terrain.

Note that at every hole of the bar, the track is altered by 50 mm on that respective side. Therefore, total alteration of the front track will be 100 mm (50 mm each side.)

There are three typical concepts of track width adjustment:

1. Infinitely variable track width adjustment, often power assisted,
2. Stepwise adjustment achieved by several modes of *rim* assembling (itself and to hub)
3. Basic track width adjustment by turning the rims.

ISO 4004 (1983) recommends the following basic track widths:

$$1500 \text{ mm} \pm 25 \text{ mm}$$
$$1800 \text{ mm} \pm 25 \text{ mm}$$
$$2000 \text{ mm} \pm 25 \text{ mm}$$

Further popular track widths for standard tractors are: 730 mm (vine yard), 1000 mm (orchard), 1250, 1350, 1600, 2200 and 2400 mm. Problems with small values typically relate to limited horizontal clearance of the tyres, mainly at the front axle for high steering angles. Problems with high track widths can arise due to traffic regulations (which limit the tractor width, e.g. 3m).

Wheel-to-hub fixing

This is a critical area regarding durability. Problems sometimes occur if a tractor is overloaded or if tyres are used with diameters above the released limits. ISO 5711 (1995) clearly favours the *flat attachment* with separated centering (in comparison with integrated centering by spherical or conical studs/nuts, as usually applied for passenger cars). ISO 5711 incorporates an informative annex for typical measures of spherical or conical studs /-nuts.

Clutch system

The normal clutch job starts when the vehicle stops. As you let out the clutch, the clutch disk surfaces rubs between the flywheel and the pressure plate until the vehicle starts to move, and then locks up when the clutch is released completely. The *durability/service life of clutches* mainly refers to *wearing of the clutch facings*. It is quite normal for clutch friction facings to be subject to wear. Slip is generated every time the clutch is engaged to compensate for the difference between the input and output speeds.

Figure 6-38: A cross- section of clutch assembly

This slip between the facings and the opposing-friction surfaces generates heat and natural abrasion/wear. Logically, the longer the slip period the more heat is generated and the higher the rate of clutch facing wear.

Figure 6-39: Facing wear worn down to rivet heads

Equally, too much grease on the spline causes the plate face slipping

Figure 6-40: Excess lubricant at the splined shaft

These are some of the factors that reduce service life/durability:

- The number of drive-offs: frequent maneuverings and stop-and-go traffic.
- The duration of clutch slip: letting the clutch slip for too long or if the drive-off engine speed is too high.
- Misuse of the clutch could cause overheating of plate due to driving or maintenance errors.

Figure 6-41: Burnt/detached facings

Brake systems

Brake device is used to slow and stop a rotating wheel and thus a moving vehicle. Brakes such as those on automobiles, trucks, trains, and bicycles use friction between a wheel and another object to slow the motion of the vehicle. The friction created by the rubbing together of two shoe pads generates a large amount of heat. A brake system must be capable of dissipating the heat as the rotating wheels slows down, because excess heat can cause the brakes to lose their grip and fail.

Types of brake systems

1. *Friction brakes*: Friction generating devices in brake system has broadly classify brakes into two types; Drum Brake and disc brake
   a. *Drum brake*: Friction is generated by the expansion of two shoe pads (lining) against the rotating drum which holds the tyre (Figure 6-42). The friction generated by this expansion slows down the vehicle.
   b. *Disc brake*: Friction is generated by two asbestos surfaces (brake pads) moving against a rotating disc to slow down disc rotation and thus vehicle movement.

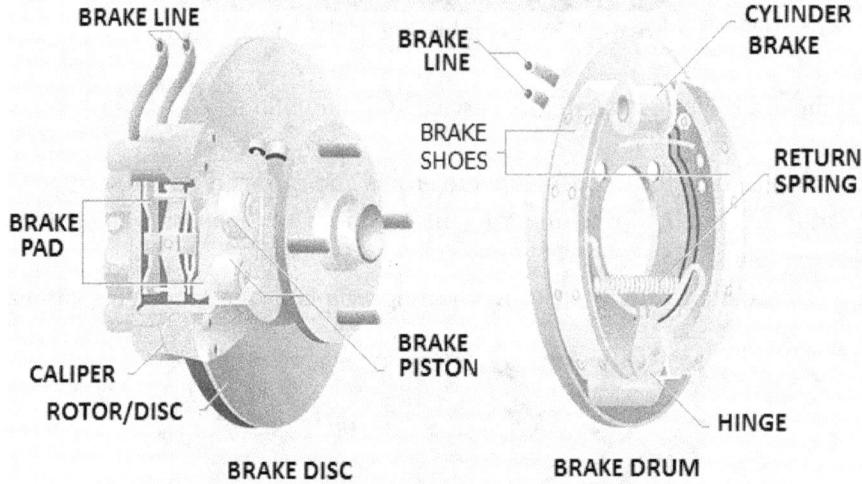

Figure 6-42: Disk and drum brake systems

2. *Mechanical brake system*: In this system, hand operated link mechanism controls the movement of the brake pads
3. *Hydraulic brake systems*: The hydraulic fluid flow at high pressure forces the expansion of compression springs which is responsible for the holding and release of the brake.
4. *Pneumatic brake system*: Tractor-trailer brakes use two types of brake systems; the first is the normal slowing and stopping system, which uses compressed air to activate the brakes.

The second system is the emergency brake system. Inside the drum brakes, the brake shoes are connected to a spring held in place by a diaphragm that is filled with air. When air pressure is applied to the diaphragm, the pressure overcomes the spring and allows the brake to be released.

However, when there is no air in the diaphragm, the spring pushes the shoes against the drum and applies the brake. As long as constant air pressure is maintained in the emergency system, the brakes remain released. But if there is a leak in the brake lines or if the air compressors fail, the air pressure inside the diaphragm decreases, and the spring-brake mechanism automatically applies the drum brakes.

Your disc brake rotors and pads handle up to 75% of your vehicle braking needs. Braking power may be severely reduced if brake pads are worn thin. Worn brake pads are not only dangerous, but will cause extensive rotor or disc damage if neglected.

Never put off brake maintenance. Brake pads normally last between 20,000 and 40,000 miles depending on driving conditions. Inspect your disc brake pads regularly and replace if needed with quality brake parts.

Performance indicators for disc and drum brakes

Energy dissipation: **Disc** brakes provide larger cooling air volume and more effective radiating surfaces than drum brake thus better heat dissipating capability.

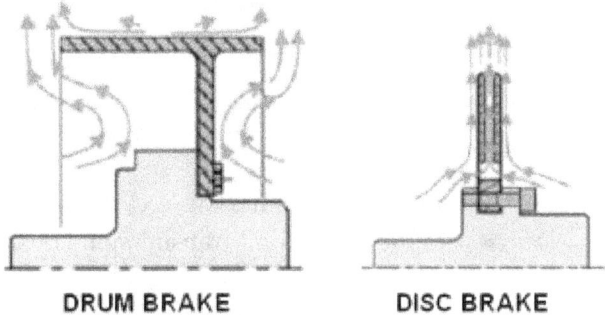

Figure 6-43: Energy dissipation in brakes

Adjustments: Disc brakes require fewer adjustments (no drum dilatation i.e. Less lining wear).

External dimensions: **Disc** brakes need less physical room and clearance for maintenance i.e. smaller motors, trolleys, couplings etc.

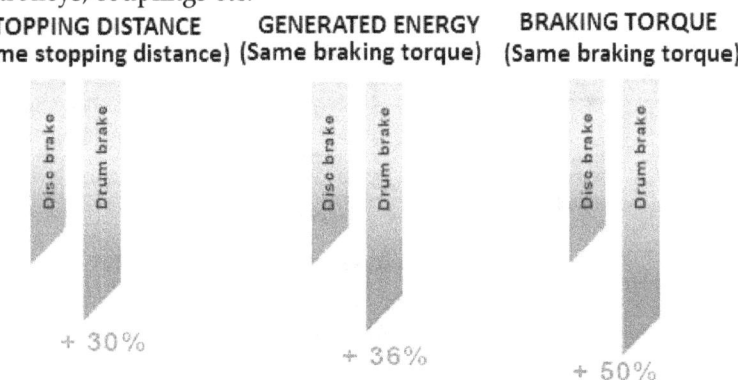

Figure 6-44: Disc and drum brake performances

Inertia – response time: **Disc** brake inertia is lower than drum brake's (60% less) i.e. Shorter response time and no need to oversize brakes

Lining/pad replacement: **Pad** replacement in disc brake is 12 times faster than lining replacement in drum brake. Replacing pads takes only 5minutes as against 60 minutes for drum brakes.

Figure 6-45: Brake lining

Tractor chassis

The concept of tractor chassis was developed around the *block design* and the *frame design*. The well-known *block chassis design* uses the engine and transmission housing instead of a frame (Figure 6-59). The block principle became the most important rule for standard tractor designs, saving initial costs and allowing the drive line elements to be perfectly encapsulated and well-lubricated.

There were some exceptions where frames were used, for instance, small Japanese tractors and German garden tractors or special vehicles. Many tractors used half frame configurations to reinforce chassis durability around the engine and simplify front attachments.

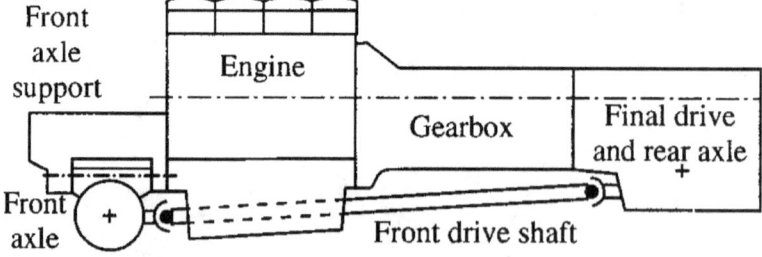

Figure 6-46: Block chassis for standard tractor

Advantages of the frame chassis in comparison to the block chassis:

1. Higher potential for bystander noise reductions
2. Improved basis for front-end loader and front hitch
3. Higher component flexibility

4. Simplified maintenance and repair

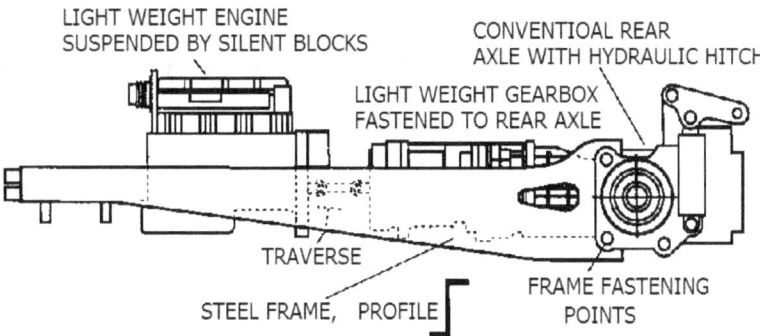

Figure 6-47: Frame tractor chassis and driveline configuration

Disadvantage

In 1992, John Deere presented two designed tractors using chassis frames, Figure 6-47. The commercial production and implementation of such design in tractor manufacture or a change from block chassis to frame chassis is not easily accepted because it requires high investment costs, as almost every component needs a new design.

Practical exercise

Students are expected to identify the various components of the following tractor systems and drives:

1. Identify the universal joint with the PTO of a tractor.
2. Hydraulic system of specific tractor in your workshop.

Prepare workbook/field report of activities and sketches.

CHAPTER 7

Tractor Drive Designs and Differential Systems

Introduction

Tractors are vehicles with specialized final drives and differential designs.

Tractor drive designs

Tractor drives are generally classified as two wheel (power tillers) drives, 4-wheel front-wheel or four wheel drive

Two-wheel (power tiller) drive: Built with the same quality design, construction, and material as the four-wheel tractors for agriculture and industry, two-wheel tractors tackle all the jobs for which larger tractors are unsuited for instance mowing slopes, manoeuvring in small areas, practicing intensive cultivation, sweeping walkways, and much more., featuring all-gear drive and an automotive-style clutch.

Figure 7-1: Two-wheel power tiller tractor

Two-wheel front drive: The two-wheel front drives have two of its four wheels controlled by the steering. Examples are common with front wheel drives in which the two front wheels control the forward and reverse movement.

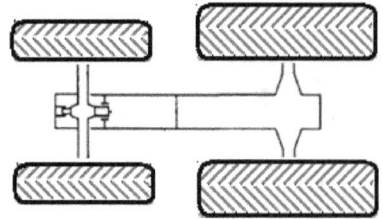

Figure 7-2: Two-wheel front drive tractor

Four-wheel drive: The four-wheel drive has a drive shaft that transmits power between the front wheels and the rear wheels. Most old two-wheel drives have manual power transmission; the newly designed two-wheelers are power drive while all the 4-wheel drives are hydraulically driven.

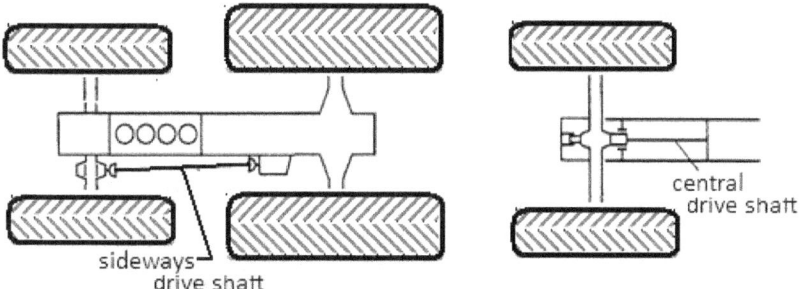

Figure 7-3: Basic front axle drive shaft concepts

Two drive line concepts are further used to classify front axle driven standard tractors, shown in Figure 7-3. This classification includes: Sideways shaft design concept and Central shaft design concept. In each case, the power typically is diverted from the gearbox output shaft, controlled by a wet clutch. This leads to a fixed ratio between rear and front axle speeds.

Comparing the two, the *sideways shaft concept* is preferred in a four-wheel drive. The *central shaft concept* is cheaper and offers more space for other components such as the fuel tank. Maximum possible steering angles are usually not influenced (except in specific cases). An offset in height at the central concept needs a universal joint but favours maximum steering angle.

The four-wheel drive offers the following advantages to designer and the farmer. Advantages for the farmer:

- ✓ Increased pull and drawbar power. The softer the soil, the higher the benefits.
- ✓ Improved tractive efficiency off-road, typically 15% resulting in fuel savings of the same level.
- ✓ Improved productivity for front-end loading.
- ✓ Improved maneuverability on slopes and under other difficult conditions.
- ✓ Improved braking performance.

The first listed benefit has often been under evaluated, when the additional pull force was derived directly from the additional vertical axle load. The true gain in traction is considerably higher due to the multipass effect and the direct compensation of the front axle rolling resistance.

Advantages to the designer:

- ✓ More freedom for adequate tyre volumes related to tractor size, power and type of mission

- ✓ Possibility of a low cost front axle brake on the front drive shaft (important for higher speeds)
- ✓ Reduced torque loads at the rear axle

The disadvantages concern mainly the additional initial costs of about 16%–20% and the higher losses for on-road operation at higher speeds.

Tractor front axle

The front axle supports the front part of the tractor and provides support for the steering arm and hydraulic cylinders. Power steering for some tractors has steering gear box mounted on the front axle. Component parts of the front axle consist of drag arm, steering arms, wheel hub, axle knee, front support etc.

Figure 7-4: Tractor front axle

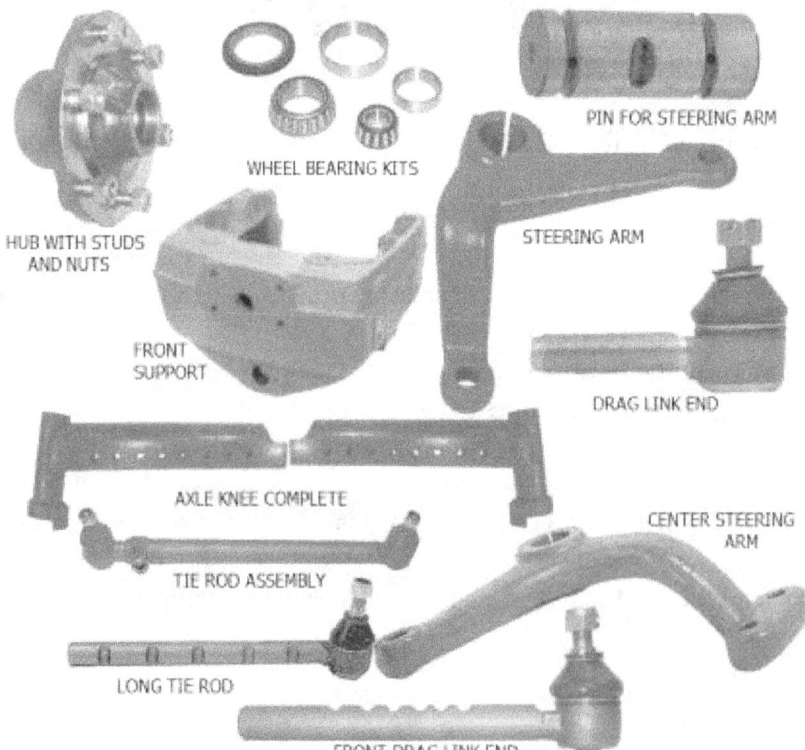

Figure 7-5: Tractor front axle components

Front axle support frames and the grilles

The front axle is attached to the engine block with set of bolts. The axle support also provides support for the front ballast and the hydraulic steering cylinder. The radiator and the battery are installed supported by the axle support and protected by the grill assembly.

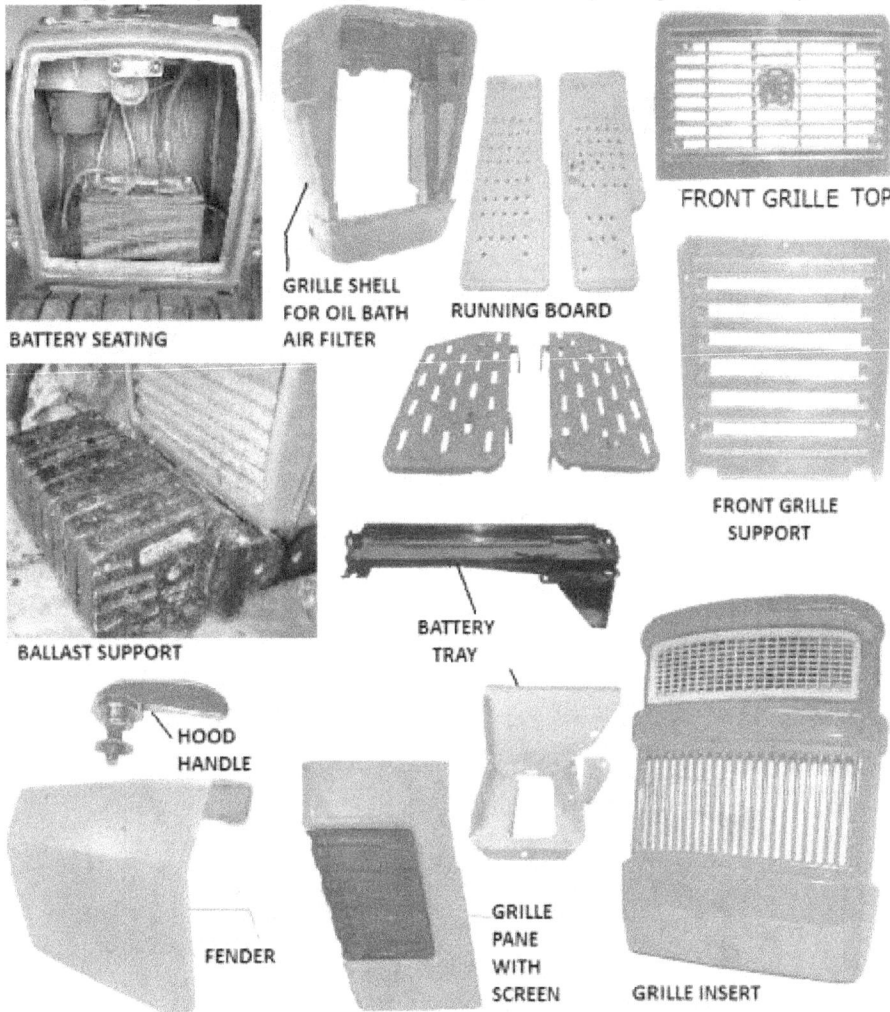

Figure 7-6: Tractor accessories

Other frame member include the running boards, battery tray, fenders, fender skin, fender brackets, hood side panels, front grilles of different designs and grille shells for oil bath air filter (Figure 7-6).

Tractor tyres

Tractor tyres are either steel or pneumatic and are in different sizes and make depending on choice of task and condition of use.

Figure 7-7: Tractor tyres

Pneumatic tyres and usage

Pneumatic tyres are made of elastic materials, natural or synthetic rubber or a blend thereof, reinforced with textile cord ply fabric carcass (or steel for some range of radial ply tyre) enclosing bead rings.

Tyre codes

Tyres may be either radial or bias ply. *Radial ply tyres* have the ply cords laid substantially at 90 to the centerline of the tread, the carcass being stabilized by an essentially inextensible circumferential belt. Radial tyres have a star marking; (*, **, etc.). Radial ply tyres are superior to bias tyres in many criteria with however little lower damping and higher sensitivity against sharp obstacles sideways.

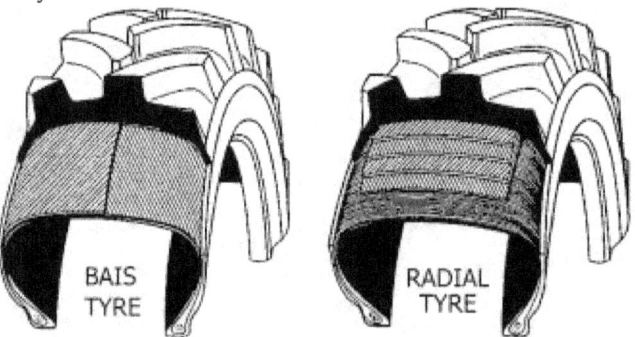

Figure 7-8: Bias ply and radial ply tyres

Typical for radials is a little larger contact area and a better internal torque transfer due to the belt concept in connection with thin and very flexible side walls. Traction coefficients and traction efficiencies are typically little higher for radials than for bias plies for the same slip but a good bias ply tyre at least can achieve the performance of a poor designed radial tyre.

Bias ply tyre (also called *diagonal ply*), has the ply cords laid at alternate angle substantially less than 90º to the centerline of the tread. Bias ply tyres could have a ply rating (4-ply, 6-ply, 8-ply etc.).

Tyre standards

Tyre size codes are based on tyre design and rim diameter, for example, based on ISO, a bias ply tyre such as *16.9 - 38 8 PR*("diagonal construction") is specified as follows (example):

16.9 = nominal tyre width (¼ width in inch)
38 = nominal rim diameter (diameter in inches)
8 PR = *ply rating* (indicates strength of carcass)

According to ISO, a radial ply tyre is specified as follows (example, same tyre size as above):

$$16.9 \text{ R } 38 \text{ } 141 \text{ A6}$$

R = Radial,
141 = Load index (2575 kg at 160 kPa),
A6 = Speed symbol(30 km/h reference speed, A8 would indicate 40 km/h reference speed).

Low section height tyres are specified by the ratio of section width and height, for example 70%:

$$520/70 \text{ R } 38 \text{ } 145 \text{ A6}$$

These tyres offer, at about the same diameter, greater width (and thus more contact area) than the standard versions. Low section height tyres have become popular as an increase of width has only minor influences on the vehicle speeds and transmission torques.

Parts of tyre

Bead: The part of the tyre which is so shaped as to fit the rim and hold the tyre on to it. It has cores made of several strands of essentially inextensible steel wire with end of the plies wrapped around the cores for anchorage.

Sidewall: The part of the tyre between the bead and the tread which flexes in service.

Sidewall rubber: The rubber layer on the sidewall of the tyre and over the carcass which may include protective ribs and fitting lines to assist in centering of tyre on the rim. This protects the tyre from scuffing and damages.

Tread: This is the part of the tyre which comes in contact with the ground and through which the driving, braking and cornering forces are transmitted. It is made of special rubber compound to give good wearing properties and in conjunction with the tread pattern to transmit these forces.

Cord: Textile or non-textile strands (threads) used in various components of the tyre carcass, plies, belts, breaker etc.

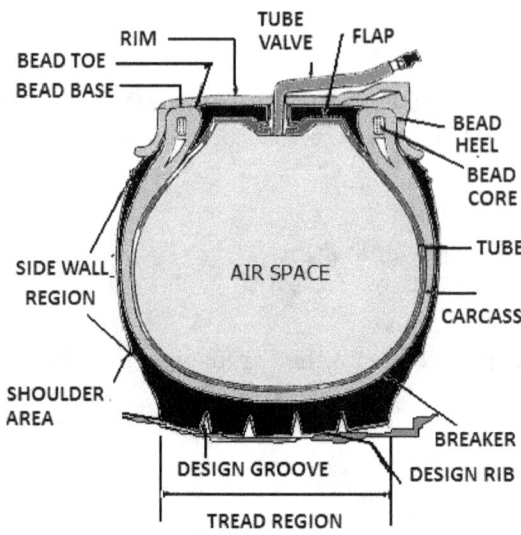
Figure 7-9: Tyre configuration

Ply: This refers to a layer of rubber coated fabric cords.

Carcass: The rubber bonded cord structure of a tyre integral with the bead, which provides the requisite strength

Tyre markings

Size designation – The size markings for the identification of tyres consists of nominal tyre width code and the nominal rim diameter code e.g. 7.5R-16.

Type of construction: For tyres of radial construction, the letter "R" replaces the dash which is used for diagonal ply.

Ply rating: The term ply rating is used to identify a given tyre with its maximum recommended load when used in a specific type of service. It is an index of tyre strength and does not necessarily represent the number of cord plies in the tyre. This also includes maximum permissible load and inflation pressure for Singles and Duals fitments, Manufacturer's Serial Number, Brand name, Country of manufacture.

Carcass ply materials – Tyres will be marked "Nylon" only if the tyre is of nylon construction. Marking of Belt material is optional, but Steel belts are usually indicated and desirable.

Tread wear indicator (TWI)

Tread wear indicator is an arrow marking on the sidewall of the tyre, to indicate the direction in which the tyre should rotate in service in the case of directional type tyres.

Tyre pattern classification

Rib tyre: These are steering wheel tyres. They can be used on all wheel position on highway application.

Lug tyre: These are designed for drive wheel positions and provide greater traction.

Semi lug tyre: These are designed for drive wheel application and can also be used as steering wheels.

Tyre selection

Selection of tyre size and ply rating shall be decided from the highest individual load on tyre on an axle determined by the GVW (Gross Vehicle Weight) distribution. The maximum GVW, which is the loaded weight specified by the manufacture for the completed vehicle depends on:

i. The kerb weight
ii. Driver and occupant weights for the designated seating capacity,
iii. Accessory weight and extra nonstandard equipment weight,
iv. Cargo load – for any uneven loading in the cargo area, the maximum cargo load must be reduced to prevent trip over and
v. Field modification to provide additional capacity, reinforcement, etc. made by those other than the original vehicle manufacturer, if permitted.

Tyre tube and flap care

Tubes: The tube and the flap MUST be of the correct size for the tyre. If in doubt, ascertain from the tyre manufacturer the suitability of a tube for use with radial ply tyre. Fitting of a new tube and a flap in a new tyre is strongly recommended without exception.

Tube failure in service may cause a tyre to fail with it. Tubes that are too large, or used tubes which have excessively worn in service, may crease or crack inside a tyre with serious risk of tyre deflation and loss of vehicle control. Tubes too small will stretch too much resulting in a quicker loss in its physical properties. Creased, cracked and weak tubes are unsuitable for service and should be replaced with new tubes.

In the event of tyre/tube repair, the tube removed from a tyre is separated and reused in the same tyre after the repair is carried out. Where tubes are fitted inside tyres, with or without flaps, the inflation pressure in tubes should be maintained at the minimum required to allow a tube to stay in shape inside the tyre.

Flaps: Whenever the use of a flap is specified by the tyre manufacturer, ensure that the correct size/code recommended for the tyre/rim size combination is used without exception. Old or damaged flaps must not be reused. Used flaps should be lightly rubbed over with dusting chalk and used without distortion in the same tyre, if possible.

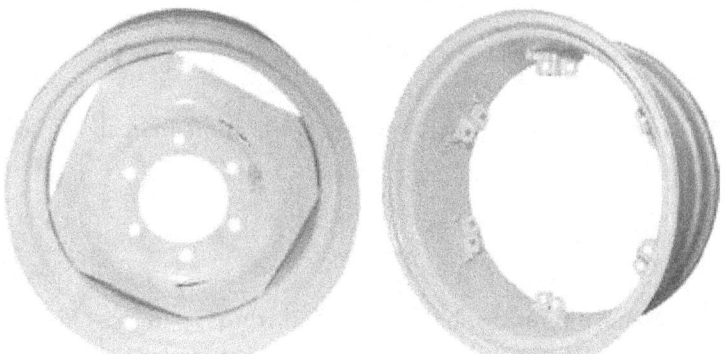

Figure 7-10: Tractor rims (6.00 x 16 and 14.00 x 30 rims)

Tyre rotation

To utilize the full potential life built into a tyre, periodic rotation is necessary. Changing the position of tyres at regular intervals is generally favoured in order to compensate for any difference in the working tread wear pattern, or uneven wear due to different working positions, thereby to obtain a longer overall tread pattern life, tyre efficiency, and stability in cornering and braking.

Standard pattern of rotation for truck / bus / LCV

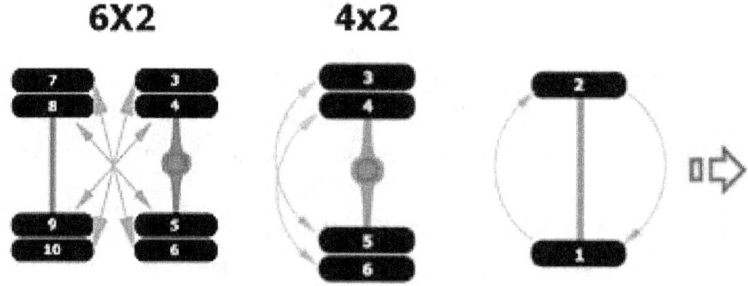

Figure 7-11: Tyre rotation for 4x2 and 6x2 axle

Direction of rotation is changed along with wheel position

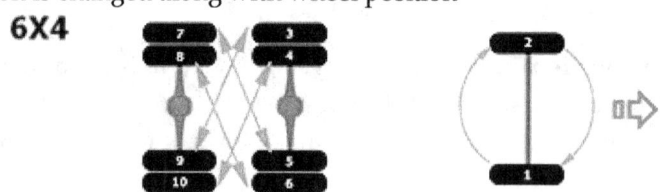

Figure 7-12: Tyre rotation for 6x4 axles

Direction of rotation is same but wheel position is changed

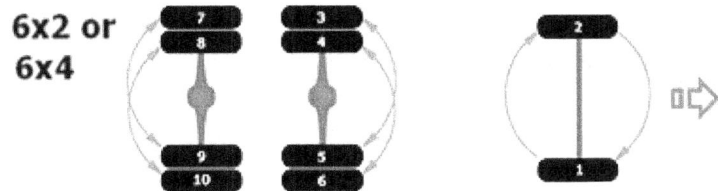

Figure 7-13: Tyre rotation by change of wheel position

Direction of rotation is same but wheel position is changed, if the tyres are not fitted on the same day in rear axle

Tyre matching

Mismatching of dual tyres imposes overload on the larger diameter tyre, causing it to over deflect and so get overheated. The smaller diameter tyre, lacking equal road contact, wears irregularly and faster than normal.

Tyre spacing

Proper spacing between two tyres (duals) is necessary to prevent the adjacent tyres getting each other rubbed at the sidewall. Such a contact generates heat & causes thinning of the sidewall rubber and may lead to separation and other premature failures.

Axle load distribution

Improper load distribution shortens tyre life. When the load is not distributed equally between the axles, the tyres on one of the axles will be overloaded(Figure 7-14).

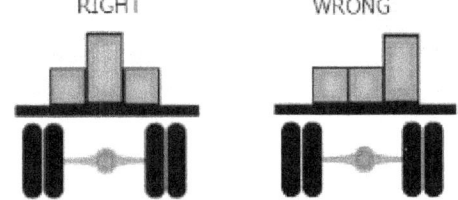

Figure 7-14: Lateral load distribution on tyres

If the tyres on one side of the truck, or trailer, are overloaded(Figure 7-15), this may affect traction, causes the driving wheels to slip at the lighter side, resulting in fast wear.

Figure 7-15: Transverse load distribution on tyres

FARM TRACTOR SYSTEMS

Tyres for agricultural uses

Efficient work on rain forest and tropical grassland and in the horticultural industry requires an extensive and varied range of machines. Land clearing machinery, ploughs, cultivation equipment, planting and harvesting, lawn mowers and horticultural machines each have a specific function, which they have to carry out under widely varying conditions. The weight of these machines also differs considerably.

Figure 7-16: Tractor tyres

As a consequence of these factors, this sector requires a wide diversity of tyre sizes and treads specifications. Tyre manufacturers have developed a diverse range of tyre sizes. Despite these diverse ranges, all the tyres share characteristics such as optimal protection of the soil structure, good stability, a sturdy and flexible carcass construction and a long life-span

Traction tyre: Crop care remains a very important factor in modern horticulture. Compact, one-axle machines, fitted with a plough, tiller or cultivator are the most frequently used tools. Traction tyres are specially developed for this type of work, and offer both traction and stability due to its open tread. It has excellent grip, perfect self-cleaning ability and are suitable for both small agricultural tractors and fieldwork equipment

Figure 7-17: Traction tyres

Features of traction tyre features include;

- Traction treads with narrow or wide tread cleats
- Special shape of the tread cleats
- Available in a wide range of sizes

Furrow wheel tyres: These has smooth contour to ensure that the tyre never becomes clogged, making it ideal as a depth wheel on ploughs. These tyres have excellent self-cleaning properties, Shakes off clinging earth and suitable for relatively heavy applications

Figure 7-18: Furrow wheel tyres

Ploughing machinery tyres: Tyres for ploughing and haymaking machines have unmistakable tread rings, guaranteed for optimal stability and good self-cleaning properties. The strong, flexible carcass is resistant to peak loads. They have relatively large contact surface, outstanding height/width ratio and strong nylon construction. The tread ring also gives extra protection against penetration by thorns and stones. It has such benefits as high resistance to punctures, good self-cleaning ability, minimal rutting and soil compaction, good stability, long life-span and easy to fit.

Figure 7-19: Ploughing and haymaking machine tyres

Tyre for horticultural uses: These are ideal solution for small horticultural and hay making machinery. They are manufactured for these types of machines with an emphasis on the smaller sizes.

Figure7-20: Horticultural tyres

They are characterized by wide lateral grooves, rounded contour and sturdy carcass. The tyre has good sideways stability, prevents damage to the turf and has long life-span.

Lawn tyres: The attractiveness of golf courses, public gardens and parks is highly influenced by the quality of the lawn. Only specially developed tyres are good enough for the strict daily maintenance schedule involved. They are very popular tyre for golf carts, equally suitable for agricultural machines, has minimal soil compaction and rutting ability, excellent shock absorption tendency and minimal damage to the grass.

The tyre for small machines and tools are used in maintenance activities on areas such as lawns and sports fields. Self-propelled lawn mowers, small driven machines and small-sized modes of transportation will perform superbly with the lawn mower tyre, without leaving marks on the lawn.

Figure 7-21: Lawn tyre

Features of lawn tyres includes

- Attractive, functional tread for use on grassland
- Relatively large and flat contact area
- Flexible sidewall construction
- Rounded shoulders

Tractor differential system

The differential is a set of gears that allow the powered wheels in a vehicle (particularly tractors) to turn at different speeds, especially when going around a curve or bend. A differential is therefore necessary to allow tractors to be steered effectively while in operation.

Figure 7-22: Tractor differential system (Valtra T series, 2002)

The differential assembly (Figure 7-22) normally includes a large spiral bevel gear that drives the differential and other small bevel gears (Figure 7-23). The larger gears (crown wheels) are the output gears which can rotate at any speed as required for turning (Figure 7-24).

Figure 7-23: Allis Chalmers B tractor differential gears

A four-wheel drive vehicle has a total of three differentials: one in the front axle, one in the rear axle, and a third between the two axles that divides power between the front and rear axles. In a *rear-wheel drive*(RWD) vehicle, the differential is located in the rear axle with metal shafts connect it to each of the rear wheels. The rear wheels receive power from the engine, via the transmission(Figure 7-24).

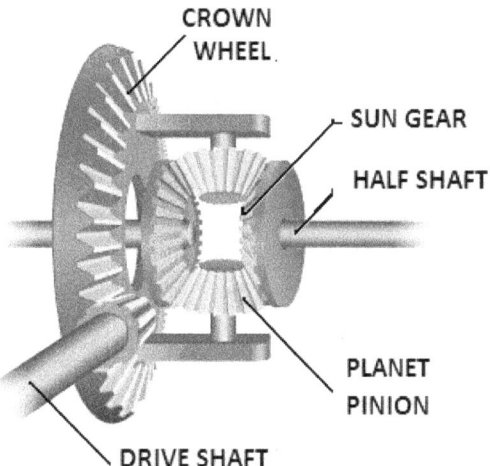

Figure 7-24: The differential gear system

The differential in a *front-wheel drive* (FWD) vehicle is contained in the transaxle, which transfers engine power to the metal shafts leading to each of the front wheels.

When a vehicle negotiates a turn, wheels on the outside of the curve must travel farther and rotate faster than the inside wheels. Without a differential, the wheels turned by an axle would not rotate at different speeds. Attempts to turn corners without a differential could damage the gears and shafts connecting the wheels to the axle, and the axle to the engine.

FARM TRACTOR SYSTEMS

With the differential system, when turning occur, power delivered to the wheels will not be the same. When one wheel is locked by braking (Figure 7-24(left shaft)), all the power is delivered to the opposite wheel (Figure 7-24(right shaft)).

Figure 7-25: Turning a corner with differential system

A differential sends most of the engine's power to the wheel with the least load or strain on it. If one wheel temporarily loses traction while the tractor is underway, a sudden surge of power to the wheel can cause the tractor to swerve dangerously when the wheel regains traction.

Figure 7-26: Differential locked together

Practical exercise

Students are expected to identify the various components of the following tractor systems and drives:

1. The front axle components
2. The tractor differential

FARM TRACTOR SYSTEMS

PART 3

TRACTOR AND ENGINE SYSTEMS MAINTENANCE

Engine tune-up, overhauling, tractor systems maintenance, repairs, and installation

FARM TRACTOR SYSTEMS

Introduction

Worn or poorly maintained engine systems have resulted to delay and drudgery of production and an increase in the running costs of farms and organizations. Sometimes malfunctioning of such engines could lead to accidents which could result to injury or death in extreme cases. Frequent breakdown of equipment has resulted in many downtime costs, economic wastes and low production output.

During the "standing" or non-use periods of a particular machine, chemical interactions between metals and fluids can cause more damage than normal wear and tear from active usage. Thus this must be considered in planning machinery maintenance.

To ensure optimum performance of vehicle systems should be regularly checked; kept clean and correctly adjusted. This chapter is designed to acquaint students with the rudimentary knowledge of performing engine tune up and overhauling in diesel equipment technology. At the end of this chapter, the students are expected to have gained knowledge of engine problem diagnosis and how to prepare to carry out engine and machinery testing, maintenance and repairs.

Learning objectives

Expected skills to learn

The students should at the end of this section be able to

1. Diagnose engine problems and prepare for engine testing.
2. Demonstrate knowledge of tune-up/adjustment of a diesel engine, and
3. Demonstrate knowledge of troubleshooting a diesel engine.

Activities

The student will be required to do the followings at the end of his training:

1. Define terms associated with engine operation and maintenance.
2. List inspections to include in a checklist during normal operation.
3. List factors to include in the procedure for stopping a diesel engine.
4. Match the causes of an engine misfiring with the corrective actions.
5. Identify causes of engine knock.
6. Identify causes of an overheated engine.
7. Identify causes of smoky exhaust.

CHAPTER 8

Engine Tune Up and Overhauling

Introduction

Tune-up is the process of making checks and minor adjustments to improve the operation of a particular machine. The objective of system tune-up is therefore to restore such a system to the performance levels intended by the manufacturer. Such systems include engine, tractor, machinery and automobiles.

Engine tune-up is the most important preventive maintenance that can be performed on any machine. However, most operators make the mistake of performing tune-up only when the engine is not running satisfactorily. This is often a costly error because, at this point, the engine may have been worn or damaged and thus requires major repair or overhaul. This can be very costly to the operator.

Basic requirements for engine maintenance work

A maintenance service practice for diesel engines consists of a complete maintenance program that is built around records and observations. The maintenance program includes appropriate analysis of the following records.

1. *Record keeping:* Engine log sheets are an important part of record keeping. The sheets must be developed to suit individual applications (i.e., auxiliary use) and related instrumentation. Accurate records are essential to good operations. Notes should be made of all events that are or appear to be outside of normal range. Detailed reports should be logged. Worn or failed parts should be tagged and protectively stored for possible future reference and analysis of failure. This is especially important when specific failures become repetitive over a period of time which may be years.
2. *Log sheet data:* Log sheets should include number of times engine starts and stops, fuel and lubrication oil consumption, and a cumulative record of the following:

    a. Hours since last oil change.
    b. Hours since last overhaul.
    c. Total hours on engine.
    d. Selected temperatures and pressures.

3. *Engine troubleshooting procedure:* Perform troubleshooting procedures whenever abnormal operation of the equipment is observed. Maintenance personnel should refer to log sheets

for interpretation and comparison of performance data. Comparison of operation should be made under similar conditions of load and ambient temperature. The general scheme for troubleshooting includes the followings.

a. *Industrial practices*. Use recognized industrial practices as the general guide for engine servicing. Service information is provided in the manufacturer's manual.
b. *Reference literature*. The engine user must refer to manufacturer's manuals for specific information on individual units.

Planning for engine maintenance

Engine maintenance is a process and must be adequately prepared for. Thus take into account the following points/precautions when planning machinery maintenance:

✓ Inspect machinery at beginning and end of each farming or harvest seasons. Repair and adjust as required. In special cases of the machinery being in continuous use throughout the year, inspection should be based on hours of operation per day, week or month. This situation is common in the tropical Africa where machinery serve a multiple purpose of hauling, field operation and other off-farm activities.
✓ Carry out maintenance work without pressure between seasons. Take out enough time for maintenance before the farming season sets in. A good time to carry out general lubrication, oil changes and filter changes is at the end of seasonal operations. First-grade quality fresh oils and lubricants at this stage provide the best protection for metal surfaces during storage periods.

Figure 8-1: Preparing tractor for inspection

✓ Follow manufacturer's recommended oil changes during working periods.
✓ Engines and hydraulic systems should be thoroughly warmed up periodically during periods of non-usage. Complete warming-up of both systems helps to keep bare metal surfaces covered with protective oil film and drive off condensed water vapour in the oil.

It is very essential to keep adequate record of maintenance works performed on particular machinery. This will enable the workshop manager to have a maintenance history and performance characteristic of individual machinery and also help to determine the rate of depreciation and reliability.

Maintenance record is particularly non-existence in most public and private organizations, a fact attributable to a high rate of emergency maintenance and the urge to get work done quickly.

Engine tune up

Tune-up exercise calls for the replacement of parts that are approaching or have reached the end of their useful life (refer to Figure 2-6), and servicing those parts that can be restored to initial "new part". Adjustments should also be made to the various engine systems so that they meet the manufacturer's specifications thus contributing effectively to the engine performance. Tune-up involves thorough checks for leakages, dirt, adjustments and probable replacement on the engine system. A tune-up should be as complete as possible.

When should an engine be tuned-up?

The intervals for tune-up may vary from 500 to 1000 hours depending upon the operating conditions. Regularity is the key to tuning up the engine so that major problems are prevented. A badly worn engine cannot be tuned up but rather overhauled (Figure 8-2). This is why the engine should first be checked to see if a tune-up will restore it, or a major overhaul is needed.

Figure 8-2: An overhauled Ford 6600 engine

FARM TRACTOR SYSTEMS

Engine tune-up chart

The following chart gives you the quick steps necessary to tune-up your engine. Check off each step that applies to your engine and make the necessary replacement of parts or adjustments as needed. Make sure you understand the function and purpose of this chart.

Table 8-1: Engine tune-up chart

No. of Step	Operations to be performed
	Engine
1.	Check cylinder head gasket for external leaks.
2.	Retighten cylinder head cap screws.
3.	Adjust valve tappet clearance.
4.	Check engine compression.
	Air intake and exhaust system
1.	Clean out precleaner.
2.	Remove and clean air cleaner.
3.	Inspect exhaust system and muffler.
4.	Check crankcase ventilating system for restrictions.
5.	Check intake manifold for leaks.

No. of Step	Operations to be performed
6.	Check air intake for leaks or restrictions.
7.	Check air induction hoses and clamps for leaks and breaks.
8.	Check colour of exhaust smoke under light load.
9.	Check colour of exhaust smoke under heavy load
	Ignition system (Spark-ignition engines)
1.	Spark plugs – Clean and adjust gap.
2.	Check spark plug wires.
3.	Distributor – Check the following items:
4.	Cap and rotor
5.	Breaker points
6.	Breaker point gap
7.	Cam lubrication
8.	Distributor timing
	Fuel systems
1.	Check fuel lines for leaks or restrictions.

2.	Clean fuel pump sediment bowl.
3.	Clean fuel strainer or filter.
4.	Check radiator for LP-Gas leaking from converter into cooling system.
5.	Drain sediment from gasoline or diesel fuel tank.
6.	Bleed diesel fuel system.
	Lubricating system
1.	Check operation of pressure gauge or light.
2.	Drain and refill crankcase.
3.	Replace oil filter.
	Cooling system
1.	Check water pump for leaks and excessive shaft endplay.
2.	Inspect radiator hoses and clamps.
3.	Clean and flush cooling system.
4.	Test the thermostat operation.
5.	Check radiator for leaks.
6.	Check radiator for air bubbles or oil indicating compression or oil leakage.
7.	Check condition of fan belt.

No. of Step	Operations to be performed
	Electrical system
1.	Check battery.
2.	Clean battery, cables and terminals.
3.	Tighten battery cables and battery hold-down clamps.
4.	Coat battery posts and cable clamps with petroleum jelly.
5.	Check specific gravity of electrolyte & add water to proper level.
6.	Check generator or alternator.
7.	Check belt tension.
	Transmission system
1.	Clutch pedal free travel
2.	Check free travel at clutch pedal.

Appropriate tools required for engine tune-up exercise

Your tools are the heart of your craft. Tools help you do your job with a high degree of quality and ease. Proper maintenance of tools and other equipment is very important. Tools can do something else, too. They can cause injury or even death! You must use the right tools for the job. Inadequate maintenance can cause equipment to deteriorate, creating dangerous conditions. You must take care of your tools so they can help you and not hurt you.

A good selection of hand tools is required for every workshop. These can be stored on a display board or in a multi-shelved toolbox so that the right one can be easily located. A set of both metric combination and imperial (A/F) tool are necessary to cover the variety of nut and bolt sizes found on equipment. Reduce your risk of injury by using the following guidelines to select hand tools:

Know your job

Note that tools are designed for specific purposes. Therefore, before you select a tool, think about the job you will be doing. Using a tool for something other than its intended purpose often damages the tool and could cause you pain, discomfort, or injury. You reduce your chances of being injured when you select a tool that fits the job you will be doing.

Look at your work space

Now look at your work space. Awkward postures may cause you to use more force. Select a tool that can be used within the space available. For example, if you work in a cramped area and high force is required, select a tool that is held with a power grip. A pinch grip will produce much less power than a power grip. Exerting force with a pinch grip means you will work harder to get the job done.

Select the tool

Over time, exposure to awkward postures or harmful contact pressures can contribute to an injury. You can reduce your risk of injury if you select hand tools that fit your hand and the job you are doing.

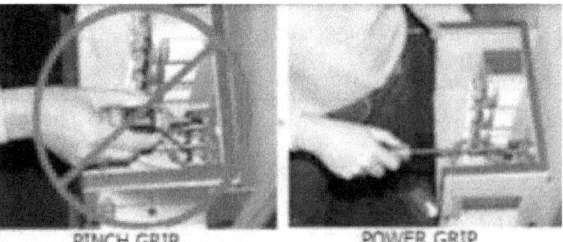

Figure 8-3: Selecting tool for work

The following tools and materials among several others have been identified as essential for engine tune-up programme:

✓ *Machine service manual*: A copy of the machine service manual/machinery maintenance workbook is required

- ✓ *Jack stands or tripod*: To support weights on the jack.
- ✓ *Battery clamp cleaning tool:* For cleaning battery, terminals and connecting battery terminals firmly
- ✓ *Masking tape:* Used for covering naked or peeled joints
- ✓ *Spark plug socket and ratchet*: This is required to remove plugs for cleaning
- ✓ *Lubrication oil and oil filter*: Required for oil change and lubrication system service
- ✓ *Inspection light*: Required to illuminate dark areas during servicing
- ✓ *Pen knife*: To cut sealed replacement parts and scrapping off of carbon from components
- ✓ *Set of Allen wrench/keys*: Required to loose bolts with Allen heads
- ✓ *Feeler gauge*: Required for gap adjustments in plugs and valves
- ✓ *Safety glasses or goggles*: Safety goggles or safety glasses shall be worn to protect against flying chips, nails, or scale. The use of safety glasses will help prevent eye injuries.
- ✓ *Pieces of hand towel/rag*: Pieces of rags must be made available to wipe off grease and oil from surfaces
- ✓ *Jack; hydraulic trolley jack*: This is required for lifting of heavy loads

Figure 8-4: Hydraulic jack

- ✓ *Diesel injector tester (DET)/petrol-nozzle-tester (BET)*: This device can be used to test the opening pressure of Diesel and petrol injection nozzles.
- ✓ *Personal protective equipment (PPE):* At every appearance in the machinery repairs workshop, students are not allowed to appear in plain clothes. Appropriate personal protective equipment, e.g. safety boots with iron nose, hand gloves, gauntlets etc must be worn due to hazards that may be encountered while using portable power tools and hand tools.

Also you need to equip yourself with torsion tools such as:

- *Set of spanners*- Various sets of ring, flat and or combination spanners are required in order to select matching ones for a particular operation.
- *Wrenches*: These includes open-end or box wrenches, socket wrenches, adjustable wrenches and pipe wrenches

Figure 8-5: Combination spanners

- *Adjustable wrenches*: Adjustable wrenches are used for many purposes. They are not intended, however, to take the place of standard open-end, box or socket wrenches. They are used mainly for nuts and bolts that do not fit a standard wrench. Pressure is always applied to the fixed jaw.

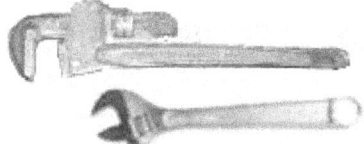

Figure 8-6: Adjustable wrenches

- *Open-end or box wrenches:* Open-end or box wrenches shall be inspected to make sure that they fit properly and are never to be used if jaws are sprung or cracked.

Figure 8-7: Socket wrenches

- *Socket wrenches*: Socket wrenches give great flexibility in hard- to- reach places. The use of special types shall be encouraged where there is danger of injury.

Figure 8-8: Socket wrench

- *Cutters and pliers:* Side cutting pliers sometimes cause injuries when short ends of wires are cut. Pliers shall not be used as a substitute for a wrench.

Figure 8-9: Cutters and pliers

- *Grips*: Griping tools include pipe tongs, pliers, special cutters, adjustable etc.

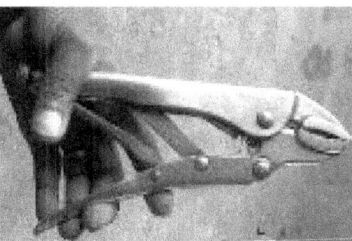

Figure 8-10: Adjustable grip

- *Screwdrivers*: Flat head, star head/Philips. The practice of using screwdrivers for punches, wedges, pinch bars, or pries shall not be allowed. Cross-slot (Phillips-head) screwdrivers are safer than the square bit type, because they have lesser tendency to slip.

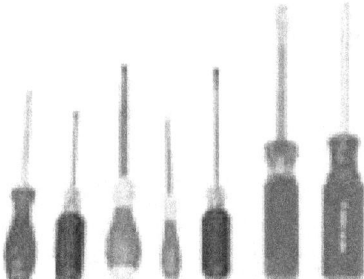

Figure 8-11: Screwdrivers

- *Hammers*: A hammer is to have a securely wedged handle suited to the type of head used. The handle shall be smooth, without cracks or splinters, free of oil, shaped to fit the hand, and of the specified size and length.

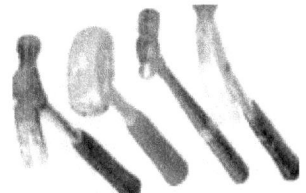

Figure 8-12: Shock/impact tools

- *Grease gun:* This is required for high pressure grease application through greasers (nipples) and in hidden parts of the machine.

Figure 8-13: Grease gun

Maintenance of petrol engine

Carburetor

Unnecessary adjustment of carburetor can results in loss of power, high fuel consumption and burned valves. The maintenance of the carburetor consists of chiefly checking nuts and bolts for tightness, removing and cleaning of jets and adjustment of idle speed and load mixture. It also includes draining of the sediment bowl to remove small particles and water condensation.

Fuel strainers

A strainer is often located in the carburetor at the end of the fuel line. Occasionally, this strainer should be cleaned to remove any sediment trapped there.

Adjusting engine idling speed

On vehicles with conventional (carburetion) systems, idling speed can be adjusted by locating the *idle set screw* at the throttle linkage on the carburetor or *anti-dieseling solenoid* in an injection fuel supply systems. Turn clockwise to increase idle speed; turn counterclockwise, to decrease idle speed.

Idling speed adjustment procedure

Here are some points to remember and procedures to follow when adjusting the idle speed:

- Be sure the engine is fully warm and off the fast idle cam when adjusting the curb idle.

Figure 8-14: Idling screw on carburetor

- If the idle is set with the air cleaner off, double check it with the cleaner unit in place.

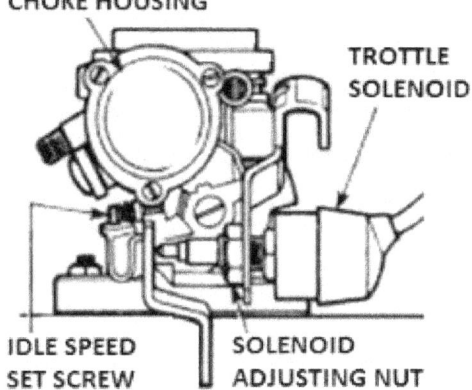

Figure 8-15: The position of idling screw

- Never try to smooth out the idle by turning up the speed; a higher idle just wastes gas and contributes to dieseling.

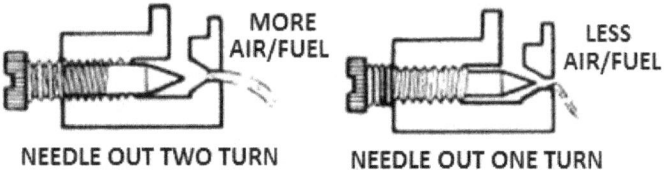

Figure 8-16: Controlling the air/fuel mixture

Distributor cap/ignition rotor should be inspected and replaced when replacing spark plugs and spark plug wires, or when a "major tune-up" is called for. The distributor cap is where the other ends of the spark plug wires connect to, and the ignition rotor in underneath the distributor cap. Some newer model vehicles do not have a distributor at all. These cars are designed with distributor less ignition systems (DIS), and therefore do not have these parts.

Fuel

Gasoline that is intended for automotive use should have an octane rating higher than 77. It is recommended that a lead-free gasoline be used because lead-free gasoline leave fewer deposits and tend to prolong valve life. It is not recommended to use any gasoline that contains alcohols such as gasohol, ethanol, or methanol. Poor fuel could cause engine knock, detonation or pre-ignition.

Engine knock and pre-ignition checks

When a part of the fuel ignites spontaneously, causing a sharp pressure rise in the cylinder, the result is an audible 'ping' sound known as knocking. Detonation or 'knocking' is thus not a specific breakdown of a component of the engine but a phenomenon which occurs following the normal ignition in spark ignition engines. As combustion pressure increases, the fuel becomes unstable and decomposes instantly, resulting in an explosion thereby producing a "knocking sound".

Detonation often occurs on the side of the cylinder opposite the spark plug. In preventing detonation, the octane rating of the fuel used must be high enough (77 and above). Octane number is the measure of a fuel's ability to resist detonation.

Pre-ignition occurs when the mixture is ignited by some abnormally hot surface such as hot plug in the combustion chamber. The burning starts ahead of normal ignition. Detonation and pre-ignition produces the same harmful effects on small engine parts. The bearings and mechanical surfaces can be distorted by excessive pressure and piston damage beyond repairs.

Possible causes of detonation

1. When the fuel is not the proper quality for the engine
2. When the engine is accelerating rapidly
3. When the engine is overloaded
4. Extreme heat and pressure created by rapidly burning fuel and air mixture
5. When the lubrication system is faulty
6. When there is lean air-fuel mixture and hot plug
7. When there is advanced engine timing
8. When there is lugging of the engine

Spark plug cleaning and setting

The ignition system of a gasoline engine consists of the spark plug, distributor and the ignition coil. These systems must be well maintained if the engine must operate well.

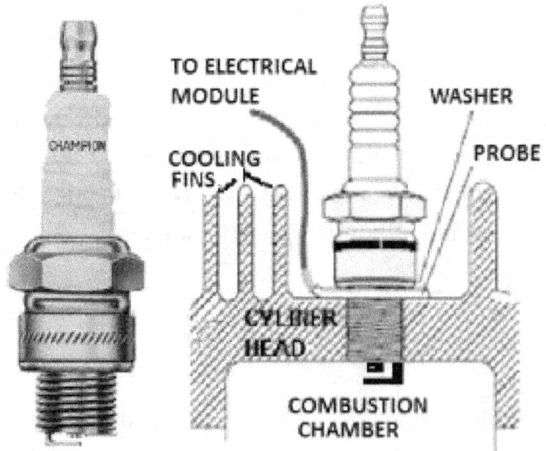

Figure 8-17: Spark plug installed on cylinder head

Spark plug problems: The following problems are associated with plug failure; when the engine fails to start, when the engine is not running smooth i.e. Engine vibration, noisy engine operation, and a smell of fuel in the exhaust smoke.

Spark plug servicing: Spark plugs should be cleaned and gapped, or replaced after 100 hours of operation. Since spark plugs operate under severe conditions, they are easily fouled.

Figure 8-18: Spark plug examination

Remove the spark plug and examine it carefully. If the electrode is burned away or pitted, or if the insulator is cracked, discard the plug and get a new one of the same designation. You will find the designation printed on the plug.

If the plug looks fine, use a penknife or a small wire brush to scrape carbon from around the electrode. If an air compressor is available, complete the cleaning with a blast of air. Solvents may also be used to complete the cleaning. A small file or wire brush is recommended to file the surface of the electrode. Do not allow any foreign material to fall into the cylinder while the spark plug is removed.

Figure 6-19: A fouled spark plug

Whether the plug is new or not, set the electrode gap to the manufacturer's specification. The recommended gap for most small engines is 0.030 inch (0.75mm). A flat feeler gauge can be used to check the gap when the spark plug is new (Figure 8-20 Briggs & Stratton Corp); however, a round-wire gauge should always be used to check and adjust worn spark plugs.

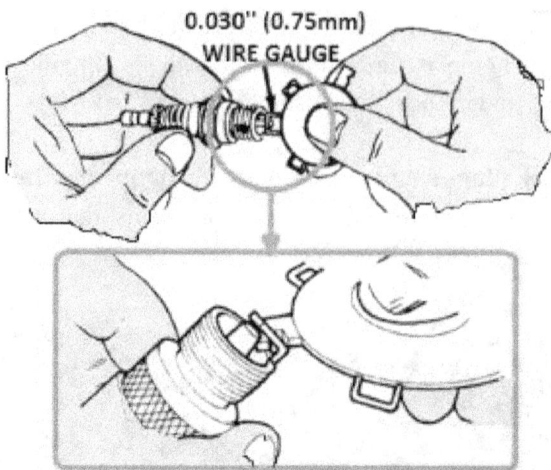

Figure 8-20: Setting spark plug gap

Worn or faulty spark plugs can cause misfire, poor fuel mileage, loss of power, and slow or extended starting time. Spark plug wires should be examined for partial contact or current leakage when replacing spark plugs. Faulty wire could lead to loss of power jerky movement etc.

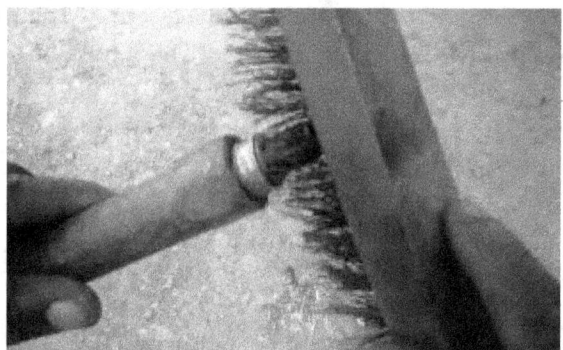

Figure 8-21: Cleaning pug with wire brush

Reading spark plug: To read a plug, you must first properly prepare it for reading, so it will give you the real reading. The best way to read the plug after removing it from the engine is to use an illuminated magnifier. With that, you can see small details and get more information. The main indicator for reading is the general appearance of the insulators.

- If it looks like it has been too hot, then it has. If it's showing some deposits, chances are it was running too cold.
- If the insulator has a nice off-white or slightly tan/gray colour with new-looking electrode shapes, this is an indication of no problems,
- If you find black fouling, it could suggest a too-cold plug, or if the plug burned white and is looking as if it melts, you may certainly change the plug, provided you have the recommended normal plug in the engine.

Figure 8-22: Spark plug gaps and electrode shapes

Spark plug wires servicing

Faulty spark plug wires should be replaced when replacing spark plugs to get maximum performance and life expectancy of spark plugs. It is advisable not to shunt current between the distributor and the coil. This could be extremely dangerous and could cause battery explosion resulting inn acid burn.

Procedure for replacing the spark plug

Spark plugs are the most critical replacement exercise of the tune-up procedure. To remove spark plug without damage, it is necessary to adhere strictly to this procedure before removal is attempted,

- Make sure the engine is warm (hand touch after cooling down). Do not remove plugs when the engine is extremely hot or cold soaked. This increases the chance the threads could be damaged.
- Remove the coil-on-plug assemblies by carefully grasping the boot and twisting it back and forth and thoroughly blow out the spark plug wells and surrounding valve cover area with compressed air.

Plug removal:

Proceed as follows:

Turn out the spark plugs, no more than 1/8 to ¼ of a turn. Work one plug at a time to avoid mixing up the wires. NOTE: Protect your eyes by wearing safety glasses or goggles when performing this process.

- Label each plug with masking tape so that it is reinstalled on the correct cylinder.

FARM TRACTOR SYSTEMS

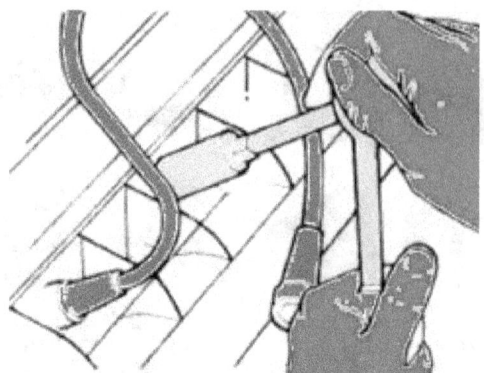

Figure 8-23: Changing spark plug

- Use a spark plug socket and ratchet to remove the plugs. If they are difficult to reach, you might also need extensions and universals. Caution ⚠ Be sure to center the plug in the socket. If held at an angle, the socket could break the plug insulator.
- Thread the plugs into the holes by hand. Finger tighten them another 1/8 of a turn, using the socket wrench and torque to manufacturer's specifications. Do not over tighten. On most small engines the torque specification will be from 18 to 22 foot pounds. Do not over tighten!

Removal of broken spark plug

Spark plugs could break in two different modes, even after following the General Spark Plug Removal Procedure:

Mode 1: The ground electrode shield is left behind thoroughly blow out the spark plug wells and as an empty shell.

Mode 2: The porcelain center and ground electrode shield is left behind and only the upper jamb nut comes out. In this case more soaking is required and long-reach nose pliers should be used to grasp and remove the porcelain center from the ground electrode shield.

Ignition system/armature maintenance

To ensure peak performance of the small engine's ignition system, the coil, armature, and flywheel magnet must remain dry, free from corrosion, and properly adjusted. To remove any deposits of oxidation on the flywheel magnet, use emery cloth, or an abrasive tape. To adjust the armature, follow the steps below as indicated in Figure 7-27.

Armature adjustment procedure

Step 1: Turn the magnet away from the armature

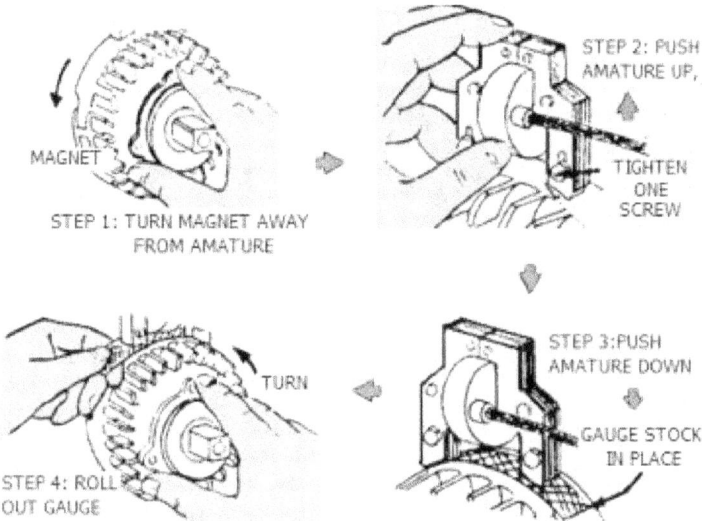

Figure 8-24: Armature adjustment procedures

Step 2: Loosen the armature mounting screws slightly and push it up and tighten one screw.

Step 3: Insert the gauge and push down the armature. The air gap between the flywheel and the armature should coincide with the manufacturer's specifications, but a good rule of thumb is to use a business card or a shim as a gauge. Simply place the card or shim between the flywheel, (Figure 8-24 step 4), and the armature, making sure that there is no space left between the three, and re-tighten the armature.

Step 4: Roll the flywheel to remove the gauge.

Replacing the distributor cap and rotor

Remove the distributor cover and clean the inside (Figure 8-25). If the terminals are corroded or if the cap is cracked or shows signs of carbon tracking (lines running from one terminal to another, replace the cap.

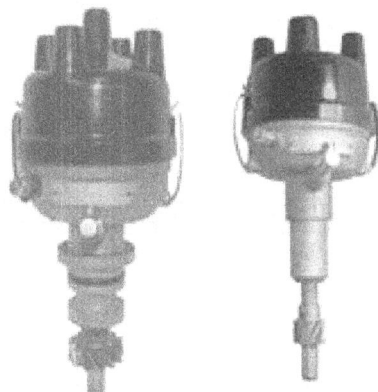

Figure 8-25: Distributors

Remove the rotor, Figure 8-33. Examine it for cracks, chips, and carbon tracking. If the rotor is cracked or if the tip is burned or corroded, replace it.

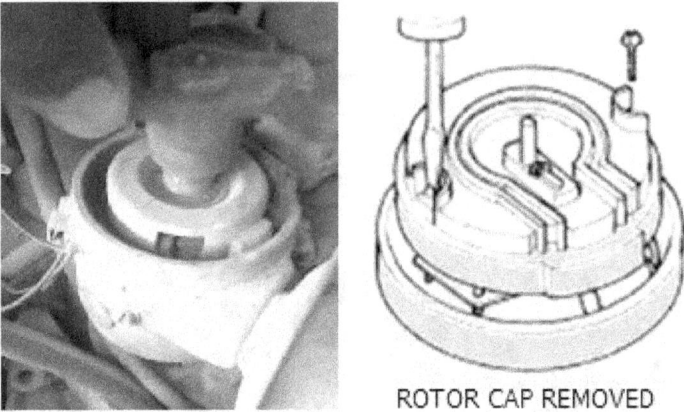

Figure 8-26: Removing the rotor and cap

Maintenance of Diesel (tractor) engine systems

A regular maintenance on the following tractor engine systems is required for efficient operation. The process, likely problems and solution that could be encountered are discussed.

Air intake system

Air filter life can be extended by reducing the amount of dirt getting to the element with the use of a pre-cleaner such as ingestive type "Visibowl" or ejective type "Sy-Klone". The pre-cleaner should need to be emptied regularly so that it does not become clogged and restrict the airflow to the engine.

Figure 8-27: Air filtering system

Maintenance of air intake system

The maintenance of the air cleaner consists of removing the oil cup/bowl cleaning it and refilling it with oil. Do not over-service air cleaner elements. Reducing the frequency of filter removal also lessens the chance of dirt dropping into the clean side of the air intake. Ensure that engine intake system is airtight and all hoses, clamps and fittings are in good condition and secure.

Procedure for cleaning wet element filters

Note: In every dusty territory, bowl oil level should be checked more frequently. If bowl sediments are considerable (much), wash lower filter element and clean filter at shorter intervals.

1. Remove the oil cup and wash it with diesel fuel, kerosene or approved solvent.
2. If there is water in the cup, check for a leak in the air tube.
3. Wipe the cup clean with a dry cloth

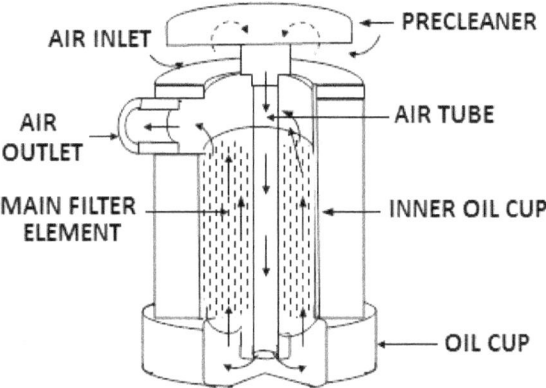

Figure 8-28: Wet- type air-filtering systems

4. Clean the tray and screen above the cup.
5. Keep the filter dry in the sun
6. Replace the tray and the filter
7. Fill the cup with clean oil to the level indicated with oil level mark.
8. Clip the cup back in place

Procedure for cleaning dry element filters

- Remove the filter element cover and bring out the filter

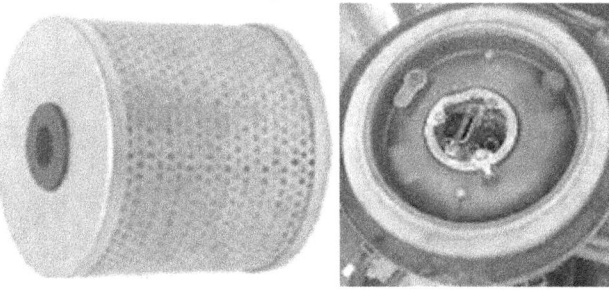

Figure 8-29: Air filters

- Hold filter in one hand and tap gently against your hand. Blow the dirt out of the filter from the inside with a compressed air. Never blow compressed air against the outside of the filter.

Figure 8-30: Cleaning the filter

- Do not wash a dry element filter in water or oil or gasoline. Replace filter if it gets oily or smoked from the engine backfiring.
- Check the safety filter element for dust too. Replace the safety filter at least once a year. Never try to clean the safety filter.
- Replace the filter.
- Couple the components back in reverse order

Tractor air filter should be checked whenever associated red light (indicator) comes on. Every year or whenever cracks are detected, replace the outer cartridge. Do not wash or blow inner safety cartridge but replaced every three times outer cartridge is cleaned, or every 400 hours.

Maintenance of the crankcase breather

The crankcase breather is often neglected because it does not require attention as often as the air cleaner. The breather is very important because it ventilates the crankcase and prevents dust from being drawn into the engine. Failure to keep the breather crankcase clean will force the lubricating oil past seals and promote excessive oil consumption. Lack of ventilation will promote the formation of acid and sludge in the crankcase.

Figure 8-31: Crankcase breather

There are many kinds of crankcase breathers. Some are located in the plate covering the push rod chamber (arrow in Figure 8-31). A common type is located in the cap on the valve cover or oil filter tube. Mist breathers contain an element of crumpled copper which is oil soaked. Dirt in the air passing this element is trapped in the oil and must be removed by washing the element. The element should be cleaned as often as necessary.

Maintenance of fuel supply system

The fuel supply system: The fuel supply system is designed to transform liquid fuel into an atomized spray, or an air-fuel mixture. After creating this air-fuel mixture, it is the job of the system to transport the mixture into the cylinder head where it will be compressed and ignited. To ensure high engine performance, the fuel supply system must effectively deliver fuel into the combustion chamber. To achieve this, all its components must be maintained.

The most important item in maintaining the fuel system is the use of clean fuel. Water is always a problem in a fuel system and may come from any of several sources – water already in the fuel, water that has leaked into the storage tank or into the tractor fuel, and water that condenses out of the air inside the tank. Condensation is the most common sources of water in tank.

Problems associated with fuel supply systems: Problem of fuel supply are caused by faulty carburetor or injection systems, dirty tanks, clogged filters, presence of water in fuel, fuel leakage, air trap in the fuel line, use of poor quality fuel and excessive heat buildup in the system. These problems often results in poor engine performance, detonation or knocking, sudden engine stoppage in mid-operation, fire outbreak, etc.

Fuel tank: The fuel tank may need to be cleaned occasional. If sediment and water condensed often in the sediment bowl, it indicates dirt in the tank and the tank will need to be drained and flushed with clean fuel. Fuel tank *should* be filled at the end of each day's work to avoid moisture from night air entering and condensing in tanks. Fuel tanks should be maintained full during storage periods to prevent interior surface corrosion from moisture laden air.

Fuel sedimenter: Some tractors could be equipped with sedimenter to completely drain off water from the tank when filled with diesel. With fuel tanks full, back off lower screw on the sedimenter to drain the water. To complete the draining, loose the upper screw on the sedimenter. Tighten the lower screw and when diesel oil started flowing out without air bubbles, tighten the upper screw.

Sediment bowl and fuel strainer: Small particles of dirt and water are trapped in the sediment bowl to prevent clogging of the small jets in the carburetor. As often as dirt or water accumulates in the bowl, it will need to be removed and cleaned. Use a metal or heat resistant sediment bowl when there is danger of dirt and leaves collecting around the bowl and catching five while in operation. Ordinarily if the glass bowls break, fuel may feed the fire.

Figure 8-32: Sediment bowl on primary filter

Fuel pump: If the system uses a fuel pump, it will be found between the tank and the carburetor or between the injection pump and the tank in tractors. There is not much maintenance that is required on the fuel pump other than to check to see that it is operating well or replaced when it fails.

Figure 8-33: Tractor fuel lift pumps

A ruptured diaphragm in the pump may permit fuel to be pumped into the crankcase. Anytime the crankcase oil is diluted excessively with fuel, or the oil level rises above normal, the fuel pump should be checked for leaks.

Fuel pump failure: Failure of the pump will be indicated if:

- The engine fails to start,
- There is loss of power at high speeds or under load,
- There is high petrol consumption, or
- There is a spray of fuel coming from the top of the fuel pump.

Fuel lines and strainers: Fittings at the ends of the fuel lines may occasionally leak as a result of vibration or impact. Usually a slight tightening of the fittings will stop the leak and this should be done as soon as the leak is detected in order to prevent further damage to the fitting.

FARM TRACTOR SYSTEMS

Figure 8-34: Primary fuel filter

Servicing the fuel supply system

The maintenance and or servicing of the fuel system is essentially that of inspecting, checking and making repairs or adjustments on the fuel tank, the carburetor, the fuel line, the fuel filter, fuel pump, injection nozzles etc. as they are needed. If clean fuel is used, only infrequently will any repair jobs need to be done.

Fuel filters

The secondary fuel filter should be replaced every four times the primary filter (first fuel cartridge) is replaced. Do not replace both filters simultaneously; replace secondary filter 40 to 50 hours after the first. If you procure a new tractor with warranty, during this period, second filter must only be removed by authorized personnel. Removal of seal on second filter and on fuel injection pump automatically invalidates the warranty in some tractor makes.

Procedure for servicing/replacing the filter

The fuel system on many types of small engines and heavy duty engines like tractor and trucks is serviced by one or a combination of the following procedures:

- Remove old and/or dirty fuel and all trash particles from the carburetor and injector pump.
- If you notice water in the carburetor, glass bowl of the primary filter, drain the fuel in the tank and wash clean.
- Examine the fuel filter, if it has gone bad replace with a new one
- Remove all jets and clean them, make through all ports, and all other carburetor parts. Change injection elements of injection pump as recommended
- Ensure that the floating cork is functioning well. The nozzles must be serviced frequently
- Replace diaphragms, filter elements, gaskets, O-rings, and/or springs when necessary.

- Replace worn or faulty parts.

Precautions against water condensation in storage tank

Taking the following precautions can prevent this phenomenon:

- Fuel tanks should be filled at the end of each day's work to avoid moisture from night air entering and condensing in tanks.
- Fuel tanks should be maintained full during storage periods to prevent interior surface corrosion from moisture laden air.
- They should be drained of moisture accumulation periodically.
- Check condition of bulk fuel supplies and periodically drain condensed moisture.
- Filter all fuel from bulk storage tanks. Moisture can pass through many standard fuel filters causing engine fuel pump damage. Prevent this by fitting filters with the clear bowl as replacement for engine mounted fuel filter only.

Procedure for fuel filter replacement and installation

- Turn off fuel valves, remove old filter.
- Clean outside of filter housing and filter base surface. Be certain old gasket is removed.
- I would suggest that you fill a spin-on filter with clean fuel before installation.
- Apply clean diesel fuel or oil to filter gasket or rubber seal. A little oil on the gasket will aid a tight seal.
- Turn the new filter element in with bear hands until the seal face touches the filter base.
- Tighten the filter ½ to ¾ turn after filter gasket after the face touches the filter base.

Caution ⚠ Fractured or leaking filters may have been damaged. Replace with new filter which is matched to the operating requirement of the engine.

Bleeding air from the fuel line

In fuel injection engines (especially diesel engines), when the engine refuses to start or suddenly stopped while in operation, the major cause could be an air trap within the fuel supply system. When you try to start the engine this air acts as a lock, preventing the normal supply of fuel into the cylinder. This air must be removed through the process of "bleeding" the fuel system before the engine could start again.

Causes of air trap in fuel supply

1. When you change a diesel fuel filter,
2. Run out of fuel or disturb the fuel system,
3. Air is trapped as bubbles within.

You may need to bleed filters, fuel pump and lines to the injectors. Here are the steps to take in bleeding air from a diesel fuel system:

Tools required: General mechanic's tool kit.

Personnel required: One mechanic and one assistant

Warning ⚠ Diesel fuel is highly flammable. Do not perform this procedure near fire, flames, or sparks. Severe injury or death may result.

Note: Have drainage container ready to catch fuel.

Injector bleeding procedure

To bleed the injector line the throttle lever must be in the full power position, and then follow these procedures.

1. Make sure there is plenty of fuel in the tank.
2. Check to ensure that the cap is turn on.
3. Open the small bleed screw on top of the engine mounted primary fuel filters (see circle) and operate the fuel lift pump (primer) by hand.

Figure 8-35: Primary fuel filter and primer

4. After the fuel filter has been purged of air, close the bleed screw (*do not tight too much,*) and open the one on the fuel injection pump. Again, after the air has been purged, close the bleed screw.

The engine should now start

If the engine could not start or runs poorly; you may have to bleed the injection line. Bleed the high pressure side as follows:

1. Loosen injection lines at the injectors about one turn. The use of two wrenches will prevent the binding or twisting of the steel lines. Usually, it is enough to bleed just half of the lines at a time.
2. Crank the engine until all air is forced out and fuel is present. When the air appears to have been purged (the fuel looks clear, not whitish and the spray will not be making squishing sound), tighten the nuts firmly, close the decompression levers
3. Engine will start to pop on one or two cylinders.
4. Tighten the injector lock nut one at a time to tell by sound which cylinders are firing properly.
5. Run the engine until it runs smoothly. This will bleed the other injectors.
6. Check for leaks and clean up any spilt fuel by:

Figure 8-36: Injection pump bleeding screws

   a. Putting a rag or paper towel around a loosened bleed screw or injector pipe nut to retain the diesel that will leak out.
   b. Use a folded paper towel or toilet paper to check around joints for leaks. Leaking fuel will quickly be absorbed by the paper.
   c. Clean the engine bay, or any other affected area, with a dishwashing liquid to get rid of any diesel smell.

Note! If the *hand priming lever* does not have any resistance through any of its travel, rotate the engine crankshaft through 360deg. Use the starter motor or crank handle. Lack of resistance in the hand priming lever could be due to the internal actuating arm being on top of the cam that drives it. Rotating the engine crankshaft 360deg will turn the camshaft 180deg, the arm will now be on the back of its cam and you should feel resistance when operating the hand priming lever.

The fuel injection pump is self-bleeding and should not require attention unless it has been removed. No attempt should be made to service the injection pump or nozzles by a non-expert. These require special tools such as diesel injector tester (DET) and know how. To test injector pump and nozzle, a diesel injector tester (DET) is used as follows;

Mode of application of DET/ BET

Dismount the injection nozzle from the engine and connect it to the tester through the adaptor pipe (1) thread. Fill in Diesel or petrol into the tank (2). By activating (pumping) the lever (3), fuel will be pumped through the nozzle holder into the nozzle. As soon as the fuel pressure exceeds the spring pressure, fine sprays of fuel will drop out of the nozzle.

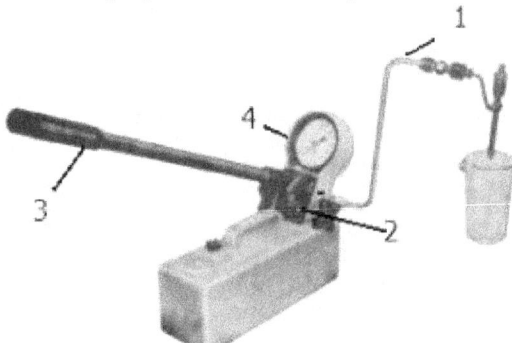

Figure 8-37: Fuel injection tester

In order to test the tightness, a pressure of about 20 bars below the opening pressure has to be built up by means of the hand pump lever. The injection valve is intact, if it does not start dropping within 10 sec. The pressure at which the nozzle opens can be seen on the pressure gauge (4). If the pressure does not correspond to the prescribed value you either have to readjust or to change the quantity of the compensating discs.

Once the pressure required for the nozzle to open is reached, the drag pointer on the gauge remains in this position so that the value can still be read later. That means, pressure gauge (4) and nozzle do not have to be kept an eye on simultaneously.

Adjusting the ignition timing

You are now ready to make the final and most important adjustment on the ignition timing. This process is highly technical and must be done under expert assistance.

Procedure for adjusting timing

- Hook up power timing light. Attach the red lead to the positive battery terminal and the black lead to the negative terminal.
- A third lead has a special connector that attaches to the number one cylinder plug wire. (The service manual will show the location of the number one cylinder pickup at the distributor cap.)
- Clean the dirt off the timing marks with solvent to see them better.

- Very carefully make a neat line with white chalk or white enamel paint over the correct timing mark. This will help you see the mark when the engine is running.

- Start the engine and let it warm. The engine must be fully warmed and idling at the proper speed to set the timing.
- While the engine warms up, check the service manual to see if there are any special timing procedures. Most engines with contact ignition systems must be timed with the vacuum advance line removed at the distributor and plugged.

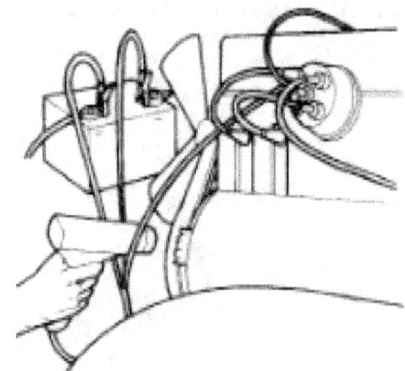

Figure 8-38 Hooking up the power timing light

- The best way to do this is by attaching a vacuum gauge in the end of the line.
- Aim the light carefully at the timing marks; not aiming the light properly is one of the primary causes of timing error. Keep the light at about a 45° angle and sight directly down the light. CAUTION: Stay away from the cooling fan blades at all times.

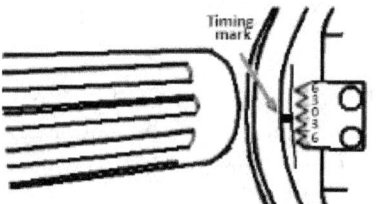

Figure 8-39 Checking the timing mark

- The light flashes when the number one cylinder fires and makes the timing marks appear to stand still. The specified mark and timing pointer will align on each flash if the engine is in time.
- If they do not line up, you will have to adjust the timing by rotating the distributor. To do this, first loosen the clamp bolt that holds the distributor in place. On older engines that bolt is easy to get to; however, on many newer engines the bolt is located behind the distributor. You might need a special distributor wrench or swivel attachment on a ratchet extension to get to the bolt.
- To adjust the timing, rotate the distributor slowly until the timing mark and pointer align. Then carefully lock down the clamp bolt and double-check to be sure the setting did not change when the distributor was tightened.

Maintenance of valve system

Valve check and examination: To begin with, before removing the top cylinder, carry out a comprehensive manual check on the valves to know which cylinders are low in compression. Examine the movement of the valve lifts before deciding whether to remove cylinder head or not. If valve motion is moving like the higher reading cylinders valves, check if the cam is worn out; if so, first fix the worn out cam. If not, proceed as follows:

- Remove the cylinder head then clean it well. Leaving the valves in their seating in the cylinder head decarbonizes the valve head by scrapping out the carbon.
- Remove the valve springs. To remove valve springs, you can do this by using a valve spring compressor or make one. An illustration of spring removal tool is shown in Figure 8-40.

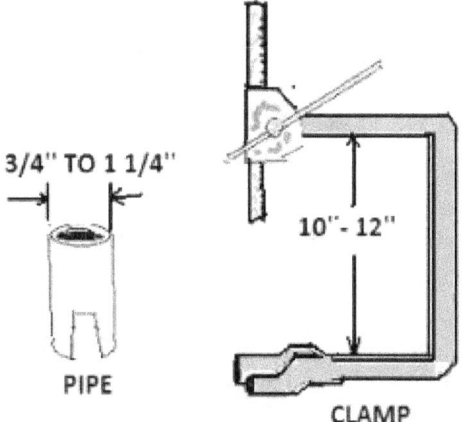

Figure 8-40: Valve spring insert

The chain is to be used with one of the two pry fork type tools made from steel bar. it will need a small hole about 3 inches from fork end to put a 1/4"screw through this attaches through the chain link itself and the chain which will be screwed snugly to the head by providing a fulcrum point to press the springs into the compressed position where you may then remove the keepers with a small magnetized screwdriver (sometimes greased) or your fingers.

Note: if the spring keeper flies off when removing the spring it may hit you in the eye so wear plastic glasses when removing the valve springs. Keep headwork on floored area, the valve keepers are hard to find if they are dropped or fly off.

- Place valve springs in box in a line maybe number the springs leave the valves in the cylinder head in the same holes (do not mix them up).
- Place cylinder head on bench or piece of wood and make a few small pieces of say 2x4x10" wood handy.
- Hook up hand drill 1/4" or 3/8" best, to a suitable plug in a drill that can run at 300 rpm.

FARM TRACTOR SYSTEMS

- Take time to clean the valves one at a time if they have carbon caked onto them. Likewise remove any greasy carbon or rust from around the immediate area of the seat, it helps a lot when grinding
- Do not clean bent valves replace them with new ones
- If the car / truck did not backfire, the intake valves may not need grinding. Used intake valves make better mileage than freshly grinded ones.
- If the exhaust valves are burned badly, they may need to be faced or changed to help with keeping efficient grinding time.

Valve adjustment: In preparation for valve adjustment on each cylinder, ensure that each cylinder is at TDC before any adjustment. Then follow the procedures listed below:

Maintenance procedures:

- Run the engine for a few seconds to ensure that the valve lifters are pumped up, (Not enough to get the engine HOT).
- If you have plug coils across the valve cover, taking them off is not necessary, tie them up to get them out-of-the-way.

  Remove ALL 4 of the plugs, (Now's a good time to change them, if you need to). If you are going to change the plugs, PLEASE, gap them at 0.032 or 0.033.

- Loosen the Allen bolts ¼ turn, at-a-time, till all are loose. (The reason we do this, is, because with a well-made metal gasket, (like the ones in our engine), loosening 1 bolt all the way out, before you get to the others, will 'possibly', put a slight, low/bow, spot in the gasket. With all the bolts loosened on the rocker cover. Take out all these bolts completely out, the rocker/head cover will lift up, and out, without the long shanks of those bolts, getting in the way.

Figure 8-41: Valve assembly of a Ford 6600 engine

- Now lift 'CAREFULLY'! You may have to 'tap' with a soft rubber mallet, to get the gasket to let go. Stick an eyeball in there, (as you are lifting), and 'watch' for the 2 dowel pins.
- Most of the time, the gasket will come off with the cover. Just be sure that you do not, kink it, or bend it.

- Got the cover off, and laying on a clean surface, Put the bolts back in to their appropriate places.
- Remove the Allen bolts holding the spacer in place, as you did the rocker cover bolts. You can actually take these all the way out, if you wish. They are all the same length.

Figure 8-42: The valve spacers and the spacer bolt

- Carefully lift up the spacer, (with gentle taping). Get the rocker and the spacer off ensuring no dowel pins is missing and the gaskets still intact and serviceable.
- With the pushrods/lifters at the bottom of their cam position, take a *'plastic'* straw, and stick it in a spark plug hole, (to 'see' if the piston is all the way up, or close). With the piston at the top, you are ready to adjust the valves. You can take the straw out, now.

Figure 8-43: The rocker valves and adjuster bolts

- Loosen the lock nut that surrounds the adjusting bolt. With the lock nut loosened, you 'should' be able to turn the adjusting nut. Back it out till there's clearance, then, slowly, turn it in, till it touches the top of the valve. Tighten lock nut.
- The adjuster, versus the other valve rocker, is a 'see-saw' effect. You can get the thinnest feeler gauge, (0.0015), you can find, and 'see' if it goes under the opposing rocker/valve. If it does, then, you have the adjuster 'touching' too much. Have to back off and try again. If you tighten down too much on the adjuster bolt, then, it 'lifts' the other rocker.

Figure 8-44: The position of the feeler gauge

- After you have got the valves adjusted, you are ready to couple back the system. Note: This is a good time to clean the outside portion of the chrome rocker & spacer covers. With the chrome cleaned, let's work on the gasket surfaces. With a 'plastic' scraper, you can get the rubber residue off, fairly easy.
- With the 2 dowel pins in place, and, the gasket in place, lower the spacer down, till it seats. Put the Allen bolts, back in their appropriate positions. Screw them in until they 'just' make contact. Tighten ¼ turn at a time, till they are all nice and 'snug'.
- Tighten the bolts away, in a criss-cross, or 'spread', pattern bearing in mind; you are 'spreading' the gasket, as you tighten.
- Get the spacer fixed and keep the gasket surfaces clean, (for the rocker/head cover). Get the chrome and the 2 dowel pins in. place the gasket placed onto the spacer.
- Place the rocker cover down on the spacer. Put the bolts back in. Tighten all bolts, till snug. Again, work in a pattern of tightening, ¼ turn at a time. Tighten/Torque them down, in a pattern.
- Once the job is finished, a 'start' of the engine is in order.

Valve grinding process

Carbon deposition on valve head and valve seating could cause gas leakage from the combustion chamber leading to loss of compression and power. If there is any cylinder with very low compression, it indicates that the valve is not well seated and such valves must be removed and grinded.

Two steps are required in grinding of valve as described below:

Step 1: Getting ready to grind valve heads

General directions for grinding valves on most engines without expensive tools and with the know-how to accomplish a very good job will require a few simple tools and materials such as:

- Hand drill (300-400 rpm) is best (Maybe 2 hand drills can speed grinding)
- Glass or bowl of water
- Cleaner for parts (carburetor spray helps)
- Valve grinding compound
- Oil or grease
- Paper towels or cotton cloth (rag)
- Possibly a sheet of wet / dry sandpaper
- Probably new valve stem seals
- Safety eye lenses or goggle

Step 2: Get down to grinding

- Keep your valve keepers out of the way in a corner where they will not spill and then begin next to grind valves, cleaning and facing of exhaust valves if necessary. Ensure you have the valve grinding compound (either water based or grease based). If it is water based, get a small bowl of soapy water nearby and a paper towel or cotton rags, water based compound is simpler and do not smell bad.

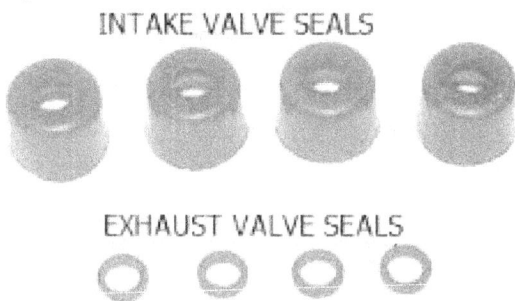

Figure 8-45: Valve keepers or inserts

- If you have the grease based grinding compound then you will need some charcoal lighter fluid or lamp oil or some petroleum (like kerosene) with a small art brush to dab it.

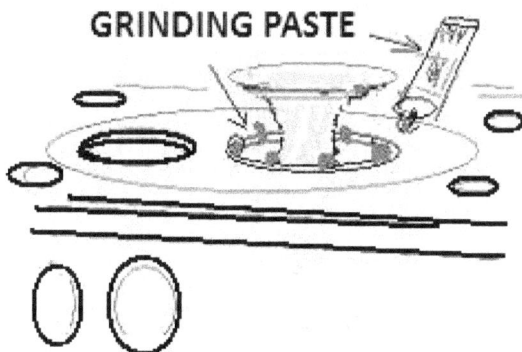

Figure 8-46: Apply grinding paste/compound

- Then put the short piece of the neoprene onto the dowel push it on say 1 inch (do not use a drill bit to turn the hose! Use a dowel or metal shaft only. Using wood you must be careful not to over tighten.
- Then push the other end onto the end of a valve (start at one end) or just do the exhaust or do all in order you decide.
- Pull the valve with the drill back into the cylinder head until the valve just touches the seat. Start the power drill slowly at first keep the valve pulled to the seat for the first 30 seconds then let the valve lift for an instant then repeat the same if possible stay around 100 rpm for a minute, after say 2 minutes then began to let valve lift off (just 1/16" to 1/8"th maximum).

FARM TRACTOR SYSTEMS

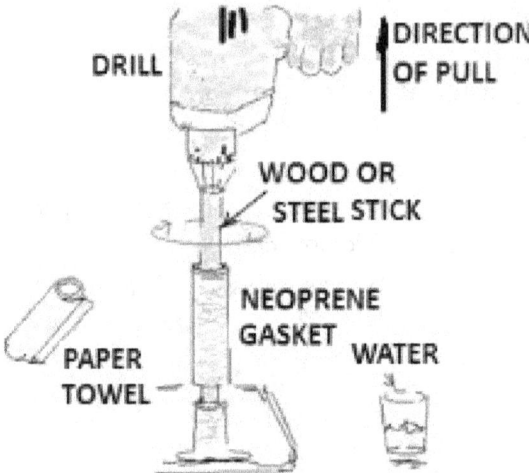

Figure 8-47: Grinding the valve

▶ Grind for 5 minutes then look at the valve lift if it is enough or out by taking the hose off the stem wash to see if the valve is showing a perfect grey white circle around the middle of where the seat is contacting the face. When it is smooth and say 1/32 to 1/16" an inch (wide) for intakes valves and or 1/16-1/8"th and inch for exhaust valves, clean both surfaces of the used compound with towel and appropriate liquid.

Figure 8-48: Grinded valve showing grey white circle

Manual grinding tool

If there are no electric hand drills, an improvised tool can be used as shown in Figure below. The tool has a hollow end and an adjuster welded to the hollow pipe to grip the valve stem.

Procedure

- Insert the valve is into the cylinder and allowed to seat properly.
- The valve stem is inserted into the hollow end of the tool and tightened with the adjuster.
- The tool is pulled back in the direction shown and twisted (turn) at ¼ turn for sometimes and the valve is checked at interval.

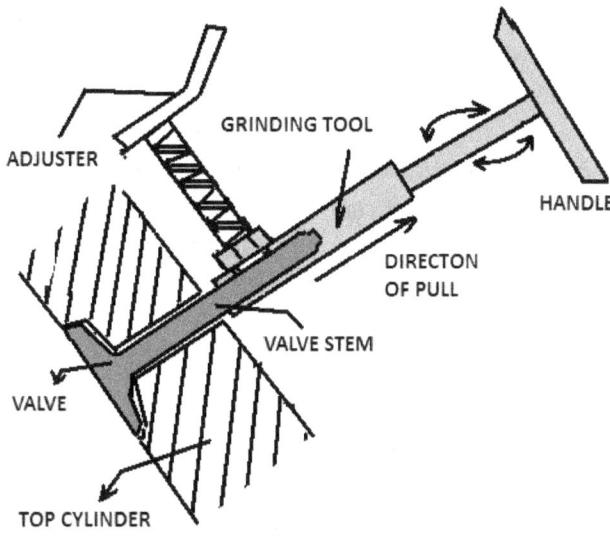

Figure 8-49: Manual grinding tool

Step 3: testing seats and valve faces quality

To check the seats of any valve suspected of being poorly fitted or how to see if your grinded valve is going to work well. This can be done as follows.

Method: When you have grinded for an interval 5 minute (valve diameter smaller then 3/4" can sometimes be ground in a minute or two!) remove the valve and wash the grinded particles off or wipe it clean enough to see if you think it is good enough you will decide this depending on the amount of tiny pits still remaining on the two surfaces (both valve face contacting area and valve seat area). When there is a smooth grey white line (circle) extending around the total circumference of the valve and no major pits that completely cross that line you may then clean and put it into the head and proceed to the next valve.

Note: If there is only half a grey white (50%) circle present showing on the valve face (not seat) then you should know that valve is bent or warped and you might have to get a replacement. Same goes for the seat face; there should be a complete circle of grey white metal showing the full circumference (not halfway) of the seat if not then you will most likely have to do one of two things; if it is a cast iron seat (not inserts) you might just need to grind more if it's hardened seats mentioned else you may need to perform the additional task of squaring the seat.

To test how good a freshly grinded valve is, you may perform a *pencil test* as follows:

- Perfectly clean and dry the grinded seat and valve, take a dull pencil and mark lines every (like 20 degrees) across the grey white grinded line space on the seat radially
- Put the proper valve into the head and push it down till it touches the seat

- Twist the valve only a short twist say 1/4" or 25 degrees back and forth. In this limited fraction of motion, it is necessary to push downward as hard as possible without loosening the slight twist control coordinate (25 degrees). Fingernails help to hold the valve edges tight in this slight back and forward motion and the valve is removed.
- Examine the lines you drew on the seat and observe the following "ifs".
  - If every line has been smeared evenly by the valves contact, the seat and valve are perfect, go on to the next one. A good pencil line smear test looks like bottom example shown in Figure 8-50.

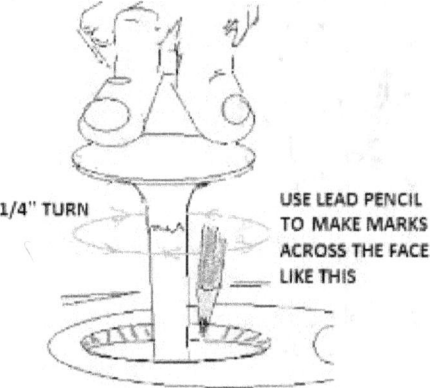

Figure 8-50: Valve ground pencil test

  - If only half the lines are smeared, check the valve to see if it is bent and retest.
  - If only 3/4 to 7/8 of the lines are smeared, the seat is probably crooked or carbon worn on one side you will have to perform an additional procedure to level that seat (carborundum square method) or the valve will still leak very bad (this is unacceptable). By this test you can find out which valves might need more work but if you know the cylinders are normal to begin with (because you measured compression) then this will not be the case with those particular valves so it helps to know which are the leaking ones.

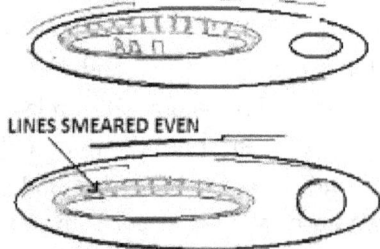

Figure 8-51: Smeared lines indicating good grinding

  - If you are repairing a head that *jumped timing* on the belt or chain, then you will be compelled to buy new valves because the exhaust valves in question are most likely (all or half) to be bent from timing jump. Those heads usually will not need this pencil test but the new replacement valves still have to be lapped (ground) into the seat by the same grinding method.

Valve spring compression

To compress valve springs, it is possible to place a small jack under a car on top of the pipe (in Figure 8-52) or use a length (say 4-5' feet) 2x4 wooden plank and press down onto the valve spring with the pipe tool, you will need something pretty heavy to hold one end of your pressing (i.e. 2x4 wood) device such methods will require a suitable helper.

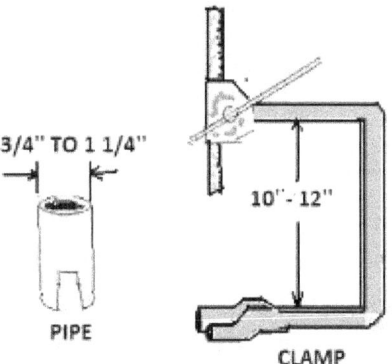

Figure 8-52: Valve spring insert

The small steel pipe is for valves set down in the head casting and you have to use a screwdriver with grease to hold the keepers when trying to put them back onto the valve stems when working on those heads where the valves are set low into the head.

Procedure for checking the valve pushrod

In maintaining the valve pushrod for enhanced performance, the following processes are involved:

- Check if the pushrods are bent irrespective of it being new or used. Using any piece of glass roll one pushrod at a time across flat surface it will roll smooth, if it is bent you will know it and change if necessary.
- Wash pushrods with soap and hot water air blowing them thoroughly. Dry inside and out.
- Hold all pushrods ends to a light inspecting holes to be completely clear.
- Lubricate lifter side of pushrods and guide plate where pushrods will rub.

Cylinder head inspection

To inspect the cylinder head, needed tool include 8S-6691cylinder head stand set. Then proceed as follows:

- Remove the cylinder head from the engine.
- Remove the water temperature regulator housing.
- Remove the valve springs and valves.

- Clean the cylinder head thoroughly. Make sure that the contact surfaces of the cylinder head and the cylinder block are clean, smooth and flat.

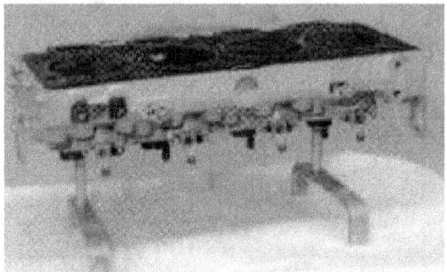

Figure 8-53: Cylinder Head on the head stand set

- Inspect the bottom surface of the cylinder head for pitting, corrosion, and cracks. Inspect the area around the valve seat insets and the holes for the fuel injector nozzles.
- Put the cylinder head on the 8S-6691 Cylinder Head Stand Set (A).
- The cylinder head for leaks at a pressure of 200 kPa (29 psi).
- The cylinder head for flatness. Cylinder head flatness can be measured in three ways; (1) Side to side (2) End to end and (3) Diagonal as indicated in the graphic illustration in Figure 8-54.

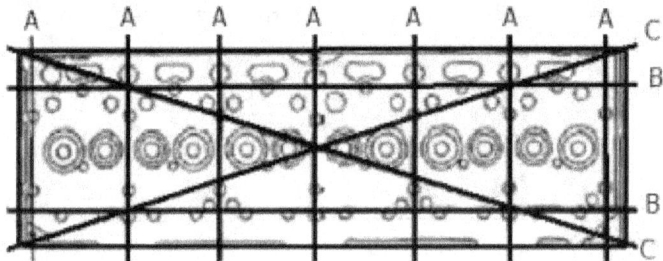

Figure 8-54: Measuring the cylinder head for flatness

1. *Side to side*: Measure the cylinder block from one side to the opposite side (A). [(1) Straight edge(2) feeler gauge].

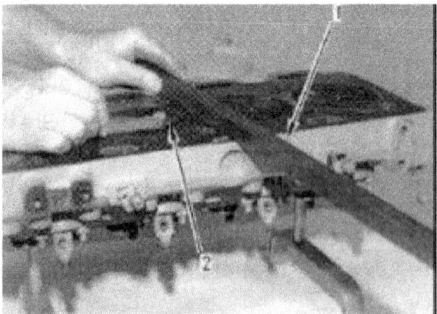

Figure 8-55: side to side.

2. *End to end*: Measure the cylinder block from one end to the opposite end (B) and
3. *Diagonal*: Measure the cylinder block from one corner to the opposite corner (C). Figure 8-56 [(1) Straight edge(2) Feeler gauge]

Measure the flatness with a straight edge (1) and with a feeler gauge (2). Refer to Specifications, "Cylinder Head" for the requirements of flatness.

Figure 8-56: Diagonal measurement

Resurfacing the cylinder head

The cylinder head face can be resurfaced by removing metal from the face if the following conditions exist:

- The cylinder head face is not flat within the specifications
- The cylinder head face is damaged by pitting, corrosion, or wear.
- The thickness of the cylinder head must not be less than 102.48 mm (4.035 inch) after the cylinder head has been machined.

If the cylinder head is resurfaced, the recesses for the valve seat inserts must be machined to the dimensions which are given in specifications, "Cylinder Head Valves". The valve seat inserts must be ground on the side which is inserted into the cylinder head. Grinding this surface will ensure that no protrusion exists above the bottom face of the cylinder head. Grind the outer edge of the cylinder head side of the valve seat insert to a chamfer of 0.90 to 1.30 mm (0.035 to 0.051 inch) at 30 degrees to the vertical

Cooling system maintenance

The cooling system performs two major functions in the engine:

1. Prevent the engine from overheating, and
2. Keep the engine at the proper operating temperature for efficiency and long life.

Cooling system maintenance is frequently overlooked. As much as forty percent of engine problems emanate from inadequate cooling system maintenance. Engines running too cool or too hot are not running correctly and this situation can lead to severe damage. Today's high tech engines demand coolant-containing glycol with the additive package and de-ionized water already premixed to avoid mixing mistakes.

FARM TRACTOR SYSTEMS

Maintaining the air cooling system

Maintenance of an air-cooling system consists of cleaning dirt, chaff and dust from the fins and intake screen and keeping the shrouds in place and repaired so the air is directed to the fins. This cleaning will have to be done as often as dirt accumulates. Failure to keep the fins and screen clean will permit hot spots to occur in the cylinder walls with damage to the rings and walls as a result of high temperature and localized burning in the combustion chamber.

Maintaining the water cooling system

Correct coolant maintenance is the key to reducing operating costs, minimizing engine downtime and ensuring better reliability and durability from engines and cooling systems alike. The regular and proper maintenance of an engine's cooling system is usually considered to be of minor importance. This however, can be an extremely costly mistake.

Research has shown that 40% of all engine problems in heavy-duty diesel engines are directly or indirectly related to improper cooling system maintenance. This figure is a very clear indication that regular cooling system maintenance is vital.

The following maintenance procedures for easy maintenance of water cooling system have been recommended to reduce engine downtime by 40% to 60%:

- ✓ Check the radiator at least once daily, refill with clean soft water. Mix the correct amount of coolant or antirust. In filling the radiator, use good quality water (rainwater or soft water) to correct level and change coolant totally every two years.
- ✓ Check systems for leaks, fins clogging, correct mechanical functioning of pumps, belt and pulleys and pressure caps.
- ✓ Check for correct thermostat operation. In the tropical climate, the operation of thermostat is non-essential because of the constant high temperature environment.
- ✓ Engines running too cool or too hot are not running correctly and this situation can lead to severe damage. To prevent this, the following procedures are necessary.

Step 1: Inspect the cooling system – Fix any leaks. Check circulating pump, fan, belts, pulleys, hoses and clamps, radiator and radiator cap. Ensure thermostat is operating properly. Repair any faults.

Step 2: Start cleaning process – As contaminated cooling systems cannot transfer heat; clean the cooling system. Flush the engine with a safe organic cleaner.

Step 3: Select coolant – Use a fully formulated coolant that meets manufacturer's specification.

Step 4: Install a coolant filter – Install the appropriate coolant filter to suit your engine. Coolant filters keeps the system clean and also provide a means of replacing the appropriate amount of coolant additive back into the system (liquid additive is available for cooling systems not using additive filters).

Step 5: Stay clean – Stay clean by topping up the system with the selected coolant only. ⚠*Do not add water* alone to systems with filter. Replace the coolant filter at the recommended service period or alternatively add liquid additive.

Step 6: Test- coolant – Test the coolant twice a year as a minimum and when major coolant loss occurs or is suspected. As a quality check, testing of the coolant is recommended at the normal oil service interval.

Procedure for troubleshooting overheating engine

The possible causes of overheating are numerous; you can start a troubleshooting process by checking the followings:

- Find out if the engine really overheats. Temperature gauges often fail or provide faulty readings. Check with a separate thermometer. Insert the thermometer into the radiator and check the reading.

Figure 8-57: Thermostat

- Open the cap and check the tank if there is enough coolant in the system. If not, it may be you have a leak, so start looking for it. While opening the cap, turn the cap through a quarter turn first to relieve the pressure due to compressed steam in the radiator. Then turn to remove the cap.
- Do not open the cap with bear hands; you could get burn by scalding.
- Allow the engine to cool, and then add ordinary fresh water as a temporary coolant slowly so that it does not cool any castings so quickly that they crack. Check water hoses and radiator for leakages.
- If coolant is vanishing, check the header (expansion) tank cap (if it is loose or defective coolant may escape there), the weep hole in the circulating pump, and all the system fittings.
- If you find coolant in your engine oil or gear oil or if you have white, steamy exhaust, the problem is likely to be corroded oil cooler, head/top gasket, or damaged internal seals.

- Check if the coolant is circulating. You can watch it swirling in the header tank (once the engine has warmed and the thermostat has opened). If not, the circulating (water) pump has malfunctioned, or the passages are clogged.
- Is the thermostat operating correctly? On most engines they are easy to remove and inspect.

You can test by putting it in a bowl of water with a thermometer and heating the water until the thermostat opens, which should begin at between 165 and 180 degrees F, depending on model. In new engines, the thermostat opens between 190 and 205 °F.

- If you have a heat exchanger, check if it is the system working properly. Look for leakages, check the strainer, or loose v-belts could be the problem.
- Did the overheating condition appear after replacing the coolant? The problem could be an air lock in the system. This is particularly likely where there are long hose runs, such as to water heaters and cabin heaters. Bleed the system by starting at the water source and progressively opening fittings until any air is vented and water comes out.
- If overheating developed after modifying the hull, changing prop, adding weight, increasing injector size, or adding power takeoffs to the engine, the problem could be an overloaded engine. Does it reach its rated wide open throttle speed? Black smoke indicates rich fuel setting, which raises temperature. The solution lies in engine adjustments or reengineering the system.

Cooling system problems

Corrosion: More and more engine manufacturers are using aluminum to reduce weight in cooling systems. But you have to pay the price for weight reduction, because aluminum is the most sensitive material in the system and corrosion is its biggest problem. Not many people are aware of it, but corrosion is not a question of age, it can start to attack an engine as early in its working life as 2,000 hours or 400,000 km. This happens when:

- Oxygen in the cooling system reacts with metal parts in the engine = oxidation rust,
- Water pumps, radiators, thermostats and manifolds are affected,
- Radiators get blocked,
- Engine overheats thereby increasing repair costs and downtime.
- Leaks appear.

Cavitation or liner pitting: The pistons in your engine move up and down about 2,000 times a minute. While they move vertically, the crankshaft is performing a completely different movement by rotating horizontally. These contradictory movements will cause your engine's liners to vibrate a lot. Although the outer wall of the liner is surrounded by cooling fluid, its inertia creates tiny vacuum pockets, causing bubbles of vapour to form on the liner walls.

When the liner vibrates back, these bubbles collapse under an enormous pressure of about 1,000 pressure bar and take small chunks out of the liner.

Acidity: Corrosion in an engine can equally occur when the pH value of your cooling fluid is lower than 7. Your cooling fluid becomes acidic due to the degradation of antifreeze and sulphates entering the cooling system. That leads to general corrosion of your liners, cylinder blocks and heads, and in the waterways and hoses.

A very high pH value could also be a bad news as you risk damage to your gaskets and to the softer metal components. Therefore, the pH value in an ideal cooling system always needs to be between 8 and 10. To achieve that, you need buffers in your cooling fluid that neutralize the formation of acids or alkalis.

Scaling: As the engine functions, the heat causes the formation of scale shell on the hot surfaces. The scale shell acts as an insulator, preventing the coolant liquid from absorbing the heat of the engine. The detrimental effect of scale takes place in the hot spots of your engine, just as it does when you boil water in your kettle. These hot spots are the liners and the cylinder heads. The consequences are worn piston rings, higher oil consumption and, in the worst cases, total engine failure.

Procedure for cleaning radiator

- Loose all hose clips and radiator mountings, and remove to a clear space
- Lose the water tank and thoroughly wash with detergent.
- Brush, blow or wash dirt, dust and chaff out of the radiator fins as it accumulate. This can be done by directing a jet of air blast from a compressor or water jet from a water tank directly to the fins. Be careful not to break fins from the tubes or to puncture the tubes.

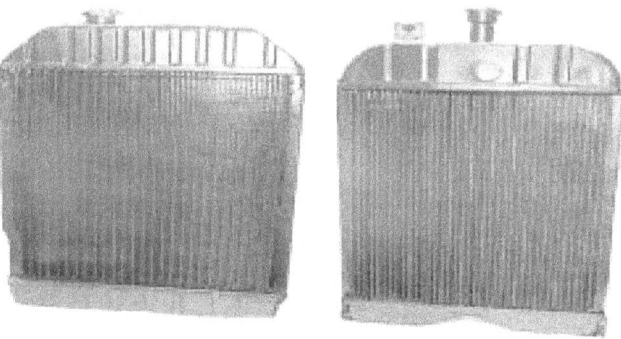

Figure 8-58: Radiators

- Wash trapped dust, bugs or shave off the outside of radiator. Those stocked can be handpicked from within the fins.
- Run water through the radiator to check for leakages along the water jackets and houses and repair them as soon as located.

- Keep the radiator tilted to at an angle to drain off all water from the fins.
- Coupled the radiator and hoses then run the system to check for further leakages.

Troubleshooting engine lubrication system and servicing

To troubleshoot the engine lubrication system, begin by gathering information on the problem. Ask the operator some questions; analyze the symptoms using your understanding of system operation. Then arrive at a logical conclusion about the cause of the problem.

The four problems most often occur in the lubrication system are as follows:

- High oil consumption (oil must be added frequently)
- Low oil pressure (gauge reads low, indicator light glows, or abnormal engine noises)
- High oil pressure (gauge reads high, oil filter swelled)
- Defective indicator or gauge circuit (inaccurate operation or readings)

When diagnosing these troubles, make a visual inspection of the engine for obvious problems. Check for oil leakage, disconnected sending unit wire, low oil level, damaged oil pan, or other troubles that relate to the symptoms.

High oil consumption

If the operator must add oil frequently to the engine, this is a symptom of high oil consumption. External oil leakage out of the engine or internal leakage of oil into the combustion chambers causes high oil consumption. A description of each of these problems is as follows:

External oil leakage – detected as darkened oil wet areas on or around the engine. Oil may also be found in small puddles under the vehicle. Leaking gaskets or seals are usually the source of external engine oil leakage.

Internal oil leakage – shows up as blue smoke exiting the exhaust system of the vehicle. For example, if the engine piston rings and cylinders are badly worn, oil can enter the combustion chambers and will be burned during combustion

Note: Do not confuse black smoke (excess fuel in the cylinder) and white smoke (water leakage into the engine cylinder) with blue smoke caused by engine oil.

Low oil pressure

Low oil pressure is indicated when the oil indicator light glows, oil gauge reads low, or when the engine lifters or bearings rattle. The most common causes of low oil pressure are as follows:

1. Low oil level (oil not high enough in pan to cover oil pickup)
2. Worn connecting rod or main bearings (pump cannot provide enough oil volume)
3. Thin or diluted oil (low viscosity or fuel in the oil)
4. Weak or broken pressure relief valve spring(valve opening too easily)
5. Cracked or loose pump pickup tube (air being pulled into the oil pump)
6. Worn oil pump (excess clearance between rotor or gears and housing)
7. Clogged oil pickup screen (reduce amount of oil entering pump)

A low oil level is a common cause of low oil pressure. Always check the oil level first when troubleshooting a low oil pressure problem.

High oil pressure

High oil pressure is seldom a problem. When it occurs, the oil pressure gauge will read high. The most frequent causes of high oil pressure are as follows:

1. Pressure relief valve struck open (not opening at specified pressure)
2. High relief valve spring tension (strong spring or spring has been improperly shimmed)
3. High oil viscosity (excessively thick oil or use of oil additive that increases viscosity)
4. Restricted oil gallery (defective block casting or debris in oil passage)

Oil indicator or gauge problems

A bad oil pressure indicator or gauge may scare the operator into believing there are major problems. The indicator light may stay on or flicker, pointing to a low oil pressure problem. The gauge may read low or high, also indicating a lubrication system problem. Inspect the indicator or gauge circuit for problems. The wire going to the sending unit may have fallen off. The sending unit wire may also be shorted to ground(light stays on or gauge always reads high).

To check the action of the indicator or gauge, remove the wire from the sending unit. Touch it on a metal part of the engine. This should make the indicator light glow or the oil pressure gauge read maximum. If it does, the sending unit may be defective. If it does not, then the circuit, indicator, or gauge may be faulty

Oil check and change

Oil is considered due for a change when it is noticed to have lost viscosity or when it is considered too black and flows easily or when the engine has run for a required operating time. A check with the dip stick will confirm this.

Figure 8-59: Checking engine oil-level

Black sooty oil and freely dripping oil from the dip stick is an indication of failed oil and must be changed immediately. Engine oil may lose viscosity or failed days after engine servicing as a result of poor quality oil. The following rules will safeguard the life of your engine;

- Use good quality oil with the right viscosity.
- Drain the oil regularly.
- Change the oil filter regularly at every oil drain.
- Keep dust and dirt out of the system.

Oil change or lubrication servicing intervals are dependent upon the various operating conditions of the engines and the sulphur content of the diesel fuel used. Oil drain intervals in all service applications may be increased or decreased with experience using a specific lubricant, while also considering the recommendations of the oil supplier.

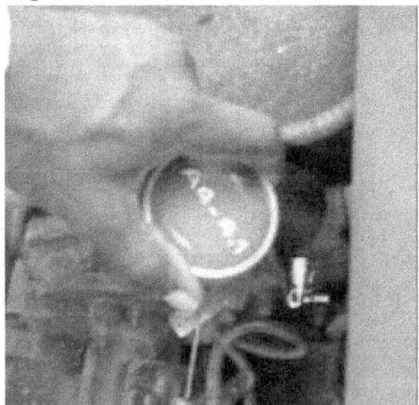

Figure 8-60: Removing the oil tank cap

 A good time to carry out general lubrication, oil and filter change is at the end of seasonal operations. First grade and quality fresh oils/lubricants at this stage provide the best protection for metal surfaces during storage periods. Follow manufacturer's recommended oil changes during working periods.

FARM TRACTOR SYSTEMS

Figure 8-61: Draining the oil (drain pan under)

Precautions ⚠ It is a good practice to drop the oil pan or (crankcase) about once a year to clean out accumulated sludge and to clean the oil pump inlet screen. This may be done when the regular tune- up is done. Ensure that the engine is cool before opening the oil drain plug. Do not drain oil from a hot engine.

Lubrication servicing procedures

In servicing the lubrication system, the following procedure must be followed:

- Position the tractor on *level* ground or inspection pit and remove the oil cap
- Drain the oil and leave for some time to drain completely
- Remove the old oil filter and dispose if it is the disposable type.
- Wash filter casing in solvent such as diesel and keep to dry if it is the replaceable type.

Figure 8-62: Gauging the oil

- Lock the drain plug when engine has fully drained of old oil
- Fix the filter and lock securely with hand to avoid leakage. Do not use wrench or locking device to tighten filter.
- Introduce the new oil with the right viscosity and specification
- Fill and gauge with the dip stick to the right level. In checking the dip stick, hold straight in your front and check the lower part for the graduations.
- Idle run the engine for some time to ensure full flow and proper oil circulation to every component.
- Re-check the oil level with dip-stick ((Figure 8-63to ensure that it is the right gauge.
- Check for leaks and make a permanent record of the hour meter reading.

Figure 8-63: Rechecking the oil level

Oil filter change

Usually the filter elements are replaced at the sometime the oil is changed. The most common filters are the spin-on filter or replaceable element type oil filter.

Spin-on/throwaway oil filters: This is replaced as a complete unit. Unscrew the filter from the base by hand or a filter wrench and throw the filter away. When replacing, wipe the base clean with a cloth and place a small amount of oil or grease on the gasket to ensure a good seal. Screw on anew filter, tightening at least a half a turn after the gasket contacts the base. Do not use a filter wrench because the filter canister could distort and leak.

Replaceable element oil filter: The filter element is removed from the filter housing and replaced. Before removing the filter, place a pan underneath the filter to collect oil from the filter. Remove the fastening bolt and lift off the cover or filter housing. Remove the gasket from the cover or housing and throw it away. Take out the old element and throw it away. Clean the inside of the filter housing and cover it. Install a new element and insert a new cover or housing gasket (ensure the gasket is completely seated in the recess). Replace the cover or housing and fasten it to the center bolt securely.

After the oil has been completely drained and the drain plug replaced, fill the crankcase to the full mark on the dipstick with the proper grade and weight of oil. Start and idle the engine. Check the oil pressure immediately. Inspect the filter or filter housing for leaks. Stop the engine and check the crankcase oil level and add to the full mark.

Oil pump servicing

Service on oil pumps is limited since they are relatively trouble-free. An oil pump will often still be operating trouble-free when the vehicle is ready for salvage. A bad oil pump will cause low or no oil pressure and possibly severe engine damage. When inner parts wear, the pump may leak and have a reduced output. The pump shaft can also strip in the pump or distributor, preventing pump operation.

To replace the oil pump, it is first necessary to determine its location. Some pumps are located inside the engine oil pan. Others are on the front of the engine under a front cover or on the side of the engine. Since removal procedures vary, refer to the manufacturer's service manual for instructions. Most mechanics install a new or factory rebuilt pump when needed. It is usually too costly to completely rebuild an oil pump in the shop.

Before installation, prime (fill) the pump with engine oil. This will ensure proper initial operation upon engine starting. Install the pump in reverse order of removal. A new gasket should be used and the retaining bolts torque as specified by the service manual.

Procedure for maintenance of lubrication system

The following procedures should be observed in the maintenance of the lubrication system:

- Position the engine well with all necessary safety rules fully observed
- Loosen all bolts holding the oil pan in place
- Remove the oil pan and the old seals. Observe if there are tiny metal fragments or sludge settlement in the oil pan. This could be an indication of metal, bearing or journal wear
- Remove the oil pump inlet screen and wash thoroughly
- If desired, the inside of the engine can be washed with clean solvent and a clean brush.
- Examine the bearings for wear and check for condition that needs attention
- Replace the oil pump screen
- Replace the oil pan with new oil seal if the old one cannot be reused
- Put fresh oil into the crankcase
- Idle run the engine for about 5 minutes and check for oil leaks.

Electrical system maintenance

Precaution: Only a qualified electrician or an electrical technician should perform electrical maintenance on tractor. Disconnect power supply before performing any maintenance. Readings with any excessive deviation from normal signals should be recorded in a log book. Such readings include:

a. Voltages and current of incoming power to enclosure
b. Current readings of all motors and
c. Current readings on primary and secondary of control transformer

Do the followings to maintain the electrical system

- Keep correct belt tension
- Keep wires clean
- Repair worn or broken wires. Do not leave any wire naked.
- Broken wire should be neatly joined and taped. Do not make thick joints to avoid local heating at those points. This is a common cause of wire burn which could cause serious damage.
- Keep battery electrolyte up to the correct level
- Avoid unnecessary wire bridging and short circuiting. This is common among the re-wires.
- Replace burnt out light bulbs, fuses etc.

Battery load

Battery "load" is how much load or drain can be placed on the electrical system before the battery begins to discharge itself. Think of it as how much electricity the battery can store before having to be recharged. It is not uncommon for batteries to go completely "flat" or discharged without ever giving a warning signal to the operator.

Figure 8-64: Batteries

Have the mechanic check the battery condition and load levels at every oil change interval if the battery is over 2 years old to ensure the battery will not leave you unexpectedly stranded. A weak battery that is not storing enough power will cause the alternator to work harder and possibly cause premature alternator failure. Dirty or corroded battery terminals can severely reduce the lifespan of the battery and alternator. A quick and inexpensive battery check and cable inspection can be done at each oil change.

The replacement battery should be the same size, have the same battery cable connections, and should be the same electrical capacity as the original battery.

Battery charging procedure

A weak or discharged battery may not be able to produce enough power to turn the engine, thus it must be recharged on a circuit charger. The system responsible for battery charging in vehicle is called alternator.

Figure 8-65: Alternators: Front mounted (L) side mounted (R)

Charging process has the following consequences; *overcharging* evaporates water from the electrolyte, when over-charging occurs, the charging system should be checked. *Undercharging* weakens battery power. This can be caused by:

1. Frequent starting and stopping the engine without letting it run enough to fully charge the battery
2. A loose alternator drive belt
3. A broken alternator or voltage regulator.
4. Loose connections between the alternator and battery.
5. Other battery problems can be caused by low battery fluid, corrosion of battery case, or cracks in the battery case.

Battery charging procedure

Observe the following charging procedures for discharged batteries: -

1. Do not smoke in the vicinity of charger batteries.
2. Remove screw-on caps; check if flaps could be removed also. Check each cell for electrolyte level and fill to the normal level.
3. Connect clips to terminals (+ve to +ve and –ve to –ve) before turning on mains at switch.
4. Charge at one tenth ($1/10^{th}$) of the amp hour (ah) rating of the battery. For instance, charging rate for a 75Ah battery should be set at 7.5.

FARM TRACTOR SYSTEMS

5. Use a hydrometer to check when battery is fully charged. Do not allow amp meter on charger to drop to zero.
6. Turn off charging mains before disconnecting clips to prevent sparks.
7. Boost charging; this uses a high rate of charge for a short period on a flat battery. This process should be restricted to about 5 minutes with strict observance of the charging rules. Sealed batteries, requiring no maintenance, are not suitable for boost charging.

Disconnecting battery

On modern tractors/vehicles with a negative earth system, disconnect the negative terminal first, then the positive terminal and replace last to prevent arcing.

Battery storage

Never store a battery on a floor, always place a wooden block beneath for insulation. When storing wet battery for a season (long period), drain the electrolyte, rinse with distil water, and store in dry place in overturn position.

Starting a run-down engine with booster battery

If a tractor battery ran down or dead, a second battery called boaster battery can be used to start the engine. Adequate maintenance of a battery and starter system minimizes the need to use jump leads. When using jumper cable use the following procedure for negative earth systems.

Step 1: Connect the first Jumper cable to the ungrounded terminal (positive terminal) of the discharged battery.

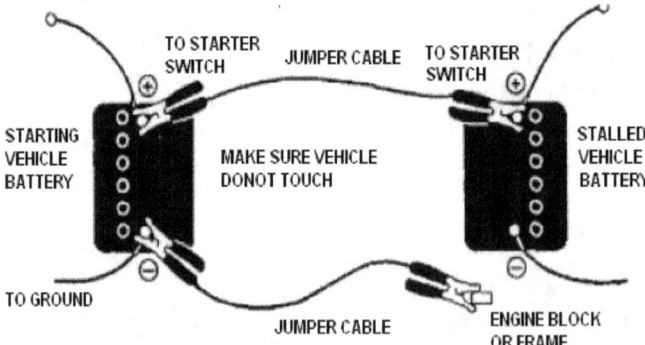

Figure 8-66: Connections for Jump-starting an engine

Step 2: Connect the other end of first Jumper cable to ungrounded (+ve) terminals of the booster battery.

Step 3: Connect the second lead to the grounded terminal (negative terminal) of the booster battery

Step 4: Connect the other end of the negative lead to the frame of machine with discharged battery – at least 500 mm from the discharged battery. *Do not* connect to the negative terminal of the discharged battery as this could cause an explosion due to ignition of gases given off by the battery electrolyte.

Step 5: Check that both leads are clear of any moving parts. Start the booster battery machine and let it idle.

Step 6: Then try to start the machine with the dead battery for short periods.

Step 7: When the engine has started, stop the engine of the donor battery.

Step 8: After the engine starts disconnect the cables in reverse order; first remove the negative from the tractor chassis, and then disconnect the others in reverse order:

$$\text{Step } 4 \Rightarrow \text{Step } 3 \Rightarrow \text{Step } 2 \Rightarrow \text{Step } 1$$

Battery and alternator maintenance

Batteries need regular maintenance in order to provide adequate power to start the engine. The following precautions must be taken to prevent battery injury and ensure safety:

- Remove all metal objects like wrist watches and rings before working around a battery-a spark or fire can occur (I have the scars to prove it.)
- Care should be taken to ensure that the jump lead clamps of a pair of grip leads connected to a battery never touch. This will lead to a heavy electrical discharge and arcing with consequent injury risk.

Figure 8-67: Battery explosion

- *Prevent battery explosions* by Keeping sparks, lighted matches, and open flame away from the top of battery. Battery can explode when connected wrongly.
- Always clean a battery about twice a year or whenever it gets dirty. If white substance appears on the battery terminals and hold down clamps, clean the battery as follows.

FARM TRACTOR SYSTEMS

- Never check battery charge by placing a metal object across the posts. Use a voltmeter or hydrometer.
- Remove the ground cable (-ve) first, then take off the second cable and lift out the battery from the seating.
- Wash the cable and clamps in a mixture of soap and water then polish the inside of the clamps with a round wire brush or sand paper rolled round your finger.
- Wash off the battery case and battery holder with soap and scrub brush. Make sure the vent caps are securely tight. Rinse the battery with clean water and dry with cloth.
- Check closely for any crack, if none; put the battery back in the machine.
- Use some grease or petroleum jelly to rub the terminal posts and cable clamps to protect them against corrosion.
- Reconnect cables, the ungrounded terminal first (+ve).

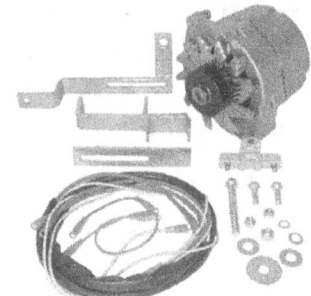

Figure 8-68: Alternator service kits

Cable ends and battery terminals should be cleaned with a wire brush and light sand paper periodically to remove any corrosion build-up. Battery terminal protecting spray coatings can also be helpful in reducing the formation of corrosion.

Alternators equally need maintenance and repairs when it fails to charge the battery. Figure 8-61 show the alternator service kit for front mount distributor.

Exhaust system troubleshooting and maintenance

In troubleshooting the exhaust system, the following component parts should be examined for malfunctioning and correction accordingly:

- *Mufflers*: Inspect area around clamps for breakage, cracks and rust through Leakage here causes:
    a. Excess noise because exhaust noise escapes before being silenced by muffler.
    b. Increase cleanup expenses as soot escapes to discolour cab or trailer.
    c. Operator discomfort as exhaust gases may escape to drive compartment causing sleepiness and even death through carbon monoxide poisoning.

Figure 8-69: The exhaust muffler

- *Elbows, stacks and exhaust pipes*: Dents or crushed portions of any tubing creates exhaust flow restriction and increases back pressure significantly. Even relatively small dents will cause decreased fuel economy and increased turbo wear. If dents are relatively large, increased bearing and lower cylinder wear will occur due to increased exhaust temperature. In such cases, significant decreases in fuel economy will result.

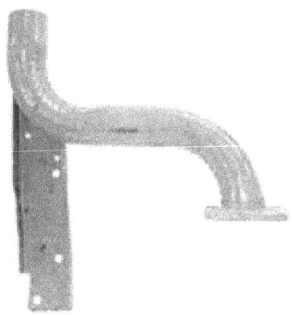

Figure 8-70: Vertical exhaust elbow

- *Raincap:* Raincap mufflers are designed to prevent rain, from passing beyond the muffler to the engine. Water entering the muffler will create slurry of soot to be blown on the trailer causing significant cleanup expense. A curved exhaust stack is an acceptable alternative to a raincap but may not be 100% effective in heavy rains.

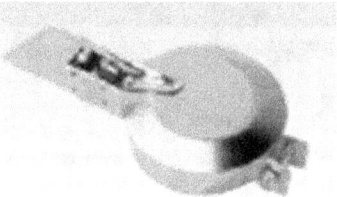

Figure 8-71: Exhaust raincap

- *Clamps*: Check band clamps for cracks periodically. Reuse of any clamp at services is not recommended.

Figure 8-72: Exhaust pipe clamp

- *Exhaust mounting*: The exhaust system should be well secured to eliminate vibration. The muffler brackets should fit securely to the muffler and to the mast or truck frame (horizontal). The muffler brackets should not squeeze the muffler body and the mast must not vibrate or wiggle.

Figure 8-73: Exhaust manifold (Allis Chalmers)

Exhaust system maintenance

- Dirty air intake may contaminate the mixture in the combustion chamber leading to unclean exhaust gasses. Thus clean or replace dry element air filters when they get dirty.
- Empty dirty oil bath air cleaners, clean the element, and pour new oil in.
- Tighten fittings on the exhaust system against leakages.
- Blowing muffler should be welded or replaced to reduce noise
- Avoid excessive exhaust vibration by tightening all guides.

Engine failure troubleshooting

The following questions and checks are recommended to troubleshoot the engine if it failed to start:

Fuel line checklist

- *Is the fuel fresh?* If the fuel is over 30 days old, replace with fresh, treated fuel
- *Is the fuel tank empty?* Fill the fuel tank; if the engine is still hot, wait until it has cooled before filling the tank
- *Is the shut-off valve closed?* Open the fuel shut-off valve
- *Is the fuel diluted with water?* Empty the tank, replace the fuel and check for leaks in the fuel tank cap
- *Is the fuel line or inlet screen blocked?* Disconnect the inlet screen from the engine and clean it, using compressed air. Do not use compressed air near the engine
- *Is the fuel tank cap clogged or un-vented?* Make sure the cap is vented and that air holes are not clogged
- *Is the carburetor blocked?* Remove the spark plug lead and spark plug; pour a teaspoon of fuel directly into the cylinder; reinsert the spark plug and lead; start the engine; if it runs for a moment before quitting, overhaul the carburetor
- *Is the engine flooded?* Adjust the float in the fuel bowl, if adjustable; make sure the choke is not set too high

- *Is the fuel solenoid (if equipped) functioning properly?* Test the solenoid (if equipped) to ensure it is functioning properly

Ignition checklist

- *Is the spark plug fouled (soothed)?* Remove the spark plug; clean the contacts or replace the plug
- *Is the spark plug gap set incorrectly?* Remove the spark plug; reset the gap
- *Is the spark plug lead faulty?* Test the lead with a spark tester, and then test the engine
- *Is the ignition switch shorted?* Repair or replace the ignition switch
- *Is the flywheel key damaged?* Replacing the flywheel key, re-torque the flywheel nut to proper specifications, then try to start the engine; if it still fail to start, check the ignition armature, wire connections or, in some engines, the points

Loss of compression checklist

Are the valves, piston, cylinder or connecting rod damaged?

Perform a compression test; if the test indicates poor compression, inspect the valves, piston and cylinder for damage and repair them as required.

Turbocharger maintenance

A turbocharger can turn in excess of 100,000 rpm and since most turbochargers rely totally on engine oil for cooling it is necessary to keep the engine oil clean. Anytime the turbo oil lines are drained of oil, it is critical that there be oil to the turbo before starting the engine. This prevents turbo damage from lack of lubrication. Damage due to lack of lubrication can also occur if oil supply to the turbo is shut off before the turbo had time to slow down. Therefore, it is normally recommended that an engine be allowed to run for a few minutes before shutting off the engine. This allows the turbo to slow down as well as to get cooled before shutting off the oil supply.

Engine overhauling

Engine overhauling is that maintenance effort (service/action) prescribed to restore an item to completely serviceable/operational condition as required by maintenance standards in appropriate technical publications. Overhaul does not normally return an item to new condition.

Overhauling kits

The Figure 8-74 show a sample of overhauling kits including pistons, sleeves, rings, piston pin, bushings, complete set of gasket, crankshaft seals etc. when ordering bearings, you order separately, also individual parts can be procure separately.

FARM TRACTOR SYSTEMS

Precautions for handling cylinder head and exhaust gaskets

1. Separate the vehicle from surrounding work areas; try to have at least 3 meters separation and avoid windy locations and cooling fans, etc.
2. Use (portable) signs to indicate that asbestos removal are going on. Wear disposable respirator (this is absolutely not observed by a higher proportion of mechanics).
3. If the gasket is damaged during separation of the components wet it with water to control asbestos fibers.
4. Keep the gasket wet and carefully remove it without using power tools.
5. Wipe down the joint faces and immediate area with a wet rag.
6. Place the waste gasket and rag, etc., into a plastic bag and seal or tie it for proper disposal.

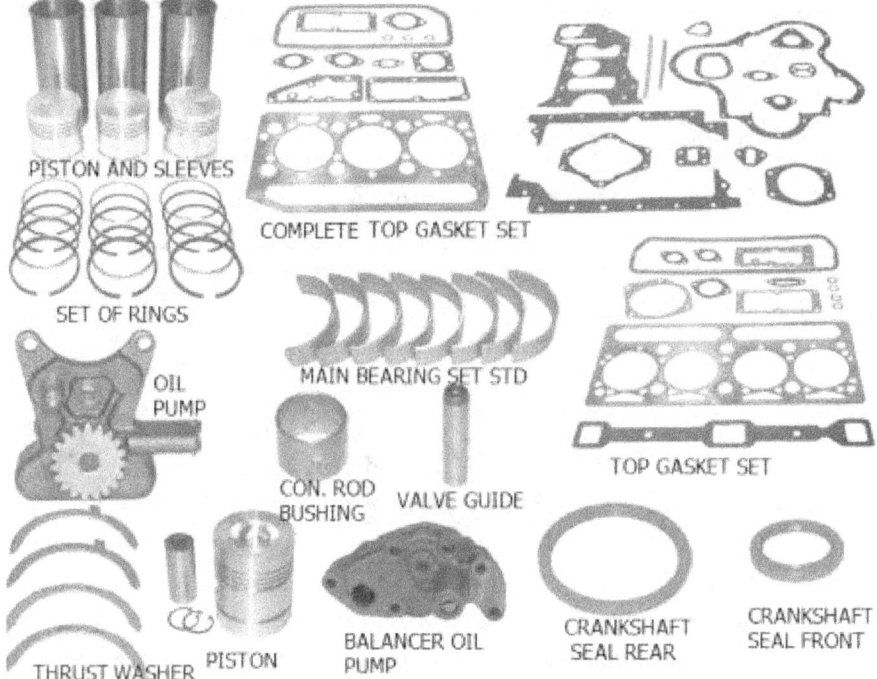

Figure 8-74: Overhauling kits

Procedure for installing new gasket

Follow the procedures below step by step to install a new gasket:

Step 1: Remove all traces of old gasket.

Step 2: Clean and degrease all sealing faces

Step 3: Check the flatness of the sealing faces (max: 0.05mm)

Step 4: Do not use any additional sealants except in exceptional cases.

Figure 8-75: Engine block showing the top gasket

Step 5: Put the gasket in installation position and fix with dowels, pins, or bolts.

Step 6: It is preferable to use new bolts and washers to reassemble cylinder head.

Step 7: Oil the threads and bolt contact areas

Step 8: Place the cylinder head and tighten bolts loosely

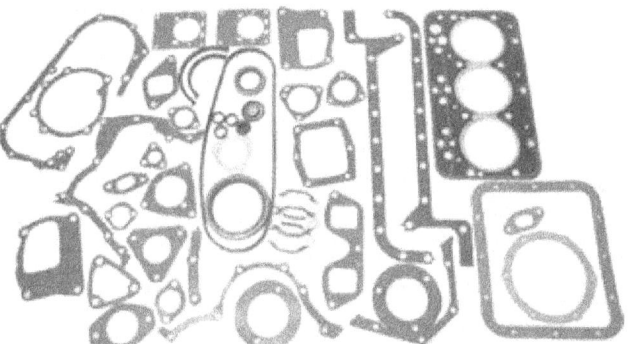

Figure 8-76: Fiat overhaul gasket kit

Step 9: Whenever re-torque is needed, first loosen each bolt by giving it a ¼ turn, and then tighten to required torque. This does not apply to angle-controlled tightening.

Cleaning of pistons and installation of new rings

Step 1: *clean the pistons*: First clean pistons thoroughly and remove all carbon deposits (Decarbonation) from the ring grooves and piston head. Use a twist drill and tap wrench or pin head needle to remove the carbon deposit from the oil drain holes. Replace cracked or deformed and worn pistons.

Step 2: *check the piston ring grooves*: If a clearance of 0.12mm or more is measured between a new, parallel-sided compression ring and the associated groove wall, this means that the piston is excessively worn and has to be replaced.

Groove clearance (mm)	Piston usage
0.005-0.01	Piston can be re-used without restriction
0.11-0.12	More caution necessary
> 0.12	Fitting of new piston necessary

Source: Motor Service Intl. GmbH (2000)

Step 3: *check the cylinder for wear*: If the wear exceeds 0.1mm for Spark ignition (SI) engines and 0.5mm for compression ignition (CI) engines, the cylinder also has to be exchanged (top ring reversal bore wear). Cylinder wear can be manually tested by shaking the piston head with forefingers while still in the bore.

Step 4: *clean the cylinder*: There is always a ring of carbon deposit just above the piston maximum travel. Remove this carbon deposits from the top area of the cylinder bore, above the piston travel zone.

Step 5: *check the ring set components*: Check the ring height by means of a measuring gauge. It is recommended to compare with catalogue data if available. The diameter may be checked by means of a measuring ring or of a reworked cylinder, the ring gap on the basis of subjective assessment or using a feeler gauge. When fitting the pistons with new rings, it is recommended in principle to change all the complete set.

Step 6: *installation of the piston rings*: Insert piston rings in the associated piston ring groove using the right assembly tools! Avoid excessive opening of the piston rings on fitting as this would cause permanent deformation and would impair the performance of the piston rings. Piston rings marked "TOP" have to be fitted with a particular side up. The mark "TOP" should point towards the piston crown so that the scraping effect is directed to the skirt lower end. If the rings are not fitted accurately there is danger of oil being pumped and the function of the ring would no longer be ensured.

When fitting rings with spiral expander the spring ends should always be positioned opposite the ring gap.

Installation of steel rail oil control rings

1. The expander spring is fitted into the grove.
2. The lower segment is inserted with a 120 degree turn.
3. The upper segment is again inserted with a turn of 120 degrees.
4. Correct: green colour mark red mark.

Hint! In the case of three-piece oil control rings the spring has a coloured dot at each end. Each of these coloured dots must still be visible after fitting the ring to the piston. This ensures that the spring ends butt up to each other and do not overlap.

Step 7: *function test/turning the piston rings*: After fitting the piston rings, make sure that they can be freely moved. Position the ring gaps with 120 degrees offset to the next ring.

Step 8: *fitting into the cylinder bore*: Oil the piston rings and the piston appropriately and use a ring compressor or a conical KS assembly sleeve in order to prevent damage to the piston rings.

Attention ⚠ Chromium-plated pistons rings must not be fitted to chromium-plated cylinder liners.

Please note! Pistons for 2-stroke engines, whose rings are prevented from turning by a small pin, must not be rotated when being introduced into the cylinder. The securing pin could otherwise slip under the ring as it springs outwards into a cylinder port. The ring would then break off at the port edge.

Ordering new piston

Over the years, engines have been replaced in many tractors, but you cannot count on ordering pistons by the tractor model number or year. The following simplified ordering process by Ford tractor manufacturer may be employed.

- Determine the diameter of the piston you are in need of, 4.2 or 4.4".
- Determine the standard bore size or oversized of your need.
- Measure the distance from the center of the pin hole to the top of the piston (see Figure 8-77) to find the value of A.
- Measure the pin diameter to find the value of B (see Figure 8-77).
- Choose the piston that matches your findings.

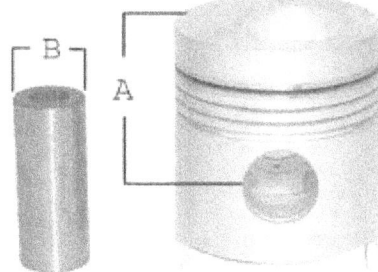

Figure 8-77: Measuring piston bore and pin

Maintaining crankshaft, gear, and main bearings

Required tools: Tool kit, arbor (puller kit and press frame); sleeve, 36mm I.D.; sleeve, 30mm O.D.;

Materials/parts: Solvent, engine oil, diesel fuel; abrasive cloth,

FARM TRACTOR SYSTEMS

General safety instructions and warnings:

1. Severe burns may result in handling of heated bearing race.
2. Protective gloves and tongs should be used for retrieval of bearing race, and gear handling.
3. Severe burns, illness, or death may result if personnel fail to handle diesel fuel properly.

Cautions ⚠

Observe the following precautions.

1. Do not inhale vapour.
2. Work in a well-ventilated area.
3. Do not use fuel of any kind near open flame; sparks, or excessive heat.

Procedures for maintaining the crankshaft

Crankshaft, gear, and main bearing maintenance task includes repair and replacement of parts. Repair of the crankshaft consists of the replacement of the rear main bearings Figure 8-78(3) and the crankshaft gear (5).

Replace the rear main bearing (3) as follows:

- Using snap ring (nose) pliers, remove the retaining ring (1) from the seal housing (2).
- Using arbor press and a 54 mm diameter pressing rod, press the outer race of rear main bearing (3) from the seal housing (2).

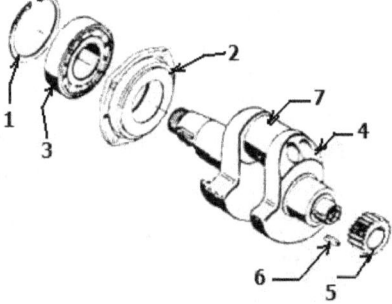

Figure 8-78: Single cylinder crankshaft assembly

- Clean crankshaft (4) in dry cleaning solvent and dry thoroughly.
- Using arbor press and 78 mm diameter pressing rod press the outer race of bearing (3) into seal housing (2), seating fully against the shoulder.
- Use snap ring pliers to install retaining ring (1) in groove of seal housing (2).
- Slide inner bearing race over tapered end of crankshaft (4) with bearing lip leading, and press the bearing (3) onto crankshaft (4) with arbor press and 36mm I.D pressing sleeve seat bearing (3) fully against shoulder.

Replacing the crankshaft gear

Replace the crankshaft gear (5) as follows:

- Clean reusable components in dry cleaning solvent and dry thoroughly.
- If crankshaft (4) is reusable, but has slight imperfection on bearing journals, remove them cloth wet with diesel fuel. Work wet cloth evenly around the circumference of crankshaft (4) at bearing journals until surface is polished smooth.
- Install key (6) in key slot of crankshaft (4).

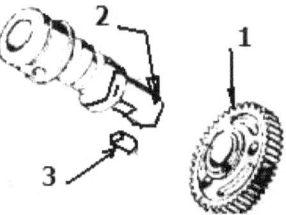

Figure 8-79: Replacing the crankshaft gear

- Heat crankshaft gear (5) on a hot plate, or equivalent, to a temperature of 195 – 210°F (90 – 100°C).
- Position crankshaft (4) in arbor press and align crankshaft gear (5) key slot with key (6) installed in crankshaft. With a 30mm OD pressing sleeve, press gear onto crankshaft until seated against shoulder.

Overhauling the governor system

The Governor is lubricated by an oil line connected to the oil filter, and drives a proofmeter (mechanical tachometer and hour meter) through a cable. There are two bolts that retain the Governor housing to the engine.

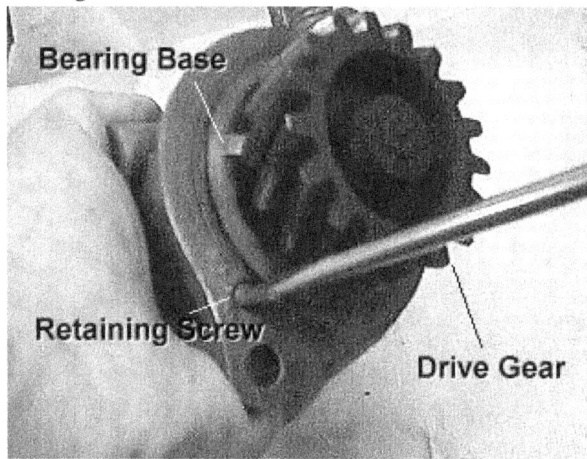

Figure 8-80: Removing the governor system

Mark the linkage between the governor and the carburetor and the dash-mounted speed control lever. Use masking tape and a marking pen to clearly show how they are to be reconnected after overhaul. Swapping the linkage will make your tractor not to run or very poorly. One machine screw retains the drive gear and shaft to the Governor housing.

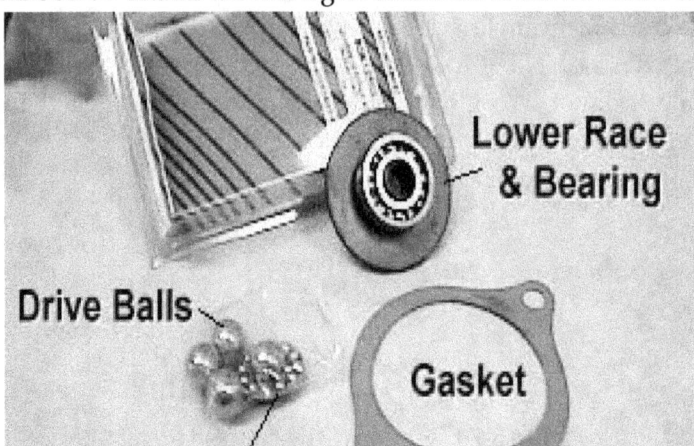

Figure 8-81: Governor Service kits

The overhaul kit consists of new steel balls, new lower race and bearing. New thrust ball bearings and gasket. The Governor fails by developing flat spots on the steel balls and corresponding wear marks on the lower race. These flat spots cause the balls to 'stick' in one position at the wear spots on the lower race rendering the governor unresponsive to engine speed changes. With the retainer machine screw removed, extract the Drive Ball Assembly from the housing. Clean the housing mating surface free of any gasket material.

Drive ball assembly

The drive ball assembly consists of the drive ball shaft, gear, fiber spacer, lower race, thrust ball bearings, washers and clip.

Figure 8-82: The Drive ball assembly

Remove the steel balls. They just fall out - nothing is holding them in at this point. Discard the thrust ball bearings. Clean the lower race of any varnish buildup.

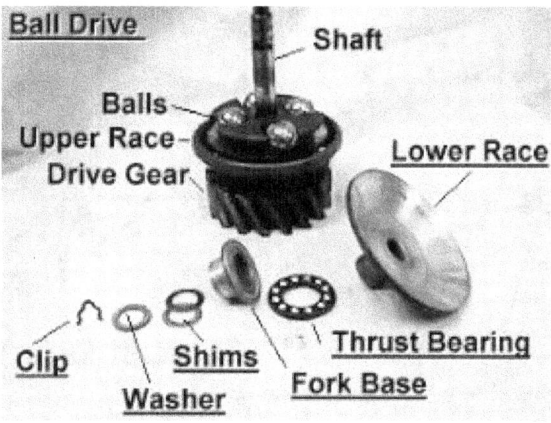
Figure 8-83: The ball drive

The drive gear is pressed on the ball drive shaft and needs to be driven off in order to gain access to the lower race for replacement. Rest the drive ball shaft in a vise with the bearing base resting on the opened jaws of the vice. With a suitable size drift, carefully drive the shaft out of the drive gear.

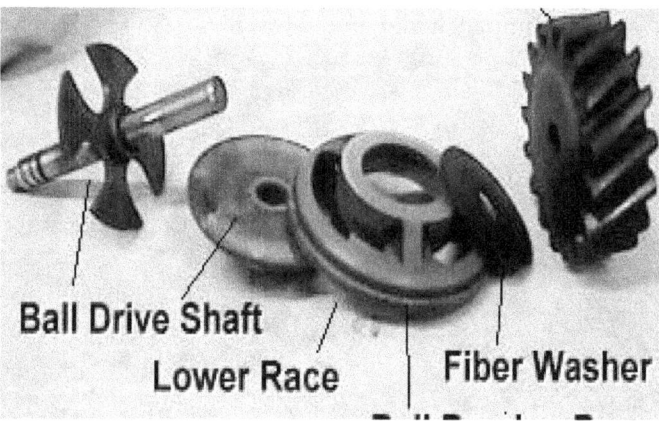
Figure 8-84: The drive gear assembly

Pay attention to which way the drive gear mounts to the shaft. Paint a small dot on the side of the gear away from the bearing base to assist in reassembly. Remove the fiber spacer, lower race and bearing base from the shaft when it has separated from the drive gear.

Reassembly is pretty much the reverse. Slide the new lower race and bearing from the rebuilt kit onto the ball drive shaft. Note the orientation of the lower race to the bearing. Fit the bearing base on the shaft up next to the lower race engaging the lower race bearing. Slip on the fiber spacer. Stand the ball drive shaft on the Proofmeter drive end. Position the drive gear on the shaft with the painted dot you put on it during disassembly facing away from the bearing base.

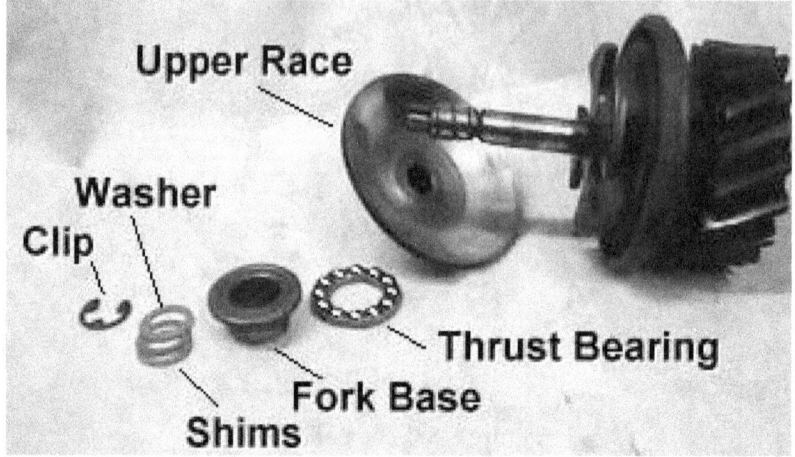

Figure 8-85: Rebuild kits

With the ball drive shaft seated in the housing, rotate the bearing base until the retaining screw recess in the bearing base line up with the recess in the housing. Install retaining machine screw. Install one or two gaskets (depending on how successful you were in getting the housing mating surface completely flat) and reinstall the Governor onto the engine.

Figure 8-86: Coupled governor assembly

Tighten the two bolts. Attach the oil line and reconnect the linkage from the throttle and from the dash-mounted speed control lever following the marking you made on the tape.

Preparing engine for end of season storage

During the "standing" or non-use periods, chemical interactions between metals and fluids can cause more damage than normal wear and tear from active usage. This must be considered in planning seasonal machinery maintenance and storage. Inspect machinery at end of season or operation, repair and adjust as required. Carry out maintenance work without pressure between seasons.

A good time to carry out general lubrication, oil changes and filter changes is at the end of seasonal operations. First quality fresh oils and lubricants at this stage provide the best protection for metal surfaces during storage periods. Engines and hydraulic systems should be thoroughly warmed up periodically during periods of non-usage. Complete warming-up of both systems helps to keep bare metal surfaces covered with protective oil film and drive off condensed water vapour in the oil.

As much as forty percent of engine problems emanated from inadequate cooling system maintenance. Today's high-tech engines demand coolant containing glycol with the additive package and deionized water already premixed to avoid mixing mistakes. Check systems for leaks. Correct mechanical function of pumps, belt and pulleys and pressure cap.

Fuel tanks should be filled at the end of each day's work to avoid moisture from night air entering and condensing in tanks. Fuel tanks should be maintained full during storage periods to prevent interior surface corrosion from moisture laden air. They should be drained of moisture accumulation periodically. Check condition of bulk fuel supplies and periodically drain condensed moisture. Filter all fuel from bulk storage tanks.

Engine pre-storage procedure

If the engine is to be kept out-of-service for a long period of time, the following pre-storage procedure is recommended:

1. Drain fuel from the tank and carburetor; start and run the engine to remove all gasoline from the fuel system. If you do not wish to do this, you may add a fuel stabilizer to the fuel to maintain fuel quality.
2. Drain oil from the crankcase while the engine is still warm. Flush the crankcase with light weight oil such as SAE 5W or 10W. Refill with the proper grade of fresh oil. Another method is to refill the crankcase with the recommended oil; then run the engine until it reaches the full operating temperature.
3. Clean the exterior of the engine to remove all traces of oil stains and service the air cleaner.
4. Remove, clean, and gap the spark plug.
5. Pour a tablespoon of oil into to spark plug hole; crank the engine slowly by hand, and replace the spark plug.
6. Paint or spread a light coat of oil over any exposed surfaces of the engine which are subject to rust and/or corrosion.

Practical exercise

1. On an engine-powered machine: perform any of the following service checks in readiness for field use. Grease all fittings, remove and clean plugs, change oil and oil filter, clean air cleaner, flush cooling system, clean radiator fins, drain fuel tank and replace diesel fuel filters. Report your activities.

2. Engine lubrication and servicing:: Carry out oil change in a tractor vehicle, change the oil, and clean the air filter
3. Fuel supply system servicing: locate the fuel supply system, identify the components and service/replace the fuel filter
4. Servicing the spark plug: Remove and inspect the plug, identify the condition and advice appropriately
5. Engine cooling system servicing: Check the water in radiator, clean radiator fins, and flush the cooling system
6. Bleeding fuel injection line: On a tractor with air-bubble trap in the fuel line, with the help of the instructor, remove the air trap from the fuel line.

CHAPTER 9

Tractor Maintenance and Servicing Schedule

Introduction

People have been killed and seriously injured while carrying out maintenance and repairs to farm tractors. Major hazards can occur when tractors are jacked and wheels are removed without following appropriate safe working procedures. These risks are magnified when carried out on bear soils. Regular maintenance of farm tractor systems and trailed implements can prevent hazardous incidents in the field.

The tractor comprises of three major parts; the engine, the transmission and the differential. Essentially, the following systems makes up the tractor systems; the steering system, brake system, clutch system, the differential etc.

Maintenance report/record

All maintenance reports are a combination of repair record, lubrication and grease record, and tyre repair and replacement record. Maintenance records make it possible to know the overall condition of the equipment and also make it possible to forecast future maintenance requirements. Records also allow you to plan for 'seasonal' maintenance. For instance, a tractor's hydraulic system must be in peak condition for ploughing, but may not be needed at other times. Maintenance may be planned accordingly.

Performance skill expected: To record and maintain hours of service requirements

Work to be performed: Given a tractor, the driver will adhere to the requirements of the hours of service regulations and complete an operator's daily log and logbook summary.

Performance criteria:

1. Comply with hours of service requirements.
2. Maintain a complete, neat, and accurate operator's daily log and logbook recap.
3. Perform all necessary calculations correct to the closest quarter (1/4) hour.

Sample tractor/ vehicle maintenance report

An example of Tractor/ Vehicle maintenance report is shown below:

FARM TRACTOR SYSTEMS

Tractor/ vehicle maintenance report

Instructions: YOU MUST submit this report monthly to:
The Workshop Manager, Department of Agricultural Engineering Technology, Federal College of Agriculture, Ishiagu, 480001, Nigeria

Maintenance report needs to be completed once every month and returned to us not later than five days after the end of the month. The reports have to be kept on file for at least two years.

Repair record: List all faults, repairs and replacement parts completed during the month. This also includes minor items as lights, reflectors, vulcanizing etc.

DATE	TRACTOR/ VEHICLE MAKE	FAULT/ PROBLEM	REPAIRS (List all repairs made to all parts/or equipment installed)

Lubrication and grease record: All lubrication and greased jobs (transmissions, differentials, joints, machinery parts etc.) must be listed.

DATE	TRACTOR/ VEHICLE MAKE	MILEAGE	LUBE/OIL /FILTER	*DIFF	*TRANS	OTHERS

*TRANS: Transmissions *DIFF: Differentials

Tyre patches or replacement record: List all tyre and tube patches, replacements, tyre switching, rotations, etc. as follows.

Unit/Section: _____ Number of Tyres: _____

TRACTOR/VEHICLE MAKE: _____ Tyre Size: _____

Model/ Reg. No.: _____ Tyre Ply: _____ Serial No.:_____

DATE	REPAIRS (List all repairs, replacements, rotations, etc.)

Note: Maintain tyre pressures at recommended settings during non-usage periods. Under-inflated tyres take permanent deformation and develop cracks at sidewalls when not attended to.

Procedure and assessment criteria:

- Interpret hours of service regulations correctly.
- Use hours of service regulations correctly.
- Keep time accurately.
- Perform arithmetic calculations necessary to recap and apply totals to the hours of service regulations.
- Determine driving hours remaining on a particular day or tour of duty.
- Keep pick-up and delivery record.

Tractor servicing and maintenance schedule

Maintenance schedule has to do with a procedure necessary for maintaining your tractor in peak efficiency. Maintenance should be made as flexible as possible. Regular maintenance should be scheduled whenever required. Maintenance can be scheduled as daily, weekly, monthly or yearly as required.

Daily maintenance operation (10 hours)

Engine:

- Check engine oil level.
- Fill fuel tank at the end of each day's work
- Drain water and impurities from sediment bowl and filters
- Activate air filter dust discharge valve
- Check coolant level

Clutch:

- Check clutch function- tractor in motion

Transmission, rear axle and hydraulic system:

- Clean transmission and final drive breathers.

Electrical system:

- Check instrument and functionality of indicator lights

Miscellaneous:

- Check miscellaneous items such brake control, clutch, transmission-gear shift lever and throttle.

FARM TRACTOR SYSTEMS

Weekly maintenance operation (50 hours)

Engine:

- Check the oil level of oil bath filter.
- Check fan belt condition and tension- adjust if necessary
- Check engine for fluid leaks
- Check the lubricant level in the power train

Clutch:

- Adjust pedal position if necessary

Transmission, rear axle and hydraulic system:

- Check the fluid level in the hydraulic system
- Check transmission oil level.

Brake system:

- Check brake pedal free travel and perform braking test
- Check parking brake locking function.

Electrical system:

- Check the electrolyte level in the battery

Miscellaneous:

- Check wheel nuts tightening.
- Check tyre pressures
- Perform 10-hour maintenance.

Bi-monthly maintenance operation (100 hours)

- Change the crankcase oil and filter
- Perform 10-hours and 50-hour maintenance

Monthly maintenance operation (250 hours)

Engine:

- Clean and re-gap or replace spark plugs
- Clean the battery
- Clean the fuel sediment bowl.
- Adjust the carburetor on spark ignition engines

Clutch:

- Adjust the clutch pedal free travel
- Lubricate the clutch-release bearing.
- Adjust steering free play on manual steering equipped tractor

Transmission:

- Check tension of drive belts

Miscellaneous:

- Perform 10-50, -hours maintenance.

Two-month maintenance operation (every 500/600 hours)

Engine:

- Check idle speed and adjust where necessary
- Remove and clean sump breather tube
- Service the distributor on spark ignition engines
- Check ignition timing on spark ignition engines
- Service the starter and generator
- Replace or clean diesel fuel filters

Transmission, rear axle and hydraulic system:

- Change the filter and oil of independent steering system
- Clean oil pump strainer.
- Change transmission and hydraulic system oil filter
- Change rear final drive oil.

Front axle (2WD & 4WD):

- Remove king pin arm free play if necessary (2 WD only)
- Change final drive oil.
- Clean filter or replace where necessary

FARM TRACTOR SYSTEMS

Power steering system:

- Change oil
- Clean filter or replace.

Electrical system:

- Check headlight and adjust

Miscellaneous:

- Check tyre pressures
- Perform other 10-50-100, and 250hours maintenance

Seasonal or yearly maintenance operation (1000/1200 hours)

Engine:

- Service the oil-bath cleaner
- Drain and refill the power train with lubricant
- Adjust valve tip clearance.
- Retighten inlet and outlet manifold bolts.
- Check general engine function: temperature, pressure etc.
- Drain, clean and refill tanks
- Test fuel injector nozzles.

Transmission, rear axle and hydraulic system:

- Adjust the second stage of the live type PTO clutch.
- Drain and refill the hydraulic system with hydraulic fluid.
- Retightened bolts between engine and gearbox and rear axle.

Front Axle (2WD & 4WD):

- Check 2WD hub bearing pre-load.
- Clean and repack front wheel bearings
- Adjust bushing terminal and joint play, if necessary.
- Check universal joint condition.

Table 9-1: Format of tractor service schedule chart

TRACTOR SERVICE SCHEDULE	
10 Hours 1. _____ 2. _____ 3. _____	**250 Hours** 1. _____ 2. _____ 3. _____
50 Hours 1. _____ 2. _____ 3. _____	**500 Hours** 1. _____ 2. _____ 3. _____
100 Hours 1. _____ 2. _____ 3. _____	**Yearly** _____ _____ _____

Engine Tune-up

		DATE								
TOTAL ESTIMATED HOURS OF OPERATION	5									
	10									
	15									
	20									
	25									

		DATE								
RECORD SERVICE	50 hour Service									
	100 hour Service									
	250 hour Service									
	500 hour Service									

Electrical system:

- Check wiring harness connection and fixing.

FARM TRACTOR SYSTEMS

- Check battery earth cable and connections.
- Check alternator and starter motor function.

Miscellaneous service:

- Check tyre pressures
- Perform other 10-50-100, and 250 hours maintenance

A typical 10 hours routine servicing schedule for New Holland 70-56 tractor is shown in the Ford chart shown in Figure 9-1.

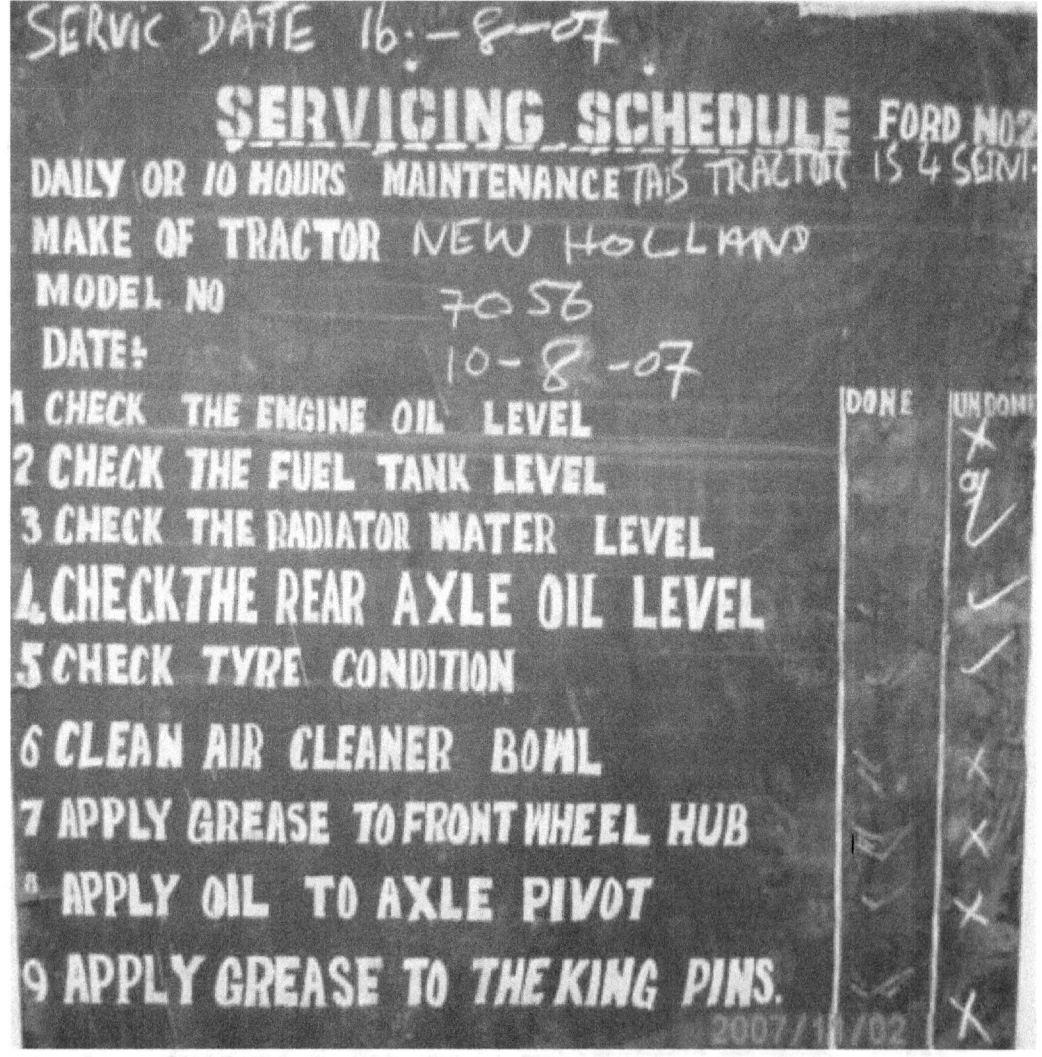

Figure 9-1: Sample chart of tractor service schedule

Operation-hour measurement and computation

Most tractors have hour-meter that records the hours the engine runs. You can figure out hours used after maintenance as follows:

$$Z \text{ (hrs)} = x - y$$

Where

Z = Time since last maintenance (hrs)

x = Present hour-meter reading, (hrs)

y = reading when maintenance was last done, (hrs)

Generally if a machine is operated 24hrs a day, do the 10-hour maintenance twice a day. If it is used less than 1000 hours each year, perform all the 1000 hours maintenance items once a year. If the machine is operated in very dusty and muddy conditions it may require more frequent maintenance.

As required:

1. Check inflation and conditions of tyres
2. Adjust brakes
3. Clean air filters in operator's enclosure
4. Tighten all loose nuts or belts
5. Repair all worn or damaged parts
6. Adjust headlights.

These checks do not override the daily routine checks before taking out your equipment in the morning. The records of the hourly operations can be documented as shown below so as to have the history of such machine.

Machinery maintenance records

Maintenance records make it possible to know the overall condition of the equipment and forecast future maintenance requirements. Records also allow you to plan for 'seasonal' maintenance. For instance, a tractor's hydraulic system must be in peak condition for plowing, but may not be needed at other times. Maintenance may be planned per hour of operation as indicated below.

Windshield wiper blades check?

Inspection: Check wiper blades for wear and washer fluid level during a regular oil and filter change. Worn out wiper has the tendency of giving your wing screen marks. Do not make the mistake of never thinking about replacing or inspecting the wiper blades until you really need them. Check fluid level in the windshield washer reservoiur and top if necessary.

Replacement: Some wiper blades are different lengths for driver and passenger side. Measure the old blades before replacing with new ones.

Windshield washer fluid

Check fluid level in the reservoir regularly. Water repelling additives and detergents can be added to the washer fluid reservoir. Not only will washer fluid aid in removing dirt from the windshield, but also it will act as a lubricant to prolong the life of the wiper blade.

Paint preservation and wash (exterior care)

Regular machinery wash can remove air borne chemicals through "acid rain" that get deposited onto the paint surface, and dull the layer of "clear coating" that is meant to protect the paint and help promote shine and luster. Car wash soap should `3eeeeeeeeeeaabe used and not dish or household soaps, as their chemical makeup can damage the clear coat. Semi-annual waxing of the exterior paint surface will help to protect this important clear coat.

Guide to preparing tractor for maintenance

1. Before inspecting or working underneath a tractor, ensure that the operator has alighted from the tractor.
2. Ensure that all movable attachments are lowered to the ground and/or safely blocked.
3. Stop all power sources to pulleys before removing or replacing belts. A common practice with flat belts is to slip belt into running pulley. This is absolutely wrong and dangerous!
4. If the wheel track is adjustable set the wheels as wide apart as practicable.
5. Stop all hazardous machinery and secure it before any work is undertaken.
6. Allow the engine to cool before removing the radiator cap, and be careful of escaping steam.
7. When jump-starting the tractor, connect the jumper leads as specified by the manufacturer, to avoid damage to the electrical system and the possibility of a battery explosion.
8. When removing and refitting tractor tyres, first remove the valve core to allow air to escape and make the tyres flexible.
9. Maintain a good grip on the tyre lever and stand to the side of the tyre when removing the tube from the rim.
10. While inflating a tyre, continually check to ensure the locking ring (O – ring) is properly seated and locked. The tyre should be inflated to its correct pressure, according to the tyre manufacturer's load/inflation specifications. Always stand to the side when inflating a tyre. An inflation cage should be used when inflating large tyres.
11. The ballasting of tractor tyres should be done in accordance with manufacturer's recommendations.

12. Keep open flames, open lights, lighted cigarettes etc. away from the refueling operation. During refueling, maintain some form of contact between the metal outlet of the refueling hose and the fuel tank opening to reduce the risk of an explosion or fire due to a discharge of static electricity. Always refuel in a well-ventilated area.

Tractor transmission component maintenance

1. *Maintenance of friction drives and systems*

Belt installation

When the pulleys have been correctly positioned on the shafts, the belts can be installed to complete the drive. The drive center distance should be reduced prior to the installation of the belts so that they may be fitted without the use of force. Under no circumstances must belts be forced into the grooves. Using sharp tools to stretch the belts over the pulley rim can easily damage belts and pulley grooves.

The installation allowance for all types of belt is given in belt tables. The take-up allowance given in the table should be added on to the calculated center distance to allow for belt stretch/bedding in.

Pulley maintenance and troubleshooting

Pulleys can be made from such materials as wood, cast iron or steel. They are in different configurations and sizes. Pulley can be single, double or multiple grooved depending on the amount of energy to be transmitted and application. The pulley is usually keyed to the shaft.

Pulley problems

Pulley groove wear can cause rapid belt failure. Check grooves for wear with a groove gauge. Before assembling the drive, check the pulley grooves to ensure the surfaces are free from scours or sharp edges, and all dimensions conform to the relevant standard.

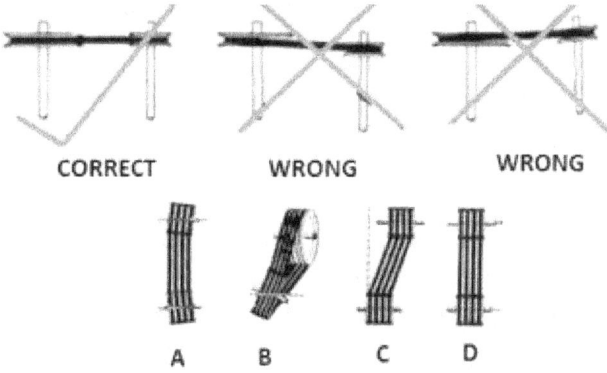

Figure 9-2: Types of belt misalignments

[A) Angular misalignment, B) Composite misalignment C) Axial misalignment, D) Correct installation - both shafts and pulleys are parallel and in alignment]

Good alignment of pulleys is important to avoid belt flank wear. Figure 9-2 shows some of the common alignment faults. Pulley misalignment should not exceed ½° angularly and 10 mm/meter drive center distance, axially.

Align pulleys with a straight edge to conserve belt life and eliminate unnecessary noise.

Table 9-1: Pulley problems and troubleshooting

Problem	Possible cause	Remedy
Frozen or stuck pulley	Damaged bearings	Replace bearings with same type
	Bent shaft	Replace shaft with same material, diameter and length
	Pulley not locked to shaft	Tighten set-screws in pulley hubs
Pulley travels or slips On shaft	Conveyor not installed square	Square conveyor frames
	Conveyor not level	Adjust to proper elevation
	Belt not square to center line	Re-cut belt, square ends and re-lace
	Lack of lubrication	Lubricate bearings
Noise	Obstruction	Remove obstruction
	Pulley moved to one side of unit	Center pulley and tighten set screws
	Damaged pulley or pulley hub	Replace the pulley
Eccentric pulley or Wobble	Bent shaft	Replace shaft with same material, diameter and length
	Loose bearings	Tighten or replace bearings
	Foreign material on pulley face	Remove foreign material

Caution ⚠ Pulley alignment may change when adjusting variable pitch pulleys. Check tension before start-up, after every pulley adjustment and regularly thereafter.

Belt maintenance and troubleshooting

Before starting maintenance

Maintenance functions on belt assembly are to be performed while the system is completely off. The main power switch to the system should be locked in the off position. This will prevent anyone from applying power to the system while maintenance personnel are at work.

Never work on a belt or conveyor while it is running, unless maintenance procedure requires operation. When maintenance performed must be while conveyor is in operation; allow only properly trained maintenance personnel to work on the conveyor.

During the maintenance

1. Do not wear loose clothing while performing maintenance on operating equipment.
2. Be aware of hazardous conditions, such as sharp edges and protruding parts.
3. When using hoists, cables or other mechanical equipment to perform maintenance, use care to not damage conveyor components. Misaligned parts are dangerous as conveyor is started after maintenance is completed.
4. Keep the environment clean. Clean up lubricants and other materials before starting conveyor.

After the maintenance

Before starting any conveyor after maintenance is completed, walk around the equipment and make certain all safety devices and guards are in place, pick up tools, maintenance equipment and clear any foreign objects from equipment. Make certain all personnel are clear of the conveyor and made aware that the conveyor is about to be started. Only authorized personnel should be permitted to start any conveyor following maintenance or emergency shut-off.

Belt drive maintenance

Heed this warning ⚠

- Do not attempt maintenance on any conveyor while it is in operation!
- Store belts in cool, dry place. Keep long belts in natural loops when storing them in pegs
- Relieve tensions on belt drives when not in use.
- Never force a belt onto or out of a pulley – this result in premature belt failure.

Belt tension check

Proper belt tension and alignment is essential for quiet operation and bearing life. With the belt grasped as shown in Figure 9-3 below, total deflection of 1" (1/2" on each side) should be easily attained.

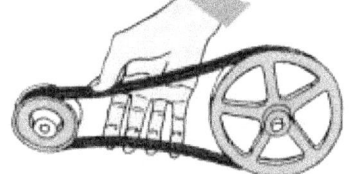

Figure 9-3: Checking belt tension

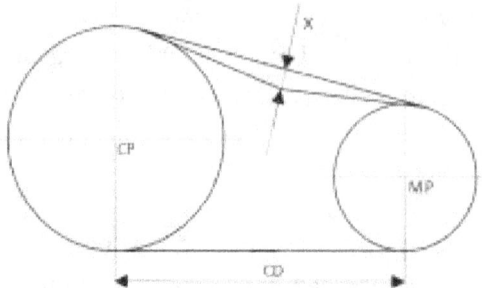

Figure 9-4: Belt tension parameters

Where
    CP = Driven pulley,
    MP = Motor/driver pulley,
    CD = Centre distance,
    X   = Deflection per 1000 mm center distance (16 mm)

In adjusting belt tension, the following approach could be employed;

Rigid base motors (RBM) *approach:* This belt arrangement releases the tensions from the belt ensuring there is no slack, measure the distance between shaft centers and add 1% to the shaft center distance and adjust the shaft centers until that value is obtained. Example: The untensioned shaft centers on a fan model measures 357.2mm (25-9/16"), Tensioned centers = 25-9/16 x 1.01 = 260.76mm (25-13/16") (6.35mm (1/4") extension).

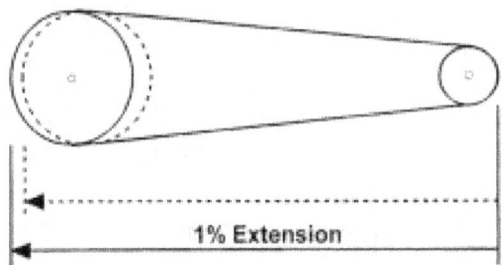

Figure 9-5: Measuring distance between centers

Tension gauge: In using a tension gauge, apply 1.81kg (4 lbs) of force to the center of the belt and adjust the tension until a deflection of 1/64" for every inch of shaft center is obtained. Ideal belt tension is the lowest value under which belt slip will not occur at peak load conditions.

Problems associated with belts

Belt slip – This is slight losses in speed between the drive and driven pulley of belt device.

Belt creep – slight stretching of the belt itself as it runs over the pulley. When belt is under load, there is a tight and a slack side to the belt. The tight side stretches, while the slack side contracts

Belt alignment: this is not quite evident with V-belts. It causes belt roll over, cord stretch, or breakage due to the load being carried on one side of belt.

Belt tension: Too little tension will cause slippage or slip and grab, causing power transmission loss, belt break or excessive belt overheating, and burned spots on the belt. Too much tension will cause belt heating and excessive stretching as well as damage to the pulleys and shaft.

Oil and grease –All belts should be kept free of oil, grease and fuel. These cause belts to become soft and slip.

Belt heating – Belts operating under temperatures of 49^0c are not usually affected by heat. Belts harden at high temperatures and cause it to crack and stretch.

Belt wearing

When there is a *damaging wear* on drive roller side of belt, one of the following problems could be responsible:

- Drive roller is slipping: Increase take-up, lag drive or increase arc of contact on drive roller with snubber oil.
- Loading end of belt has buildup of material, which is being, ground between belts and drive roller: Improve belt-loading procedures. Install plows or scrapers in front of drive roller on return run. If leakage through fasteners, use belt with flap or vulcanized splice.
- Idlers are sticking: Increase maintenance and lubrication.
- Bolt heads protruding above lagging
- Replace worn lagging on pulley.
- Tighten bolts,
- Cement lagging to pulley.

For *excessive edge wear problem*, one of the following could be responsible

- Folding of belt edge on edge guard or frame: Consider using more stable construction. Provide more clearance. Smooth any rough areas on
- Side loading: Load in direction of belt travel.
- Buildup of material on drive roller: Install scrapers to eliminate buildup from pushing belt against frame.

Belt stretching

- *Belt too tight*: Reduce take-up tension.
- *Belt not heavy enough for desired application*: Switch to a higher grade belt.
- *Frozen idlers or buildup of material on drive rollers and idlers*: Improve maintenance and clean-up. Lube or replace frozen idlers.

Belt shrinking

The following conditions could lead to belt shrinkage and possible solution

- *Bad edge on belt due to rubbing*: Improve alignment of idlers; drive rollers and tracking of belt. Consider narrower belt.
- *Belt damage by abrasives, chemicals, mildew, acid, heat and oil*: Specify belt with proper resistance for material being handled.
- Improper adhesion. Check maintenance procedures.

Belt cracking

- *Abrasion, chemical, acid or rot damage*: Specify belt with proper rubber compounds for material being handled.
- *Belt too tight*: East tension on belt and use some type of compensating take-up. Consider lagging drive and using snub oil.
- *Small radial cracks on belt side and base:* Generally caused by slippage due to insufficient belt tension, but excessive heat and/or chemical fumes can also cause the same problem.
- *Belt swelling or softening:* Caused by excessive contamination by oil, certain cutting fluids, water or rubber solvents.
- *Whip during running:* Often caused by incorrect tensioning, particularly on long center drives. If a slightly higher (or lower) tension does not cure the problem, there may be a critical vibration frequency in the system, which requires redesign or use of banded belts.

Practical notes on maintenance of flat belt/conveyor

- Pulleys need to be crowned to prevent belt from wandering off. Belts tend to move to tightest position
- Tension required enabling belt to operate. Tensions in belt are normally set by adjusting centre distance between pulleys to ensure some stretch of belts (say 2%).
- Best drives result from belts with high flexibility, low mass, and with surfaces engineered to provide a high coefficient of friction

Practical notes on maintenance of Vee-belt

☞ Setting the belt tension is readily achieved by jacking the pulleys apart and measuring the transverse distance the belt can move.
☞ Higher shaft torsion loads are handled by using multiple belt pulleys.
☞ A jockey pulley can be installed to increase the angle of contact and allow transfer of more power. It can be mounted on either the tight or loose belt side and adjusted inwards to provide more angle of contact.

Table 9-2: Belting problems and troubleshooting

Problem	Possible cause(s)	Remedy
Belt slips on drive pulley	Take-up pulley not adjusted properly	Adjust each take-up screw in small increments
	Face of drive pulley or pulley side of the belt is slippery	Replace lagging if worn smooth. If objects in the lagging cause slippage, clean by scraping with a wire brush. Do not use belt dressing, thinners, oils, gasoline or solvents. These items could impregnate a belt or pulley lagging and cause premature wear.
	Snubber roller is misaligned	Realign snubber to increase wrap on the drive pulley
Belt lace pulling great	Out belt tension too much	Reduce belt tension by adjusting the take up
Wear on Pulley Side of Belt	Belt slipping on drive pulley	Adjust each take-up screw in small increments
	Pulley or roller bearings sticking	Check alignment, damage and lubrication
	Misaligned or damaged beds	Check slider bed for smoothness and alignment
Top surface of belt damaged	Obstruction	Inspect conveyor for obstructions and remove
	Damaged idler or snubber roller	Check return idlers and snubbers for foreign material
Belt travels to one side of unit	Rollers upstream not square	Adjust roller on side shifted, to belt travel direction
	Bed frame structure not square	Check bed with level and adjust to proper elevation
	Foreign material on roller or pulley face	Clean material off rollers and pulleys and check for Freeness and alignment
	Belt travels off drive or tail pulley	Check alignment of drive pulley, snubber roller, return rollers or the slider bed. Adjust tail pulley to increase tension on the side of the belt which has traveled

Choice and maintenance of Vee-belts

When using Vee-belt it is important to consider the followings:

- In multiple belt drives all belts must be a matched set. A matched set of belts means each belt is of equal length so they all take equal loads. In an unmatched set the shorter belt is more heavily loaded and fails soonest.
- Note that life span of Rubber belt is limited and they can be further deteriorated by oxygen in the air, heat, dirt and oil.
- Dirty, dusty conditions destroy both belt and pulley. Dust settles on the pulley and imbeds into the rubber. Eventually the belt starts to slip thereby polishing the pulley and reducing the friction between belt and pulley. It may be necessary to install a ventilated, dust proof enclosure around the equipment. This is often the case for belt driven air compressors in dusty locations.
- Belt life is limited and they will fail by stretching, snapping or slipping when their working life limit is reached. Drive belts and pulleys should be inspected on a preventative maintenance routine or replaced prior the likely end of their working life. Replace a stretched, cracked or shiny belt immediately.
- Provide pulleys of generous diameter so the belts are not excessively flexed. Flexing around a small pulley produces high stress gradients across the belt.
- Determine the optimum number of belts for a drive so they are loaded to the manufacturer's recommendations.
- The separation distance of the pulleys determines the angle of contact the belt makes with each pulley
- When the pulley Vee faces become polished the pulley must be replaced. If it is possible reclaim the pulley by lightly machining the groove.
- As more belts are added to the drive and tensioned to the required load the driven and driver shaft bending loads increase.
- This will raise the radial loads on the shaft bearings nearest the pulley and act to reduce their working life. Insure the shaft bearing type selected can accommodate high radial loads.
- For safety reasons the belts and pulleys require a guard over them. The guard must be well ventilated and not allow heat from the motor or working equipment to be trapped and overheat the belts. Provide plentiful access and make removal of the guard easy.
- Belt tensioning testers are available to check the amount of tension applied to the belt meets manufacturer's recommendations.

Problems associated with gear

The major problem associated with gear is the wearing of the teeth. The major types of gear tooth wear and failure include:

- *Normal wear* – Normal polishing of the gear teeth as they operate.

☞ *Abrasive wear* – surface injury caused by tiny particles carried in the lubricant or embedded in the tooth surfaces.
☞ *Overload wear* – contact surface is worn but smooth.
☞ *Rolling and peening* – Overloading and sliding leaves a burr on the tooth edge which results in rolling. Peening occur as a result of back lashing.
☞ *Rippling* – This is a wavy surface on the teeth at right angles to the direction of slide. Surface yielding due to lack of lubrication, heavy load or vibration scaring may cause rippling – This is caused by temperature rise and rupture of the lubricant film arising from too heavy loads.

Table 9-3: Motor and gear reducer troubleshooting

Problem	Possible cause	Remedy
Motor hard to start, stalling out or running hot	✓ Lack of Lubricant ✓ Stiff pulley ✓ Stiff roller ✓ Overloaded ✓ Electrical fault	✓ Check oil level in gearbox, verify if vent plug is open ✓ Inspect all pulleys and bearings, replace if any is faulty ✓ Inspect all rollers, replace if faulty ✓ Remove load and possibly increase horsepower ✓ Check wiring, circuits and take amp readings
Excessive noise	✓ Lack of lubricant ✓ Damaged gears ✓ Faulty bearing	✓ Check oil level in reducer & add if needed ✓ Replace unit ✓ Replace bearings

☞ *Pitting* – very minute or micro pitting can occur as gray surface, which may advance slowly to an actual, pitted condition. A condition often associated with thin oil film due to high oil temperature.
☞ *Scratching* – Caused by particles of metal flaking off the gears which are larger than the abrasive particles.
☞ *Corrosion* – Corrosive wear results in an erosion of the tooth surfaces by acid formed when moisture and lubricant reacts together.
☞ *Burning* – usually caused by complete failure or lack of lubrication.
☞ *Cracking* – Most heat cracks in gears are extremely fine and may not show up until gears have been used for some time. These failures are manufacturer's fault.

Maintenance of chain drive

Chain tension and alignment

Chain tension should be adjusted to allow ½" to 1" of movement between the sprockets. Replace Chain Guard after adjusting chain tension. Use a straightedge to align sprockets. Make sure the setscrews are tight when finished.

FARM TRACTOR SYSTEMS

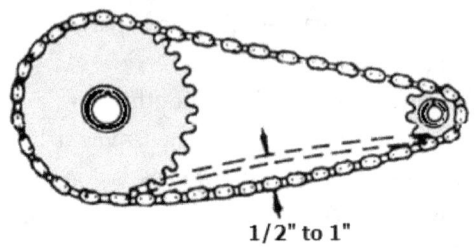

Figure 9-6: Chain tensioning

The major maintenance exercise carried out on chain drives is that of cleaning and lubrication.

Cleaning of chain

Periodic cleaning is good economy under even the best operating conditions. To clean chains follow these steps:

1. Remove chain from sprocket
2. Wash clean in diesel fuel
3. Drain fuel and soak chain in oil
4. Hang chain to drain off excess lubricant. Remove chain from sprocket and coat with heavy grease, wrap in heavy grease resistant paper and store in a dry area.

Chain lubrication

Lubrication reduces wear, protect against corrosion and prevent galling or seizing of pins and bushings. The importance of lubrication increases with higher chain speeds. Lubrication can be done in the following ways:

1. *Manual lubrication*: oil is applied with a brush or spout can once every 8 hours of operation and under all condition.
2. *Drip lubrication* – Oil drops are directed between the link plate edges from drip lubrication.
3. *Bath or disk lubrication* – the lower strand of chain runs through a sump of oil in the drive housing. With disk lubrication, the chain operates above the oil level and a rotating disk picks up oil from the sump and deposits it onto the chain by means of a trough.
4. *Oil stream lubrication* - a pump that sprays oil onto the chain through nozzles supplies the lubricant.

Table 9-4: Chain and sprockets troubleshooting

Problem	Possible cause	Remedy
Abnormal wear	Excessive chain tension	Reduce the chain tension
	Misaligned sprockets	Align sprocket faces with straight edge
	Chain not lubricated	Lubricate with proper lubricant
	Damaged chain or sprocket	Replace damaged component

	Misaligned chain guard	Adjust as required
Excessive noise	Loose chain	Adjust chain tension
Excessive noise	Chain not lubricated	Lubricate with proper lubricant
	Misaligned sprockets	Align sprocket faces with straight edge
Pulsating chain	Improper chain tension	Adjust chain tension
	Overload	Inspect for obstruction causing drag and remove
Broken chain	Frozen pulley, sprocket or shaft	Inspect and replace damaged items
	Worn or damaged chain	Replace damaged chain
	Obstruction	Inspect conveyor for obstruction and remove
Sprocket loose on shaft	Loose set screws	Align sprocket faces with straight edge and tighten set screws
	Worn or damaged key	Replace key and inspect shaft keyway for damage
Chain slack	Normal wear	Adjust chain to proper tension

2. *Hydraulic transmission systems servicing*

Maintenance of hydraulic and transmission is essential for safe and smooth operation of the tractor. It is generally recommended that you drain and flush the transmission, differential and final drive when it is found to be badly contaminated and refilled with correct grade fluid once a year or after 1000 hours of operation. If this is carefully done, it should prevent any breakdown in this part during the life of the tractor.

Note: In the earlier tractor models, the hydraulic fluid lubricates the transmission and hydraulic systems.

Servicing procedure

1. Drain the transmission, differential and final drive cases.
2. Flush the transmission, differential and final drive cases.
3. Refill all cases with the right amount of the correct weight or viscosity oil.
4. Make sure the hydraulic filter is changed as recommended and the system is drained and refilled regularly.

In maintaining those parts of the tractor requiring pressure gun grease, it is a good practice to watch carefully as the periodic greasing is done to see that the fittings are in place and taking grease.

Tips for hydraulic system maintenance

1. Look after hydraulic hose connectors.
2. Keep fittings tight and repair leakages
3. Do not run equipment with unplugged connectors. Dirt entry here may go straight through a pump or control valves, causing severe damage.
4. Inspect hydraulic rams for leaks at glands. Replace defective gland packing. If oil can leaks out, dirt can be drawn back into the hydraulic system and again cause damage and reducing filter service life.

Figure 9-7: Power steering add on kit

5. Maintain hydraulic systems at correct level and keep breathers clean. Each time a single acting hydraulic ram travels full stroke, an equivalent amount of air moves in or out of the hydraulic tank
6. Keep the hydraulic oil clean and change at recommended intervals.
7. Change filters, and keep oil reservoir filled to the right level.

Figure 9-8: MF 435 Tractor hydraulic system

Hydraulic fittings connect lines or hose to other parts such as trailer tipping and braking devices (Figure 9-8). They must be tightened enough not to leak. To work on the hydraulic fitting,

1. Turn off the engine and secure it properly

2. Relieve the system pressure by moving the control levers back and forth.
3. Loosen fittings slowly and carefully
4. Replace the damaged fitting.

Figure9-9: Hydraulic actuator for trailer tipping device

Hydraulic pumps

The following can damage hydraulic pumps:

1. Dirt, rust, metal, particles, and even bubbles can damage hydraulic pump
2. Overloading the pump.
3. Lack of oil, or bad hydraulic oil that is too thick, thin or has wrong additives.

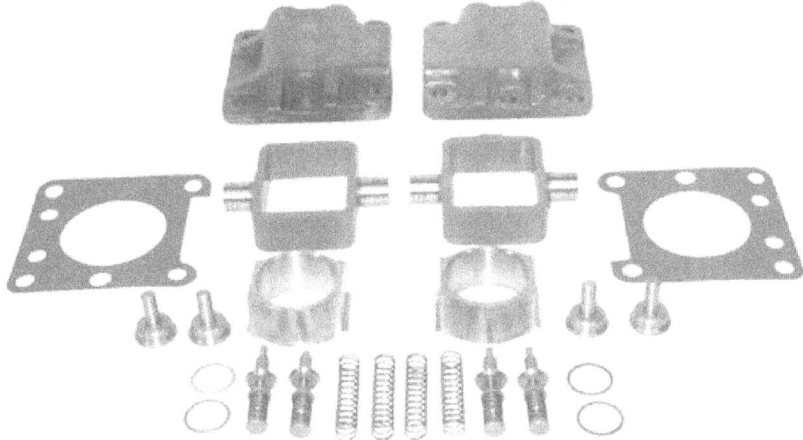

Figure9-10: Repair kits for hydraulic pump

Protect hydraulic systems from dirt by:

a. Storing hydraulic cylinders fully retracted so that dirt cannot stick to the exposed rod .
b. Look after hydraulic hose connectors. Keep fittings tight

c. Do not run equipment with unplugged connectors. Dirt entry here may go straight through a pump or control valves, causing severe damage.

4. Inspect hydraulic rams for leaks at glands. Keep vents wiped clean. Replace defective gland packing. If oil can leak out, dirt can be drawn back into the hydraulic system and again cause damage and reducing filter service life.
5. Maintain hydraulic systems at correct level and keep breathers clean. Each time a single acting hydraulic ram travels full stroke, an equivalent amount of air moves in or out of the hydraulic tank.
6. Keep cylinders aligned
7. Clean fittings before you hook an implement up to the tractor hydraulic system.

Clutch system maintenance

A clutch is a wearable item just like tyres they wear out. A tremendous amount of heat is generated in the clutch, pressure plate, and flywheel. This happens each time you start moving and to a lesser degree, when you changes gear. As time goes by, the friction surface wears thinner, and it will get to a point where it is too thin to transmit enough torque to move the car normally. The clutch is now slipping and probably smells – really bad! At this point, a clutch repair is inevitable.

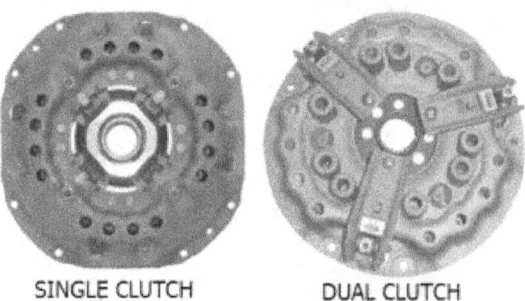

SINGLE CLUTCH   DUAL CLUTCH
Figure 9-11: Clutch disc assembly

Clutch maintenance helps maintain the maximum service life for clutch components. The clutch system should be properly adjusted and kept working as the manufacturer intended. The process begins with the clutch linkage adjustment. Some vehicles have hydraulic clutch master and slave cylinders, which require no periodic adjustment after initial installation. Other cable-operated systems have a self-adjusting mechanism built into the clutch pedal mechanism to eliminate the need for periodic adjustment. But the majority of cable-operated clutch systems do require periodic adjustment.

Changing and bleeding the clutch system

Normally, brake/clutch fluid is changed manually by creating pressure to avoid air bubbles in the system. One technician keeps pumping the brake pedal and other bleeds the brake/clutch pipes. This process is time consuming with multiple manpower operation. Yet, the job done will not guarantee 100 percent braking efficiency due to air and humidity locks in the system. To overcome this deficiency, it is recommended to use brake/clutch maintenance device while changing brake/clutch fluids. Following are some various symptoms and possible causes.

Table 9-5: Clutch symptoms, possible causes and solution

Symptoms	Causes	Solutions
Clutch chattering or grabbing	Incorrect clutch adjustment	Adjust clutch
	Oil, grease or glaze on facings	Disassemble clutch and troubleshoot replace
	Loose "U" joint flange	Tighten or replace "U" joint flange
	Worn input shaft spline	Replace input shaft
	Binding pressure plate	Replace pressure plate
Clutch chattering or grabbing	Binding clutch disc hub	Replace clutch disc
	Unequal pressure auto repair plate contact	Replace worn/misaligned components
	Loose/bent clutch disc	Replace clutch troubleshoot disc
	Worn pressure plate, disc or flywheel	Replace damaged components
	Broken or weak pressure springs	Replace pressure plate
	Sticking clutch pedal	Lubricate clutch pedal & troubleshoot vehicle linkage
	Incorrect clutch auto repair disc facing Engine loose in chassis	Replace clutch disc Tighten all mounting bolts
Clutch slipping	Pressure springs worn or broken	Replace pressure plate
	Oily, greasy or worn clutch facings	Replace clutch disc
Clutch slipping	Warped clutch disc or pressure plate	Replace damaged components
	Binding release auto repair levers or clutch pedal	Replace release components
Clutch squeaking	Worn or damaged release bearing	Replace release bearing
	Dry or worn pilot or release bearing	Replace bearing assembly

Symptoms	Causes	Solutions
Clutch squeaking	Pilot bearing turning in crankshaft auto repair	Replace pilot bearing and/or troubleshoot cars crankshaft
	Worn input shaft bearing	Replace bearing and seal
	Incorrect transmission alignment	Realign transmission
	Dry release fork between pivot	Lubricate release fork and pivot
Clutch pedal sticks down	Binding clutch cable	Replace cable
	Springs weak in pressure plate	Replace pressure plate
	Binding in clutch linkage	Lubricate and free linkage
Noisy clutch	Dry release bearing	Replace release bearing
	Worn input shaft bearing auto repair	Replace bearing
Transmission clicking	Weak springs in pressure troubleshoot cars plate	Replace pressure plate
	Release fork loose on ball stud	Replace release fork and/or ball stud
	Oil on clutch disc damper	Replace clutch disc
	Broken spring in slave cylinder	Replace slave cylinder

Clutch plate repairs and replacement

1. Separate the vehicle from surrounding work areas; try to have at least 3 meters separation and avoid windy locations and cooling fans, etc.
2. Use (portable) signs to indicate that asbestos removal are going on. Wear a disposable respirator (this is absolutely not observed by a higher proportion of mechanics).
3. On separation of the gearbox from the engine, wet wipe inside the bell housing (banjo) and around pressure plate.
4. On removal of pressure plate and clutch plate, vacuum/wet wipe flywheel, housing and components.
5. Place used rags and removed components, etc., in a plastic bag and seal or tie it.

Figure 9-12: Clutch pressure plate

Hydraulic fluids inspection

Power steering fluid: Inspection- oil level check: Power steering fluid can either be pink or clear in color, usually only a very small amount is needed to top off fluid level. If more than 2 fluid ounce* (0.06 liters) is needed to top the fluid level, have the system checked for leaks or wear. Note: * 1oz = 0.0284lt.

Fluid replacement- Power steering fluid just like any other fluid becomes dirty and contaminated and should be replaced with clean fluid periodically. Dirty power steering fluid can cause the power steering pump or the power steering gear assemblies to fail and can cause premature wear to occur.

*Transmission fluid: Inspection-*Usually the transmission fluid level is checked with engine running, hot and in park. Check your owner's manual for proper fluid type and proper fluid level inspection procedures. Automatic transmission fluid is usually pink in color. Most standard shift transmissions will have a drain plug to service the fluid. Some stick shift transmissions use engine oil as a lubricant; consult your owner's manual when servicing. It could also be a good idea to have the replacement fluid type information available for the repair shop.

Replacement- Consult the owner manual for proper fluid type and service intervals. If applicable, replace the internal automatic transmission filter or clean the re-usable screen when changing the transmission fluid.

Differential fluid: A rear differential is only found on rear wheel drive cars, trucks and tractors.

Inspection- The rear differential fluid or grease should be checked during each routine oil change and topped off as needed with the fluid prescribed in the owner's manual.

Replacements- Drain and flush the rear end fluid periodically to remove any metal filings that have normally accumulated in the differential housing. Replace the differential cover gasket and add any recommended supplemental additive prescribed in the owner's manual.

The differential system maintenance

The differential is located at the rear of the transmission case and is supported by two taper bearings, which are held in place by the rear axle housings. The rear axles and housings are located on each side of the differential assembly, with a planetary reduction at the outer end. The right axle housing includes the differential lock assembly, and both housings contain a brake mechanism.

Figure 9-13: The differential housing (Mustang fever)

The rear axle planetary transmission components use straight mineral SAE 90 oil. Figure 9-9 shows a cross section of the differential, pinion and rear axle assembly.

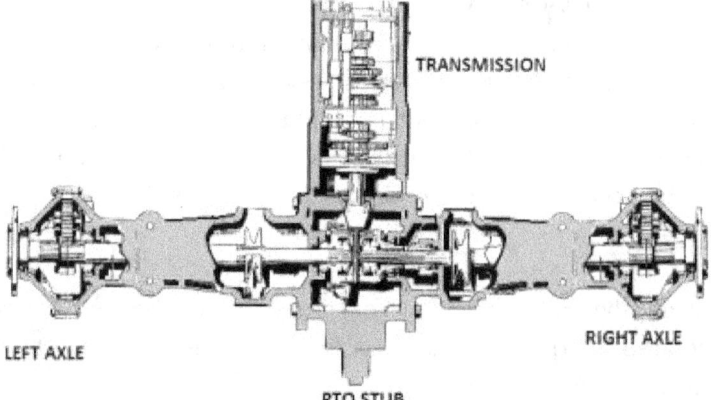

Figure 9-14: Tractor transmission and differential

Removing the differential

In order to remove the differential assembly for repairs and component replacement, it will be necessary to remove the lift cover which is done by first removing the mounting cap screws holding the cover to the transmission housing and then lift the cover. Place a wedge on each side between the front axle and the engine, support the transmission on a jack or stand and remove the rear wheels.

Figure 9-15: Differential housing and the planetary unit

Remove the housings and the planetary units (Figure 9-15). Remove the differential assembly through upper part of transmission case.

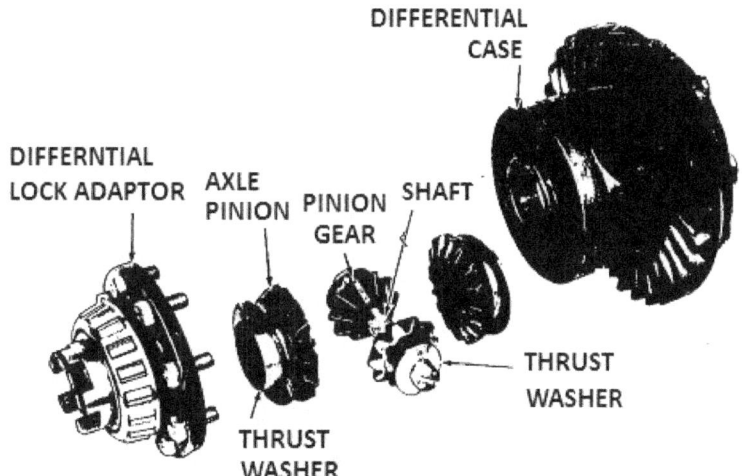

Figure 9-16: The components of the differential

Mark the position of the differential lock coupling on the differential housing. Remove the bolts holding the assembly together, drive out the pins holding the differential pinion shaft.

Figure 9-17: The components of the differential

Remove the shaft, pinions, the trust washers and the axle shaft pinion gears and their trust washers. Remove the cap screws holding ring gear to the differential cage, and remove the ring gear. Reassembly follows the reverse order above.

The rear axle shaft and housing

The planetary gear reduction unit is located at the outer end of the axle housing. The backlash between the pinion gears and the planetary ring gear is 0.003-0.007". Adjustment is made by the use of shims, 0.002-0.005" thickness.

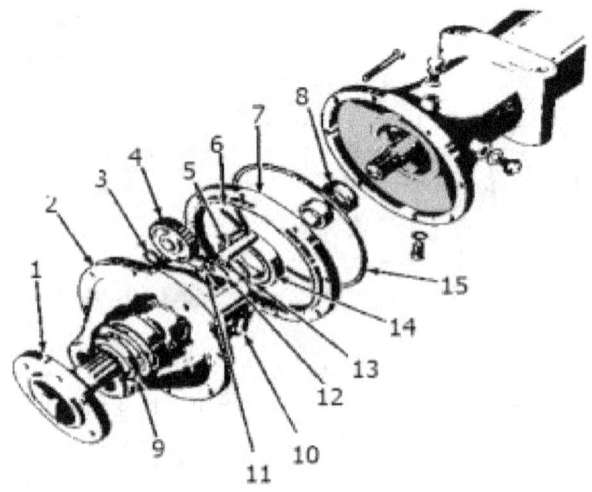

Figure 9-18: Differential pinion and rear axle

[1. Axle 2.Housing 3.Spacer 4. Pinion gear 5. Pinion gear shaft 6. Groove pin 7. Ring gear 8. Oil seal 9.Oil seal 10.Carrier 11.Needle bearing 12. Thrust washer 13. Snap ring 14. Bearing 15. Gasket]

Removing the planetary unit

Support the rear axle housing on a stand; remove the wheel and draining the planetary unit. Punch mark on the relative position of the ring gear and cover to the axle housing. Remove the bolts attaching the planetary unit, use a screwdriver in notches provided and pry the unit apart.

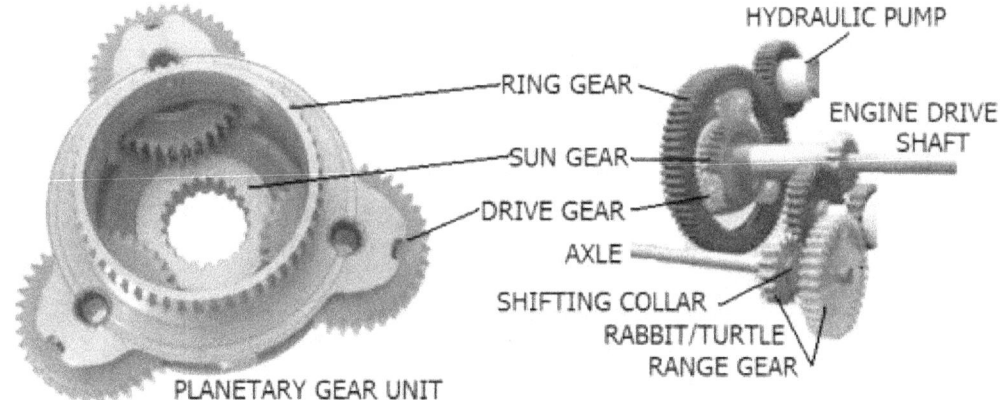

Figure 9-19: Planetary gear unit

To disassemble the shaft and the planetary pinion carrier, proceed as follows:

Drive groove pin securing the pinion shaft out. Push the pinion shafts outward and remove the pinions with their needle bearings, remove the snap ring from the axle shaft and remove the carrier assembly from the axle by the use of a press. In reassembling, clean the grease from the splines and coat with the edge of the planetary carrier splines with locktite and press into place. Replace the snap ring on the axle shaft. Install the pinion gears, bearings and shafts with the help of grease. Install the new groove pins to secure pinion shafts. Fit a bushing of inside diameter 1.693-1.692" to support the axle shaft. Inspect the bearings, install a new gasket and assemble unit into the axle housing.

Removing the rear axle and housing

To remove the axle shaft, it will be necessary to remove the axle housing as an assembly. The planetary assembly must be removed as described above. Remove the differential lock control assembly by removing setscrew and sliding shaft out of the yoke, and remove the yoke through opening at the top of axle housing. Put a slight pressure against spring and remove the remaining snap ring.

Gently shift the collar, and the spring and washer can then be removed in order to have access to the inner oil seal. Special seal protector collar is installed to protect the inner oil seal and another on the outer oil seal and unscrew driver hub from the tool.

Figure 9-20: The differential and transmission

Remove the axle shaft through the outer end of the axle housing while holding outer seal protector collar in place, and at the same time catch the brake hub as it slides off the end of the axle splines.

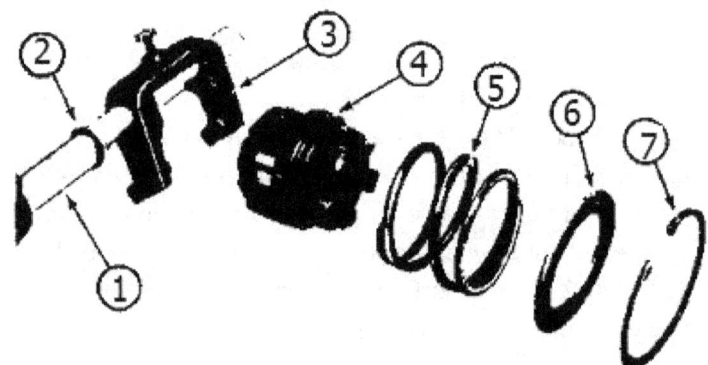

Figure 9-21: Differential lock shift assembly
[1. Shaft 2.O-ring 3.Yoke 4.Hub 5.Spring 6.Washer 7. Snap ring]

Adjusting the differential lock

Adjust the foot pedal on the differential lock so that the pedal is 1/8" off the step plate when the lock is fully engaged. To make this adjustment, it will be necessary to loosen the lock bolt on the lever assembly and run it on the shaft.

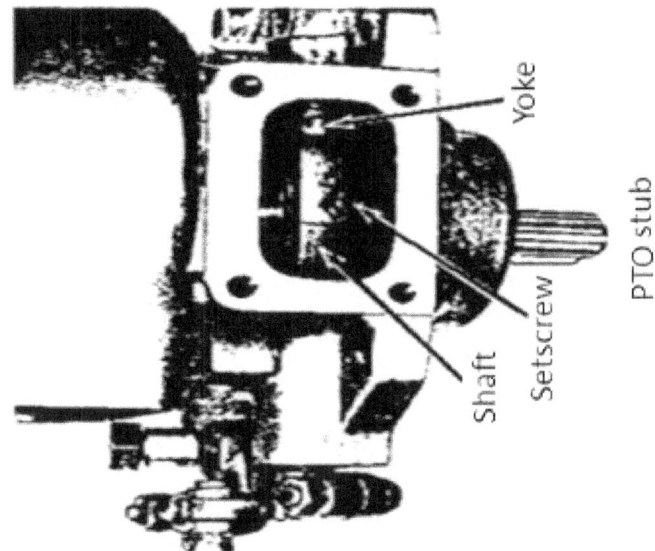

Figure 9-22: Differential lock control

Brakes system maintenance

Brake fluid inspection- Check the brake fluid level. The fluid level should only need to be slightly topped off occasionally. If more than 2 oz. of fluid is needed, the brake system should be inspected for leaks and component wears. Add only the recommended type of brake fluid as listed in the owner's manual. Do not add any other fluid to the brake fluid reservoir, and keep all foreign objects out of the fluid. The fluid in the reservoir should be clear in appearance and free of dirt and debris.

Figure 9-23: Allis Chalmer brake master cylinder

*Brake fluid replacement-*Brake fluid retains moisture and should be flushed and re-bled (remove the air from the system) to keep brakes working effectively. Check brake cylinder, for leakages or pressure failure due to valve or spring defects, check brake worn out pads and shoe wear.

FARM TRACTOR SYSTEMS

Identifying brake failure and warning signs

Worn or poorly maintained brakes and steering have resulted in many deaths on our roads. Do not wait until you hear grinding noises to have the brakes inspected. Have brakes checked periodically for wear. Some warning signs of brake problems are:

1. Noises when brakes are applied,
2. The steering wheel shakes when brakes are applied,
3. When you need to add more than 2 oz of brake fluid to the brake fluid reservoir, leakages is obvious
4. When you have a soft or squishy brake pedal, or
5. The brake pedal goes to the floor slowly while brakes are applied.

The following steps should be followed in brake assembly repair and maintenance

Procedure for changing brake pads

Precaution: In vehicle with ABS brake system, hydraulic line pressure is retained even after the vehicle has been parked for hours. Therefore it is required to always depress the pedal 10-20 times before beginning a brake job to release residual pressure.

Figure 9-24: Rear axle showing the wheel hub

Procedure

1. Slack the lug and or wheel nuts.
2. Raise and secure the vehicle on fully supported jack stands on level ground.
3. Remove the nuts and bring out the wheels.
4. Clean brake assembly with brake cleaner and inspect the brake lines for leakage.
5. Next remove the caliper fasteners but not the caliper. On some vehicles brake pads are held into the calipers with special clips. Make sure you note the original location before removing. Install new clips, if supplied, upon reinstallation.

Figure 9-25: Removing the caliper fastener

6. Remove and secure the caliper to the vehicle using heavy wire. Never allow the caliper to hang by the brake hose.

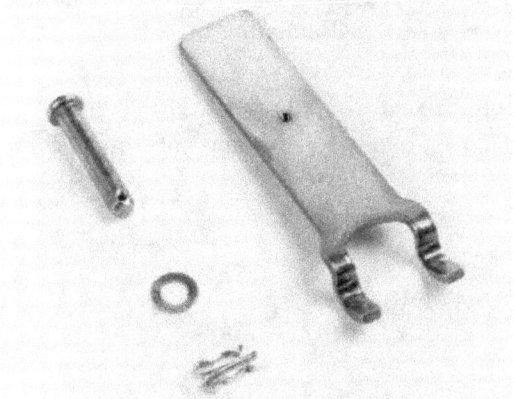

Figure 9-26: The caliper locking assembly

7. Remove the outboard pad and use a spreading tool or a c-clamp to compress the caliper piston. In some cases a long metal bar is used to depress the piston- this could be dangerous! A wedge could be placed between the caliper piston and the camber while the steering is wheeled to push the piston in.

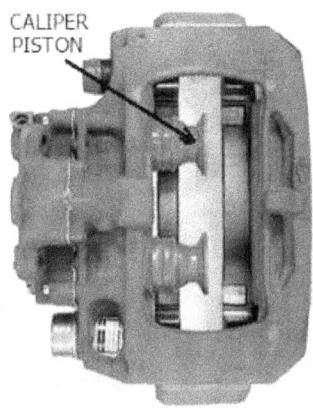

Figure 9-27: Brake caliper showing piston

8. Check the brake caliper piston for free and easy movement. With both pads removed, and the caliper piston compressed, inspect the caliper and replace if leaks are found on the caliper or piston boot. Uneven wear indicates caliper problems.
9. Inspect the rotor/disc for excessive grooves, scoring, cracks, bluing or shiny spots. If the disc is damaged, remove and replace it. Remove the caliper if disc removal is required. This is a good time for wheel bearing inspection.

Figure 9-28: Removing the wheel flange

10. Reverse the disassembly procedure making sure to keep things clean along the way. Do not forget to install fasteners, retaining clips and anti-squeal compound or shims.

Figure 9-29: Bolting the brake caliper

11. Insert the pad retainer (Figure 9-30) into the groove in the brake caliper and press down with a screwdriver until the pin can be inserted into the bore. Insert locking pin and washer and secure with spring clip. Insert a feeler gauge between tappet and brake pad back plate and adjust the running clearance to 0.7 mm by turning the hexagon drive on the adjuster.

Figure 9-30: Inserting the pad retainer

12. Replace the disc hub and tightened the lug nuts

Figure 9-31: Replacing the brake disc

Procedure for changing brake linings

In replacing a brake lining on disc drum observe the following procedures

1. Slack the lug and or wheel nuts.
2. Raise and secure the vehicle on fully supported jack stands on level ground.
3. Remove the nuts and bring out the wheels.
4. Clean brake assembly with brake cleaner and inspect the brake lines for leakage.
5. Remove the clips holding the caliper in place. Jack out the assembly and remove the piston. Ensure that you drain the fluid.
6. Remove all old linings from the caliper and clean the caliper.
7. Position each half-lining on the drum and hold in place with a brass pin. This is best done on a vice where the pin is gently riveted such that the fiber is not destroyed.
8. Repeat step 7 for the other half of the drum
9. Fix back the mechanism, cover the drum and fix the tyre.

Figure 9-32: Asbestos brake linings

Bleeding process for the brake system

Follow these procedures in bleeding air from a brake system:

1. Check the brake line for any possible leakage
2. Ensure you fill the reservoir with brake fluid before opening any bleeding nipple
3. Open one of the bleeding nipples and pump the brake until all whitish fluid is removed.
4. Hold down the pedal and lock the nipple then pump the brake until it is strong

5. Repeat the procedure for each of the brake points
6. Test the brake

Shock absorber installation

Before the installation of new shock absorber, it must be manually activated before installation on the vehicle. Open and close the polished rod through the entire stroke while in the vertical position with the rod side up. This action will get rid of any air or gas bubbles which may have formed in the pressure tube during storage. This process must be repeated several times until the shock absorber feels uniform without the presence of vacuum.

The shock absorbers must always be installed with the rod side up. When installing shock absorbers, it is recommended that the vehicle manufacturer's instructions be carefully followed with respect to tightening, gaps, components installation sequence, usage of glues, adhesives and vibration/noise insulation, as well as to checking the state of all other suspension system components for repair or replacement if necessary.

Do not grip polished piston rod with any tool, because any nicks or scratches on the rod will result in decreased service life. It is recommended that the suspension components be checked every 10,000 km. It is also recommended that the product be installed under the supervision of a specialist.

Note: Steering dampers (steering shock absorbers) do not require the activation process described above.

Tyre service and maintenance

Selection of agricultural tyres

Selection of tyre size and ply rating is based on the highest individual wheel load on each axle when the vehicle is statically weighed. Maximum load per tyre should not be greater than the load permitted for the size and ply rating. Drive wheel tyres on agriculture tractors when operating in the field must be so selected as to be able to withstand the maximum pull of the tractor under the operating conditions intended.

Maximum load in field service or haulage is to include:

i. Net weight – service with standard equipment, including maximum fuel, oil, and coolant capacity.
ii. Accessory weight, optional equipment weight and special order modifications.
iii. Tyre ballast, if used
iv. Field modifications.
v. Bin & tank load- include total weight when full

vi. Tractor Trailer – covers weight carried by the axle for haulage purpose.
vii. Cyclic loading on agriculture harvesting equipment – i.e. gradual increase in payload to maximum allowable load with unloading before off-field transport.

Proper inflation – Tyres are designed to carry loads up to the maximum specified at the inflation pressure for desired deflections, road contact, tread wear. Any neglect of inflation pressure will result in one or more of serious tyre failures or loss of tyre life potential.

i. *Over Inflation* – Load carrying capacity of a tyre cannot be increased above the maximum rated capacity, merely increasing its inflation pressure. Over inflated tyres do not flex as designed, do not absorb shocks, or impacts, more prone to cuts, concussion, snags and rapid center wear.
ii. *Under Inflation* – Under inflation results in excessive flexing of tyres, excessive heat generation, rapid shoulder wear.

Figure 9-33: Type of tyre inflation

Determining optimum tyre pressure

Tyre inflation pressure: Tyre inflation ratings are for guidance only. Such factors as tyre make, ballast and service conditions, etc., may dictate higher or lower pressures. The optimum tyre pressure configuration for a tractor depends on

1. The type and size of the tractor,
2. The type, size and number of tyres,
3. Soil type and soil condition, and
4. Draft, which depends on the type, width and operating depth of the tillage tool or other implement.

Variables such as tyre size and implement type can be controlled, but soil characteristics can vary considerably within a field. Hence, an optimum configuration must be determined for average conditions. The following procedure is a step-by-step guide to determining an optimum configuration.

Step 1: *Select a field operation* such as ploughing, field cultivation or planting and mount the implement on the tractor according to guidelines in the owner's manual, your experience or other available guidelines.

Step 2: *Weigh and record each axle separately.* Fill mounted tanks such as fuel tanks and sprayer tanks. If mounted equipment is used, weigh the tractor with the mounted equipment on a weigh bridge.

Step 3: *Determine tyre type and size* from the tyre codes on the sidewall of the tyres. See tyre code for details.

Step 4: *Check and adjust weight distribution.* Refer to the owner's manual or to Table 9-1 and adjust front and rear weights so that the weight of the tractor is properly distributed between the front and rear axles. Record the final weight of each axle.

Step 5: *Determine the weight supported by each tyre.* Divide the axle weight by the number of tyres on the axle.

Step 6: *Adjust tyre inflation pressures.* Refer to the tyre inflation pressure chart for your tyres, which is available from your tyre dealer. Locate your tyres on the chart by tyre size. Determine the minimum tyre inflation pressure listed in the chart for the weight supported by each tyre, as calculated in Step 5. Adjust tyre inflation pressures as necessary.

Step 7: *Assess tractive efficiency by measuring or observing wheel slip.* Wheel slip can be estimated by observing the appearance of the track. *If the track is scrambled* so that the lug marks are completely broken, slip is high. *If the track is well defined* and the lug marks are unbroken, slip is low.

Step 8: *Add or remove ballast to optimize slip. If slip is high or greater than 15 percent*, excessive power is being lost to wheel slip.

Step 9: *Repeat steps 4–8* if ballast adjustments are made until slip is approximately 10 percent.

Tyre inspection

A few minutes tyre inspection may disclose damage or a condition which, if neglected, could cause a breakdown. Check pressure and tread wear. Check air pressure to correspond to the prescribed pressure stamped on the tyre. Inspect tyres for uneven tread wear, punctures, bulges, or cuts in sidewall of the tyre. Avoid moving a vehicle with deflated tyre, this could cause tyre twisting or squeezing.

As a result do any of the followings:

1. Ensure that the tyres fitted are correct in size, ply rating and type for the application.

2. Ensure that the tyre inflation pressures are strictly adhered to the recommendations for the load and that the maximum cold inflation for the maximum load carrying capacity of the tyre is never exceeded.
3. Ensure that the valve stem in tube is free from damage and always use a valve cap.
4. In case of any abnormal loss of tyre air pressure, investigate and rectify the defect.
5. Check valve core / tube
6. Ensure that the rim wheel is in good condition.
7. Always fit a new tube and a new flap (wherever recommended) with a new tyre.

Figure 9-34: Twisted/deflated tyre

8. Do not change tyre size / ply rating from the original equipment for the vehicle without prior approval from the tyre and vehicle manufacturer, and the required inflation pressure ascertained.
9. Follow recommendation for tyre rotation and tyre matching.
10. Ensure timely removal of tyre for repair / retreading.
11. Consult tyre manufacturer's specialist whenever in doubt and whenever a significant improvement is not noticeable, even after the corrective actions recommended.

Tyre balancing and front end alignment

Tyre balancing: Routine rotation and balancing can greatly extend the life of your tyres. Most front-end "shake and shimmy" complaints of the tyre in motion can be attributed to out of balance, or out of round tyres. When rotating or replacing automobile tyres the best tyres should go to the front of the vehicle. Some tyres are "directional" and should be kept on the same side of the car turning in the direction indicated on the side of the tyre. Ask your mechanic if he would inspect the brakes for free when rotating and balance tyres.

Front end alignment – The front end components of a vehicle can be out of alignment, but not give any indication or warning signs. Shimmying and shakes in the front end are usually not caused by the car being "out of alignment," but by out of balance or lack of rotation with the tyres. The vehicle pulling to one side, or unusual tyre wear are the two most common "out of alignment" warning signs.

Check the alignment and all wearable parts in the front end periodically. Always have the front end aligned when replacing tyres. A front-end alignment is commonly referred to as a "four wheel alignment" these days. Some adjustments to the rear alignment are available on most new model vehicles, thus the term four-wheel alignment.

Rim, wheel and axle maintenance

Rim maintenance

1. Whenever a tyre is to be removed from a wheel or prior to mounting of a tyre, the condition of the rim/wheel should be checked thoroughly, particularly for any distortion of the rim flange or wheel disc. Any rust is to be removed by brushing off with a wire brush.
2. Damaged, cracked or distorted wheels or wheels having stud hole seating cracked or deformed or showing ovality must not be repaired or put in service.
3. Mounting faces of the hub, ball seats and flat mounting surfaces of wheels should be clean and free from foreign material or excess paint.
4. Threads of studs and nuts should be clean, free from burrs or damage.
5. On disc wheels, the nuts must be tightened in a cross sequence and to the recommended torque.
6. Nuts should always be kept tight.
7. Never permit oil or any lubricant to get into the ball seats of wheels, or on the ball faces of the nuts.
8. In order to avoid tension crack corrosion on the rim wheels, which are likely to damage tyres, an anti-corrosive protection on the wheel must be fully ensured, even on the tyre side of the rim and lock rings.
9. Always avoid vehicle or tyre/rim overloading. Do not use a higher inflation pressure than that permissible by the rim / wheel manufacturer for the size / type, to avoid over stressing the wheel. Avoid subjecting the nuts to an over torque.

Wheel maintenance

Remove hub cap, and filled with recommended grade of grease then re-install the hub cap. Lubricate the two front axle pivot points. Also lubricate the kingpins using recommended grease.

Wheel/tyre slip

Excessive wheel slip may be caused by:

1. Soil that is too wet,
2. A draft force that is too large,

3. Tyres that provide inadequate contact with the soil surface because they are too small or overinflated, or
4. A tractor with inadequate ballast.

For inadequate ballast

i. Consider equipping the tractor with larger tyres, dual tyres or radial ply tyres.
ii. Ballast may also be added to improve traction.

Increased ballast will cause greater soil pressures and increase compaction. Consider using dual tyres to decrease soil pressure and remember to adjust tyre pressures. *If slip is low or less than 5 percent*, excessive power is being lost to rolling resistance and ballast may be removed to increase tractive efficiency. Removing ballast reduces the weight of the tractor and will decrease the severity of compaction.

Note: Decrease tyre inflation pressures when ballast is removed or when dual tyres are used. Increase tyre pressures when ballast is added or when dual tyres are removed. Weigh each axle to determine the load on the tyres and use a tyre inflation pressure chart to determine the correct tyre pressure.

Measuring wheel slip by counting tyre revolutions

Wheel slip can be determined by counting tyre revolutions. This procedure requires two people and can be performed in only a few minutes without any special tools. One person operates the tractor and one person counts tyre revolutions.

1. *Make a clearly visible mark on the tyre* or plan to use the valve stem to count tyre revolutions. Chalk makes a good temporary mark.
2. *Lay out a course.* Choose a section of a field with conditions similar to average field conditions. The course should be long enough to provide extra space at each end of the course for starting and stopping.
3. *Set a flag* in the field off to the side of the course to mark the beginning of the course.
4. *Begin to* operate the tractor and implement under average or normal operating conditions and far enough behind the mark that normal operating conditions are reached before the tractor passes the flag at the beginning of the course.
5. *Note the position of the valve stem or chalk mark* as the tyre passes the flag.
6. *Walk parallel to the tyre at a safe distance* from the tractor and implement.
7. *Count 20 tyre revolutions.* Stop at the moment the tyre mark or valve stem reaches its original position for the 20th time.
8. *Mark the end of the course.* Lift the implement.
9. *Return to the beginning of the course.*
10. *Operate the tractor at the same speed* on undisturbed soil parallel to the original course without the implement in the ground.

11. *Count tyre revolutions once again* as the tractor travels the entire course. This time, instead of marking the end of the course after 20 revolutions, count the number of revolutions between the beginning and the end of the course. Because the tractor is no longer pulling a heavy load, slip should be almost eliminated and you should count fewer than 20 revolutions.
12. *Use the equation below to determine slip.* Each missing tyre revolution under no load accounts for 5 percent slip that occurred during the first pass when the tractor was pulling the implement. For example, 19 revolutions indicate 5 percent slip, 18 revolutions indicate 10 percent slip and 17 revolutions indicate 15 percent slip, etc.

Slippage of drive wheels: Travel reduction on soil surfaces, or slip (s), is calculated as follows:

% Slip = 100 x (loaded revolutions – no-load revolutions) ÷ loaded revolutions

$$\% \, Slip(s) = \frac{a_n - a_1}{a_n} x100 \ldots\ldots\ldots 9.1$$

Where:

S = slip, percent;

a_n = the advance under no load conditions per wheel or track revolution, m (ft);

a_1 = the advance under actual load conditions per wheel or track revolution, m (ft).

Example: During the first pass with the tractor and implement, 20 revolutions are counted. During the second pass with the implement raised 18 revolutions are counted.

$$\% \, Slip(s) = \frac{20 - 18}{20} x100 = 10\%$$

Motion resistance becomes appreciable when heavy implements are used in soft or loose soils. Tyre parameters and wheel loadings must be known or assumed to calculate this value. Total implement motion resistance is computed as the sum of the individual wheel values.

Ballast

Ballast is the important link between field speed and drive wheel slip. Too much ballast can create power loss from increased rolling resistance, soil compaction, and high mechanical loading on axles, bearings, and the drive train. Insufficient ballast can create too much slip, high tire wear, and lost productivity.

Ballast (tractor weight) management

Careful management of ballast and tyre inflation pressure can maximize tractive efficiency, minimize compaction, and increase tractor drive train life and increase profitability. Ballast should be used to achieve just enough traction to transmit power to the ground without excessive wheel slip. Excess ballast causes a deeper track that increases rolling resistance. Power is lost from an over-ballasted tractor. Eliminating wheel slip by adding ballast does not maximize power transmitted through the tractor drawbar. The amount of grip obtained by wheel on soil depends on three factors:

a. The area of driving surface in contact with the land
b. The amount of weight on that surface
c. Condition of the soil.

Tilled or soft soils require more ballast for traction. Ballast should be distributed between the front and rear of the tractor in the correct proportions to achieve maximum tractive efficiency and stability (Table 9-6). The location of the drawbar on a tractor causes weight transfer from the front axle to the rear axle when the tractor is pulling an implement. Weight transfer is especially evident on a two-wheel-drive tractor when the front end becomes so light that steering becomes difficult.

Table 9-6: Static weight distribution in front and rear of tractor

Tractor design / implement type	Weight distribution	
	Front	Rear
Two-wheel drive / Trailing implement	25%	75%
Two-wheel drive / Semi-mounted implement	30%	70%
Two-wheel drive / Mounted implement	35%	65%
Front-wheel assist / Trailing implement	40%	60%
Front-wheel assist / Mounted implement	45%	55%
Four-wheel drive / Trailing implement	55%	45%
Four-wheel drive / Mounted implement	60%	40%

Adding ballast to the bracket on the front of a tractor reduces the weight on the rear axle. For example, adding 100 kg on the front increases total tractor weight by 100 kg but may increase the weight on the front axle by 150 kg and reduce the weight on the rear axle by 50 kg. The rear ballast weights are attached to the outside of the rear wheel, additional weight can be bolted to the first weight.

The static (no-load) weight distributions shown in Table 9-3 should be maintained by adding and removing ballast from the front and rear of the tractor. Front-wheel-assist and four-wheel-drive tractors should have relatively more weight on the front than two-wheel-drive tractors because the front wheels also provide traction.

Figure 9-35: Tractor front and rear axle ballasts

Ballast (tractor weight) selection

The correct drive wheel ballast is the one that results in optimum slip. Therefore the following three steps will help select the proper ballast while optimizing field speed and drive wheel slip.

1. Select a relatively high field speed, above 6 mph.
2. Select an implement width that matches this speed with available drawbar power of the tractor.
3. Ballast the drive wheels to obtain a slip that is near the optimum slip for the particular soil conditions (refer to tractor operator's manual for specific values).

This selection process will help maximize efficiency of power used in tillage operations.

Wheel track adjustment

On tractors, track is measured between wheel center lines. The track can be adjusted according to operational requirements, bearing in mind consideration for the following factors:

1. Type of operation and implement
2. Type of cultivation, and
3. Type of soil terrain.

The track width is of paramount importance in adapting the tractor to the implement and the job to be done. During pulverization (harrowing), the track width should be such that the wheels pass between the crop rows, thus avoiding as far as possible damage to the crop. In ploughing (Figure 9-36), the track width will determine the cutting width of the first disc, which should be equal to the others.

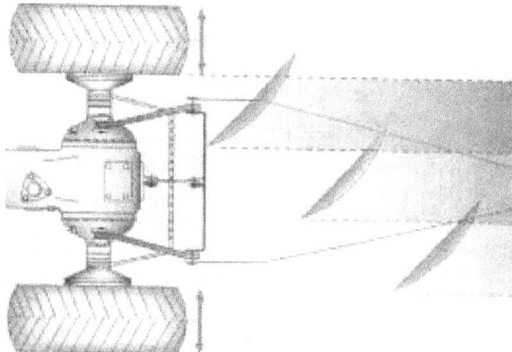

Figure 9-36: Track adjustment for ploughing operation

2-wheel drive track adjustment: Remove the bolts holding the front frame and move the whole internal wheel assembly to the desired position. At every hole of the bar, the track is altered by 50 mm on that respective side. Therefore, total alteration of the front track will be 100 mm (50 mm on each side). Avoid the use of maximum front axle track when operating on irregular land, as it transmits shock to the axle.

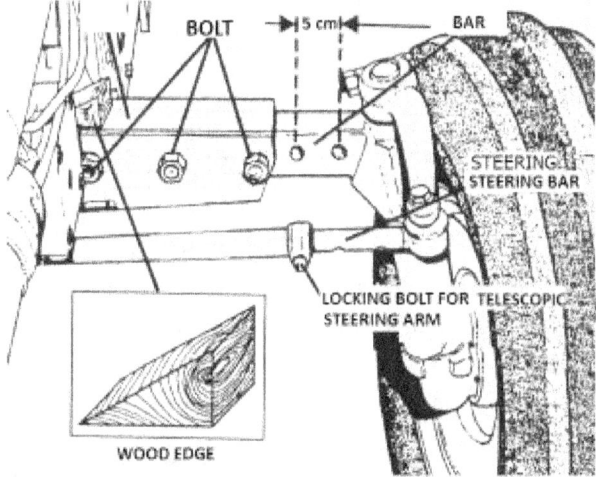

Figure 9-37: 2-wheel drive track adjustment

4- Wheel drive track adjustment: This system enables up to 8 different settings with different track widths, according to the mounting scheme chosen. To change the track, block the rear wheels and raise the front axle, supporting it on a suitable safe trestle. When necessary to invert the rims, change complete wheels: left wheel to right side and right wheel to left side. Tightened the wheels to the right torque and retightened later.

Setting the steering angle: The steering angle limitation adjustment can be made by turning two stop bolts, one on each side of the axle. Loosen the counter nut 1 in Figure 8-36, and turn the stop bolt 2 anticlockwise to reduce the toe-in and clockwise to increase it.

Figure 9-38: Front axle steering angle adjustment

The head of bolt 2 sets the maximum toe-in. adjust both bolts equally, to avoid interference in either direction with the steering wheel at full lock. Retighten the counter nuts and lower the tractor.

Greasing and adjustments

Tractors generally operate in dusty environments, paddy fields, deep water etc, and coupled with a lack of maintenance care, the intervals of service should be regulated for better performance and long life. One way of ensuring this is through grease lubrication at those grease points located all over the tractor and machinery. Grease lubrication points are always located in conspicuous areas where they can easily be reached by either oil can or grease gun.

On 4-WD tractors, lubrication points are located as follows:

Pivot pin: On central drive type tractors, the greaser point is located at the front bearing. On side drive type tractor the greaser point is located on each bushing and pins

Swivel pins: grease all the four points at approximately 600 hours of operation in normal conditions or 300 hours in severe conditions.

Universal joints: the greaser point is located in two places. Grease at every 100 hours in normal conditions or every 50 hours in sever conditions.

Drive shaft for side drive type axles: grease at every 100 hours in normal conditions or every 50 hours in sever conditions.

Front axle check and lubrication

Check and ensure that oil level is up to the marked level as indicated on the casing. If not, top up as necessary. The International specification recommends multi grade transmission oil of viscosity 20W-30 for all axle; axle casing and final drives for all two-wheel, and four-wheel drives. Check the front axle final drive and top the oil when required. On 2 wheel drive tractor, lubrication points are located as follows:

On central joints, one greaser is located

On kingpins, one greaser is located on each wheel

On the wheel hubs, one greaser is located on each wheel

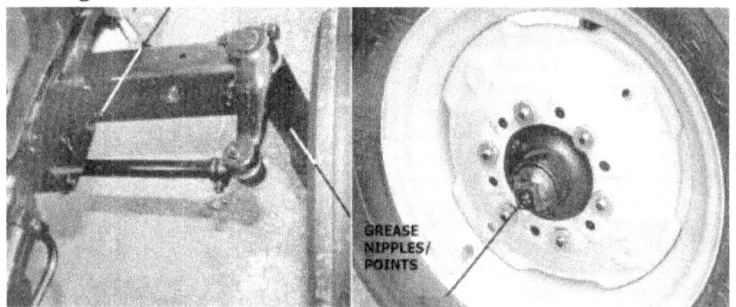

Figure 9-39: Grease points on 2 WD tractors

Hydraulic lift system: For all points, grease every 200 hours in normal condition or every 100 hours in severe conditions.

Leveling arms (1 or 2 grease points)

Leveling case (1 greaser if fitted)

Lateral stabilizers: chain type (2 greasers each), telescopic type (1 greaser each).

Upper link arm (2 greasers, if fitted)

General points on the tractor includes clutch pedal, brake pedals, clutch cross shaft (right and left), brake cross shaft (left and right), interior articulation points, steering cylinders (2 WD tractors 1 or 2 greaser points depending on tractor). Grease all points at every 100 hours of operation as follows:

Figure 9-40: Hydraulic grease points

FARM TRACTOR SYSTEMS

Tractor service access points

The following service access points have been designed on the tractor for easy accessibility for adjustment and maintenance checks:

Front grille- The front grille provides access to the headlight bulbs, battery and the air cleaner. To remove the grille, pull them by the handle; the headlight wiring can be disconnected if necessary.

Hood and side panels- Access to the top of the engine, fuel tank and radiator is gained by removal of the hood and side panels.

Front side panels- To remove the front panels, unscrew the fastening bolts, five each side and remove the panels.

Hood- To remove the hood; unscrews the twelve hood screws, and lift up the hood.

Radiator cap- To remove the radiator cap, (forward of the two caps above the hood) - press the radiator cap down and turn slowly counterclockwise to allow pressure to escape before removing it. This is particularly important if the engine is hot.

Fuel filler- to remove the fuel filter cap (rear of the two caps above the hood) - press the cap downward and turn counterclockwise.

Practical exercises

1. With the aid of workshop technician, carry out transmission system evaluation of an identified tractor and make your report.
2. Explain how power is produced or transferred in

   a. Hydraulic system,
   b. Selected power transmission system of your choice

3. Explain each step in ONE of the following maintenance procedures:

   a. Tightening hydraulic fittings.
   b. Servicing spark plugs.
   c. Lubricating a clutch-release bearing.
   d. Cleaning a work piece with a wire-brush wheel.

4. Prepare a tractor/ vehicle daily maintenance report for a period of one month.

FARM TRACTOR SYSTEMS

PART 4

TRACTOR DRIVING AND OPERATIONS

Highway codes, tractor driving tests, drive control devices, equipment hitching and field operations

FARM TRACTOR SYSTEMS

Introduction

Tractors as a utility vehicle composed essentially of all systems in an automobile and a few other systems for specialized operations. Tractor systems include all other systems apart from the engine which function together for efficient delivery of its duties. Such systems include the hydraulic systems, steering systems, PTO systems, transmission systems, and brake systems.

A good employee should maintain these systems in safe operating condition by making regular inspections and following the manufacturer's recommended servicing and maintenance procedures. Logbooks and maintenance records should be kept for scheduled maintenance, repairs and any modifications made on any system, which might affect the safe operation of the tractor.

Expected learning skills

At the end of this chapter, the students are expected to

1. Be able to identify the various systems that make up a tractor, their functions and how to maintain them.
2. Be acquainted with various field operations, tractor driving techniques etc.
3. Be able to perform 10, 50, and 100 hour maintenance
4. Be able to prepare a maintenance record, schedule and charts for maintenance operations
5. Be familiar with the basic requirements for carrying out some minor repairs and some major and more complex repairs on the tractor.

Student activities

1. Students are expected to be familiar with Nigerian and International highway codes, tractor and operation signs/signals and road marks.
2. They should be able to handle tractor and pass a driving test examination.
3. They should be able to carry out field operations.

CHAPTER 10

Tractor Driving and Operations

Introduction

Tractor driving and operation is designed for drivers and non-drivers alike over 18 years of age who want to learn how to drive and operate a tractor. The driving session covers training on identification of highway codes, road signs and hand signals in transporting equipment on the highway and on the field. Other essential trainings include the movement of tractors, coupling or attachment of the various implements and as well as basic maneuvering and field operation. Following the initial routine machinery preparations, driving the tractor to perform field operations must be learned. This chapter covers all you need to know about highway codes, road signs, hand signals, tractor driving and field operation.

List of available tractor model in Nigeria

Students are required to be familiar with the various tractor makes/models that have been used for farm operations in Nigeria. Such records will keep students informed of the type of tractor they are likely to encounter on the field. Some of the different types of tractors that had been used in Nigeria include;

1. Fordson Ferguson
2. Mersey Ferguson (Good for the Nigeria soils and largely used till date)
3. Marson Harrison (No longer available today)
4. David Brown (Used immediately after independence)
5. Ford (there are 4 types common in Nigeria)
6. Internal harvester (Three types have been used in Nigeria)
7. John Deere (5 types available, equally good on Nigeria soils)
8. Styer (2 models used)
9. Fiat (Italy)
10. New Holland (recently introduced by Fiat)
11. Urssus (recently introduced)
12. Mahindra 615H (India) and
13. Dentz among several others

FARM TRACTOR SYSTEMS

Tractor drive control systems

Performance skill: read and interpret control systems

Given the instrumentation and control panels of a typical tractor, the operator will be required to identify, locate, read, and interpret the typical tractor instruments and controls found on the tractor.

Performance requirements:

i. Drivers/operator/student must identify and locate each of the tractor driving controls and the various monitoring devices (gauges, alarms, lights, etc.) required to operate the vehicle safely and efficiently.
ii. Drivers/operator/student must read instrument/gauge accurately within ± 1 unit of measure correctly each time.
iii. Drivers/operator/student must operate control/switch correctly each time.
iv. Drivers/operator/student must supplement gauge/control information with other data.

Performance assessment criteria:

1. Identify, locate, and read/operate each of the *primary controls* including those required for - steering, accelerating, braking, and parking.

2. Identify, locate, and read/operate each of the *secondary controls* including those required for - control of lights, signals, windshield wipers and washers, interior climate, engine starting and shutdown, suspension and coupling.
3. Identify, locate, read/operate, and indicate the acceptable reading range of the various instruments required to monitor vehicle and engine speed as well as the status of fuel, oil, air, cooling, exhaust, and electrical systems.
4. Augment displayed information from other sources, given that instruments may malfunction or not be entirely accurate.

Tractor drive control systems

The followings are some drive control systems in the tractor Figure 10-18. Each control point labeled in the figure is described below:

1. The shuttle lever allows you to move from forward to reverse without stopping – a real plus for loader work.
2. Semi-flat platform rests on an isolated mount so when the tractor vibrates you do not. The uncluttered layout gives you room to spread out and let you mount or dismount from either side.
3. Scooped steel fenders with sturdy grab handles on both sides make entry and exit easier.

4. PTO hydraulic engagement lever for the independent PTO for easy operation and quick of implements.

Figure 10-1: Tractor control points

5. A retractable seatbelt and folding ROPS (Rollover Protection System) combine to provide added protection.
6. Spring-suspension seat glides forward or back to put you right where you need to be for optimal control. Get a perfect fit with height and weight adjustments too.
7. Position control lets you operate the 3-point hitch with increased precision.
8. Foot throttle gives you solid control of engine speed when your hands are busy operating other controls.
9. Conveniently located gear selector lever makes shifting a snap.
10. A range lever lets you match the ground speed to the job.
11. A brake lock lever makes it easy to engage the brakes before leaving the platform.
12. The hand throttle is perfect for when you want to maintain a set engine RPM for PTO applications.
13. Automotive-style instrumentation makes it easy to keep an eye on major operating systems without losing sight of the task at hand.

Each of these control systems are further described in subsequent sections below.

FARM TRACTOR SYSTEMS

The clutch pedal

Power transmission between the engine and the other parts of the tractor is governed by the clutch which is controlled by the pedal. When the pedal is depressed, the engine is disengaged, when released, the engine is engaged.

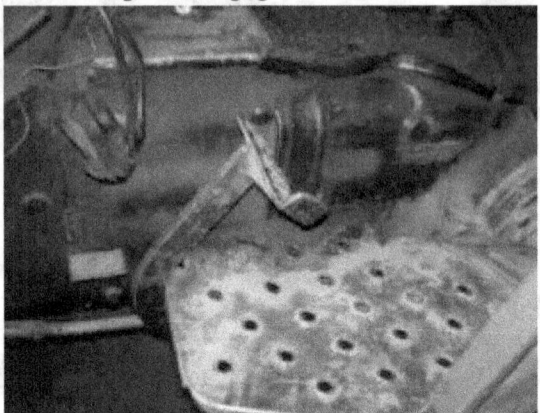

Figure 10-2: Tractor clutch pedal

Tractors with dual clutch assembly need the clutch depressed half way down to stop the wheels and fully down to stop the PTO. The clutch pedal is located on the left side of the driving seat. Clutch pedal are not to be used as foot rest, as it could cause undue clutch wear

Throttles (hand controlled lever and foot pedal)

There are two types of throttles used in tractor; the hand and foot throttle. The hand throttle is used to set the required engine speed for field work. This speed is maintained by the engine governors. The foot throttle is used for road movement and other transport operations where speed varies from time to time. The foot throttle can be used to override the hand throttle at times.

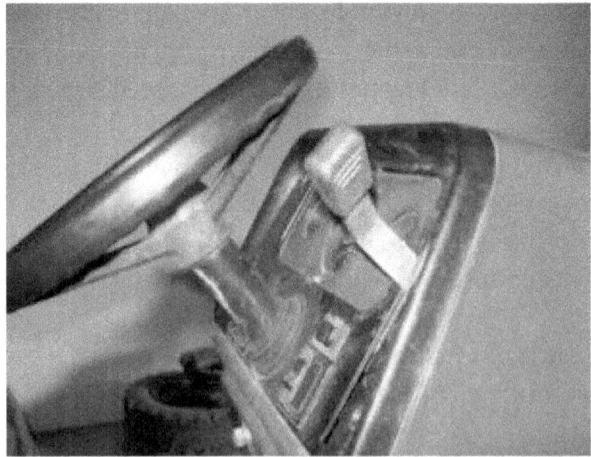

Figure 10-3: Tractor hand throttle lever

Gear engagement levers

Tractors are designed with two set of gears, four forward gears (1, 2, 3, 4) and one reverse gear. The gears were controlled by the gear selectors or levers located in front of the seat. There are two set of levers; one is short and the other long. The long lever is the selector for the range of forward and reverse gears. The shorter lever controls the auxiliary gear. The auxiliary gear has high and low range.

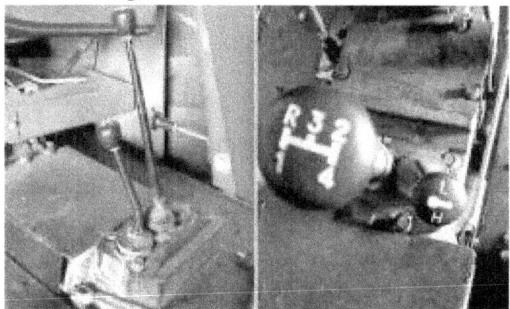

Figure 10-4: Tractor gear levers/selector (MF tractor)

A typical tractor has eight forward and two reverse gears. Some has an extra high or low ratio system operated with electrical or hydraulic power. It is control with a switch or lever on the instrument panel. This device is in addition to the two gear levers and can give at least sixteen forward gears.

The forward gears are used for increasing speed. Gear number 1 is the lowest but develops the greatest torque. This gear is designed for power at low speed while the 4th gear is designed for speed but less torque or power. The two sets of gears were designed for complementary use and a combination of these gears makes for effective farm operations. The forward gears in combination with high and low gears are used for field operations as follows

Brake control lever/pedal

The brake system performs three functions in the tractor:

1) To stop the tractor,
2) To park and
3) To assist in steering especially in confined spaces or at headlands.

There are two brake pedals, one for each rear wheels. They can be used independently in the field to help with turning on the headlands. The pedals must be locked together before driving the tractor on the road. The handbrake is used for the purpose of parking.

FARM TRACTOR SYSTEMS

Figure 10-5: Differential brake pedals

Differential lock pedal

This cut out the action of the differential. When the pedal is pressed down, it engages the differential by lo0cking together the two wheels and allows the wheels to travel in a straight line until the pedal is released. The pedal lock is used to reduce wheel spinning and improve traction when working under difficult soil conditions. The differential lock could be disengaged automatically or released by pushing down the clutch pedal or brake pedal after releasing the differential lock pedal.

Figure 10-6: Differential lock pedal

Power takeoff (PTO) lever

PTO Drive lever is located at the left side of the driver or directly under the seat in some tractor makes. *To engage the live type PTO, depress the clutch pedal and move the lever forward*. The drive can be engaged after the clutch pedal is pressed down. Some tractors have separate clutch for the independent PTO. It is not connected to the main transmission clutch. *To engage the independent PTO, this system does not require the clutch pedal to be depressed: only decrease speed, move the lever forward and select* the desired engine speed. The clutch lever is moved by hand to engage or disengaged the PTO drive. Many tractors have a choice of two PTO speeds; one is the British standard speed of 540 rpm and the other is America standard speed of 1000 rpm which is mainly used to drive equipment with a high power requirement.

Figure 10-7: PTO engagement lever (arrow)

Tractor hydraulic controls

This is otherwise known as the hydraulic system position control lever. There different hydraulic controls and each tractor manufacturers have special features in the design of the hydraulic systems. This lever is used for:

1. Operation with implements working over the surface of the field, such as pulverizing, loading platforms, undergrowth clearing devices etc.
2. Coupling of implements,
3. Transportation of implements to the working or loading place (hoist, platforms etc.)

The basic hydraulic controls are;

1. *Draft control system*, which is used for controlling soil engaging implements such as plough or cultivator, maintains a regular soil working depth without the use of depth wheels. This control system operates in the yellow (Draft) range of the quadrant and is used for depth control in soil-engaging implements. The further the lever is moved to DOWN position (Figure 9-25), the deeper the implement will penetrate the ground and the vice versa.
2. *Response control system*: This control system is located between the quadrant and the operator's seat. Position control system is used for controlling equipment used above ground level such as sprayers and mowers. It regulates the lowering speed of the 3-point hitch implements. With the selector fully down, there is a quick reaction (quick lowering speed = Hare). When completely up, the reaction is slow (turtle). Intermediary adjustments leave the selector within the intermediate band of the quadrant.

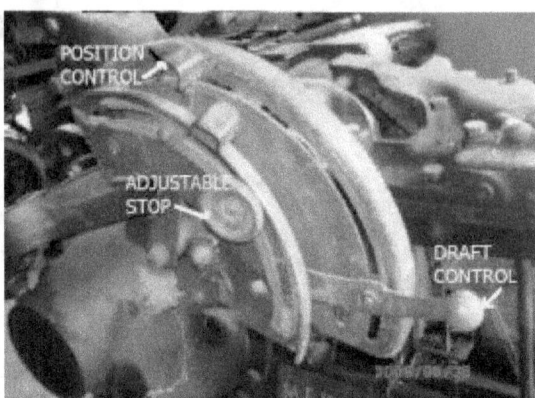

Figure 10-8: Tractor hydraulic control systems

3. *Position control lever:* This operate on the Red (Position and Transport) and the Blue (Constant Pumping) range of the quadrant. The Red sector is used to control the working height of the implements above the ground. The lever should NEVER be positioned in the Blue (constant pump position) or the hydraulic relieve valve be constantly open and heating of the hydraulic oil will occur.

Adjustable Stop (hand knobs) locks the adjustable lock in place when the desired working depth or height is reached. Moving the lever into the external position automatically selects constant pumping internally.

4. *External ram systems* are operated with hydraulic external service system. The ram on a tipping trailer is an example of the use of this system.

The following systems are areas of further developments in instrumentation and control.

Tractor instrument controls

Instruments, or gauges, on the tractor control panel tell the driver about the operating conditions within and around the tractor. All tractor drivers should know what instruments are available to indicate that the tractor is operating properly. When tractor systems are not working properly, continued operation may cause costly repairs and possible injury.

Figure 10-9: The modern tractor instrument panel

FARM TRACTOR SYSTEMS

Power and energy requirements for the agricultural field operations can be easily monitored and controlled by installed on-board instrumentation and sensors strategically mounted at various locations on the tractor. Modern agricultural tractor is equipped with on-board instrumentation monitoring and acquisition systems capable of providing the information on the engine speed (Tachometer), PTO speed, Forward travel speed, Drive wheel slippage, and Fuel consumption per hour and per hectare, Distance travel, Drawbar pull force, PTO shaft torque, etc.

Such control systems permits further research on the development of an agricultural database, especially on the aspects of energy saving and high performance.

Indicators and gauges

Instrument indicators can be warning lights, analog gauges, computer digital displays, buzzers, or standard gauges. It is important for the beginning operator to develop the habit of regularly checking the instrument panel. Check the gauges:

i. At start up,
ii. At regular intervals during operation, and
iii. When changes occur in the normal sounds of operation

Abnormal gauge readings, plus changes in operating sounds, indicate that there is a problem. You should immediately *stop the engine in a safe place*, and seek help from the owner or an experienced operator.

The following instrument controls are found on typical farm tractors:

Figure 10-10: Gauge panel

Tractor meter

The tractor meter comprises of the hourmeter and the tachometer (revolution counter). Tractor hour-meter records the number of hours worked by the engine and also measures the engine speed in revolutions per minute. Engine hours are recorded at about 2/3 of the maximum engine speed. With the engine at full throttle, the engine hours recorded will be more than those shown on the driver's watch. The engine speeds at which the PTO shaft runs at the standard speed of 540 rpm and 1000 rpm are marked on the engine revolution counter.

Figure 10-11: Tractor engine speed indicator

Some tractor meters show the forward speed of the tractor in each gear. Tractors without a speed meter have a chart on the instrument panel which gives the forward speed in each gear at selected engine speed. The temperature gauge gives the temperature of the engine coolant. Some gauges have a red section to warn the driver if the engine overheats.

Tachometers show revolutions per minute (RPM). Engine RPM must be matched to the job being done. Incorrect RPM can lead to:

- Engine damage
- Driveline and PTO damage
- Hazardous situations

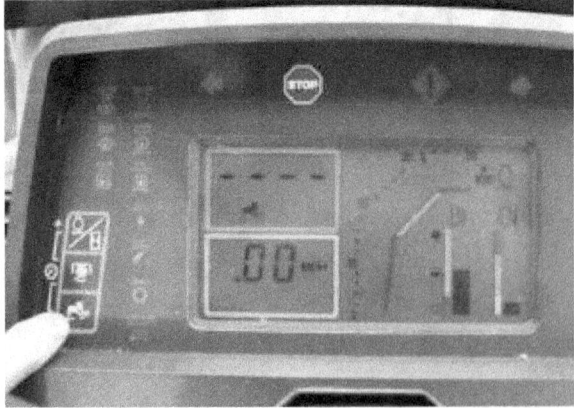

Figure 10-12: Digital display tachometer

Low engine speed while in a higher gear and beginning to pull a heavy load can stall the engine. High-engine speed with a low gear while attached to a heavy load can also create enough torque (rotational force) to tip the tractor backward. Accelerating quickly with a heavy load going up a slope can cause the tractor to rear up and tip backward. Engine speeds (rpm) must also match PTO-driven machine requirements.

Speed up the engine before engaging the PTO to operate an implement. *Low-engine speed* could stall the tractor. *High-engine speed* could shear off the implements safety pin if the pin was already under load.

Battery charge indicator

The battery charge indicator is a light bulb which show when the starting switch is turn on without the engine being on. With the engine operating, this light should be off showing that the battery charging system is working correctly. The battery charge indicator, or ammeter, shows whether the alternator or generator is charging the battery properly. Each time the tractor is started, the battery is discharged. During operation, the battery is recharged. Gauges will indicate + or - charge. Lights will show red at low charge. If the battery is discharging, find out the problem. The engine may not start the next time due to a low battery.

Oil pressure indicator (oil light or gauge)

This indicator is important to the life of an engine. It cones ON with the ignition switch and OFF when the engine starts. If oil pressure falls because of oil leak or low oil levels, the light or gauge shows you must stop the engine immediately. Never operate the engine with low oil pressure or oil levels. Oil lubricates the internal parts of the engine and prevents major repair expense. If low or zero oil pressure is indicated, shut down the tractor engine immediately to avoid costly engine rebuilds.

Figure 10-13: Oil pressure indicators

Engine temperature gauge

The engine must be cooled to prevent damage. Water-cooled engines can overheat if coolant is lost, radiators become clogged with debris, or the radiator leaks. If the engine overheats, stop the engine, allow it to cool, and then check for the problem. Never open the radiator cap while the engine is hot. Scalding from extremely hot water can result.

Figure 10-14: Danger of overheating

Fuel gauge

Check the fuel gauge before leaving for the field. Running out of fuel is inconvenient. On some tractors, running out of fuel (diesel) means time-consuming bleeding of air from the fuel lines in order to be able to start the tractor again.

Electrical controls

Electrical control system in tractor control the followings: front headlight, front auxiliary headlights, rear work light, rear red lanterns and brake light, road warning lights and license plate lights.

Other gauges

Tractors may come equipped with instruments to monitor air filter conditions, transmission temperatures, hydraulic system oil levels, and of course hours of work (hour meter). Become familiar with all instruments before operating the tractor.

Data instrumentation systems

This system is built into some modern tractors and is capable of providing information on engine speed, PTO speed, forward travel speed, drive wheel slippage, acres worked, fuel consumption per hour, fuel consumption per hectare, acres per hour, cost factor, fuel consumed, fuel remaining, and distance travelled of the tractor.

It utilizes a radar sensor to measure ground speed, a magnetic pickup sensor to measure rear wheel axle rotation and a flow meter in the engine flow line to measure fuel flow into the engine. The system digital readout is situated in the right pillar of the tractor cab.

Doppler radar

The Doppler radar unit for the tractor ground speed measurement is mounted rearward facing at 35 degree and located about midway on the left side of the tractor. It uses the Doppler radar effect from a microwave emission to generate a frequency signal that is proportional to ground speed. This radar unit could generate output signal at a frequency of 27.3 Hz per Km/hr within 3% tolerances.

Figure 10-15: Doppler radar

Data acquisition system

This system when mounted on the tractor and is capable of recording information on drawbar pull force, PTO shaft torque, drive wheel torque, and both vertical and horizontal forces at the 3 point hitches of the tractor-implement. It utilizes a locally designed drawbar pull transducer to measure horizontal pull at tractor drawbar point, wheel torque transducers to measure the torque at both tractor rear wheels, PTO shaft torque transducer to measure the torque at tractor PTO output, and a 3-point auto hitch dynamometer to measure the horizontal and vertical forces on the implement behind the tractor.

Flow meter

The fuel flow rate is measured by a micro oval flow meter that is located in the fuel line between fuel filter and injection pump of the tractor. The device is an aluminum case with a measuring chamber having two oval wheels that rotate one on top the other.

Figure 10-16: Flow meter

The fuel flow meter output pulse as a result of the repeated opening and closing of the reed switch is proportional to the flow rate of fuel passing through the two oval wheels. This factory calibrated fuel flow meter could measure the flow rate in the range from 0.025 to 40 L/hr within ±1% accuracy.

Rear wheel torque transducer

The rear wheel torques is measured by a pair of specially made transducers that are mounted on each side of the rear wheel axles of the tractor. The design of the transducer is based on an extension shaft that is securely mounted between the rear wheel axle flange and tyre rim.

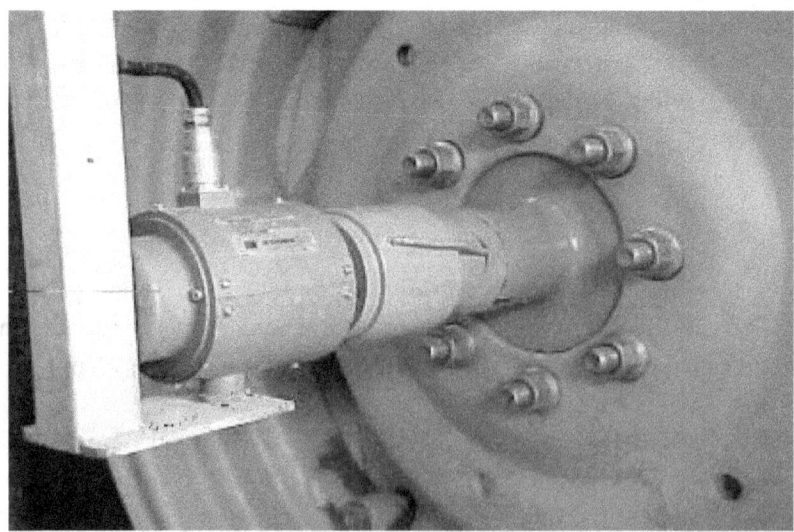

Figure 10-17: Torque transducer

PTO shaft torque transducer

The PTO torque is measured by a specially made PTO shaft torque transducer that is located between the tractor PTO and the implement PTO input. The free end of the female PTO drive shaft is provided with a raised collar, a screw bracket and a lock nut to position a slip ring while the other end is welded to a universal joint.

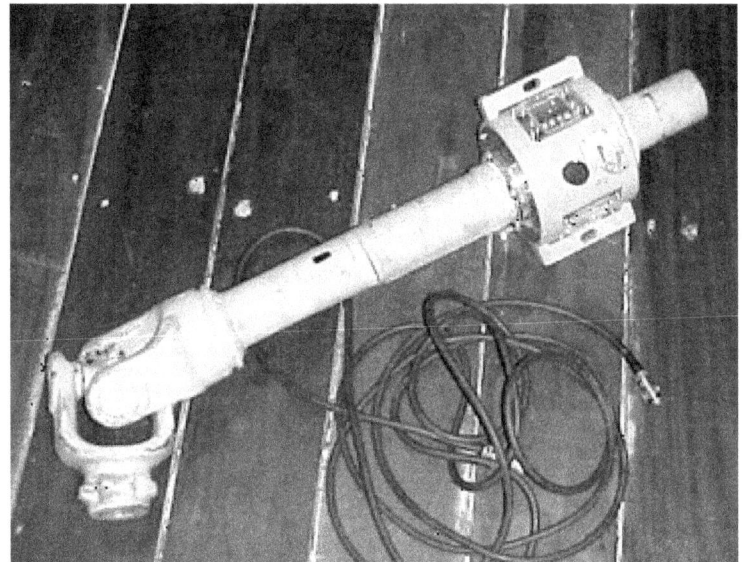

Figure 10-18: PTO shaft torque transducers

Draw bar pull transducer

The pull at the drawbar point is measured by a special made drawbar pull transducer. The thick proof ring part is made close to front pin of the drawbar pull transducer to reduce the effects of lateral and longitudinal moments on the transducer measurements.

Figure 10-19: Drawbar-pull transducer

3-point auto-hitch dynamometer

The force sensing elements comprises of three steel extended octagonal transducers that are located between the frames and hook brackets. Each transducer is designed for a maximum horizontal and vertical force.

Figure 10-20: 3-point hitch dynamometer

Tractor driving and operation

Tractor driving test

Tractor driving operation is designed for drivers and non-drivers over 17 years of age who want to learn how to drive a tractor. The driving session covers the movement of tractors, coupling or attachment of the various implements and as well as basic maneuvering and field operation. Following the initial routine machinery preparations (Figure 9-38), driving the tractor to perform field operations must be learned.

Figure 10-21: Routine tractor preparation for driving test

FARM TRACTOR SYSTEMS

The driving test is designed to find out if you:

i. Know the rules of the road.
ii. Have the knowledge and skill to drive competently in accordance with those rules.
iii. Drive with due regard for the safety and convenience of other road users.

Operator/trainee's requirements: The trainee or operator will be required to read the number plate at a minimum distance of 20.5m. If you fail the eyesight test then you will have failed the driving test.

The examiner/instructor's requirement: *The examiner will watch to see how you check and adjust your seat and mirrors (if you have them). He will ensure it is safe to start the engine and that you follow the correct procedures.*

Performance evaluation for tractor operators

The following safety information should be reviewed with the trainee before conducting the exercise:

1. No extra rider policy. Never allow an extra rider if a seat and seat belt is not available.
2. Tractors should only be started while the operator is sitting in the seat.
3. Potential locations for tractor overturn on the site where the tractor will be used.
4. PTO safety:

a. Dress safely (close fitting clothing, remove hood strings, tie, long hair back, etc.)
b. Keep all PTO shielding in place and operating properly.
c. Never step or lean over a PTO shaft. Always walk around the tractor and implement.
d. When making repairs or adjustments, first shut off the PTO, the tractor engine, and wait for everything to come to a complete stop.

Performance evaluation form for tractor operators

Evaluation instructions: Please mark each task as a pass or fail. If the item does not apply, mark the item as N/A.

Operator Name _____ Date_____

Department_____

Trainer_____

Tractors approved to operate: Supervisor or designated trainer may decide to list each tractor or general categories of types of tractors that the operator is approved to operate. If, during exercise, an employee is assigned to drive a type of tractor that he/she is not familiar with, the employee must contact their supervisor/designated trainer to obtain training prior to operating the tractor.

	Task: Ability to	Pass	Fail
1.	Perform pre-operational inspection of tractor		
2.	Wore seat belt		
3.	Start the tractor properly		
4.	Take off tractor		
5.	Demonstrate smooth acceleration and stopping		
6.	Properly hitch implement to tractor		
7.	Properly attach/detach hydraulic lines on implement		
8.	Demonstrate safe driving while on level ground or at a slight grade		
9.	Demonstrate safe procedures for lowering or blocking out implements		
10.	Demonstrate proper emergency shutdown procedure for tractor, PTO, and implements		
	Overall performance		

_____   _____
Operator signature   Evaluator signature/Date

Guide to tractor driving

Good eyesight is essential for driving. Tractor driving test requires that you read the number plate of a vehicle at a *minimum distance of 20.5m*. If you fail the eyesight test then you will have failed the driving test. You need to make sure it is safe to start and drive your tractor. The examiner will watch to see that you check and adjust your seat and mirrors (if you have them).

Mounting the tractor

When you approach the tractor after the initial routine checks, you must be cautious of your safety! Mount the machine and dismount the machine only at locations that have steps and/or handholds. Before you mount the machine, clean the steps and the handholds. Inspect the steps and handholds. Face the machine whenever you get on the machine and whenever you get off the machine. Always be seated in the tractor and never climb the tractor from the right hand side. Always climb from the left.

Maintain a three-point contact with the steps and with the handholds. *Note:* Three-point contact can be two feet and one hand. Three-point contact can also be one foot and two hands.

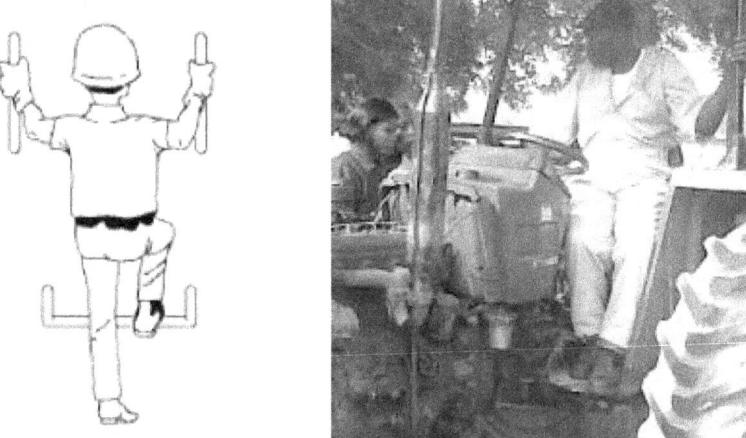

Figure 10-22: Mounting and dismounting tractor

Starting the tractor

Do not jump start a tractor on the floor. To start the tractor, be in the sitting position, carry out the following processes.

1. Check and ensure that the gear levers are in disengaged or maintained in NEUTRAL or PARK position. Most tractors has safety mechanism requiring either the gear lever to be set in neutral, the clutch pedal pressed fully down and or the PTO control lever be disengaged before the tractor could start. This reduces the load on the starter motor.
2. Set the hand throttle lever between ¾ and fully open mark
3. Put the stop control knob in the start position
4. Put the key in the ignition switch and turn clockwise. If the engine does not start within 30 seconds, release the starter switch and wait for further 30 seconds before trying again. Excessive trying of the starter without rest can run down the battery.

You should show your examiner that you understand the functions of all the controls. You should use them smoothly, correctly, safely and at the right time.

Operator's chair

The design of the modern tractor includes considerations of human factors. These factors allow the operator to perform many complex tasks with efficiency, safety, and minimum of fatigue.

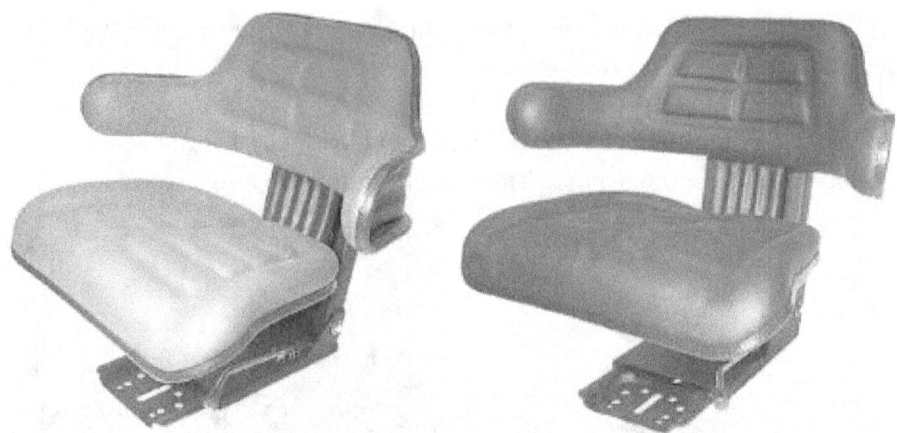

Figure 10-23: Tractor seats

Tractor ergonomic design are handled by the manufacturer based on field investigation and such factors as easy access to all control systems, pedal and levers in seating and relaxed position are important. Operator should have enough comfort on his seat during operation. Worn out seats and torn leather from seat should be replaced or mended.

Driving the tractor

Start the ignition; depress the clutch pedal to disengage the engine form the transmission. Push the short lever down to select the HIGH gear then hold the longer lever and push it to the far left then shift backward to engage gear number 1.

Note: the position of the forward gears varies according to tractor design. The position is always indicated on the selector.

Throttling/accelerating the tractor

There are two types of throttling device in use on the tractor;

i. The hand throttle lever, located at the right hand side of the dashboard just below the steering. The hand throttle lever is designed for use on the field to ensure uniform speed while in operation. This is so designed because of the uneven terrain the tractor operates. When the lever is shifted up, speed increases and then reduces when pushed down.
ii. The pedal throttle lever, located below the brake pedals depending on design. The pedal accelerator is designed for road movement.

Gently press the throttle pedal (accelerator) while you gradually release the clutch pedal for full engagement, then control the steering. Change gears from 1-2-3-4 as the tractor gain speed. The recommended maximum speed for tractor on highway is 25km/hr.

FARM TRACTOR SYSTEMS

Adjustments: Adjust your seat so that you can reach and operate the main driving controls, and the brakes. Ensure that the gear selector is in neutral, hand throttle/accelerator or engine starting controls are properly set. You should show your examiner that you understand the functions of all the controls. You should use them smoothly, correctly and at the right time.

Accelerator/throttle: Use a foot accelerator instead of a hand throttle or 'cruise control' for on-road driving. Hand throttles are normally use for off road/field work. Avoid accelerating fiercely leading to a loss of control thereby making the vehicle surge or lurch.

Footbrake: If your tractor has a split left and right braking system, use the pedals locked together while moving on the road. Your tractor may 'pull' to one side or the other slightly when pedals are in un-lock position, especially if you have been working off-road using the independent brakes. Avoid braking harshly except in an emergency. Avoid applying the handbrake before the vehicle has stopped moving or moving off with the handbrake on.

Gears: Make sure you are in the right gear range and use two-wheel drive instead of four.

Steering: Keep both hands on the steering wheel unless you are operating another control. Do not turn too early when steering around a corner. If you do, you risk cutting the corner when turning right and putting other drivers or road users at risk or striking the kerb when turning left. Do not turn too late, you could put other road users and pedestrians at risk by swinging wide on left turns or overshooting on right turns.

Moving the tractor: You should understand the functions of the following: lights, indicators, windscreen wipers etc. When moving off, make sure that you check all mirrors and over BOTH shoulders before moving away. Rear observation mirrors on some tractors may be affected by vibration. If your vehicle is affected, you will have to turn to look behind. When travelling slowly, early signals help other road users in good time; however, giving signals too early could mislead other road users.

Emergency stop: Large, chunky rear back tyres fitted to tractors can give limited grip, therefore be prepared to take immediate correcting action if you over-brake and skid. Avoid braking harshly except in an emergency. In reversing into a road side, you may be asked to reverse into a road side without a kerb. Do not let the wheels go onto the grass.

At junctions: you may need to creep slowly forwards, looking both ways until you can see it's safe to emerge. When meeting and passing other vehicles use the height of your vehicle to get a good look ahead. Do not try and squeeze through a tight gap. If you need to wait, do not get too close to the obstruction.

Driving tractor on highway

Before you operate a tractor on a highway, be sure to:

- Lock brake pedals together.
- Adjust the seat position so you are able to safely reach the steering wheel, pedals and gear shifts.
- Adjust mirrors for good vision.
- Make sure all lights and flashers work properly.
- Check tyre pressure and make sure wheel bolts are tight.
- Add weights, if necessary, to balance the tractor, especially if you are pulling or hauling a load.

The highway codes must be obeyed whenever a tractor is driven on public highways.

Nigerian and international highway codes

Nigerian and international highway codes were designed to provide the road users with relevant information on road networks, road use activities, and qualification for road use, vehicle registration, basic vehicle safety checks, road signs, signals and markings. Others include: road traffic offences, driving under special conditions etc.

The following codes were extracted from the Nigerian Highway Code, a publication of the Federal Road Safety Commission (FRSC), Nigeria published in November 2008.

Road signs and signals

An understanding of traffic signs, and road identification marks are compulsory tenet for all drivers and operators if you must handle a machine. The following signs, signals and markings are described by the Nigerian Highway Code.

Figure 10-24: Prohibitive regulatory signs and codes

Traffic signs let you know about traffic regulations, special hazards and other road conditions, construction areas, speed limits etc. *Regulatory signs* are mostly circular in shape and are of two types. Those with *red & yellow circles* are *prohibitive signs*. Those with *blue circles* but no red border are *mandatory signs*. They give positive instructions.

Figure 10-25: Mandatory regulatory signs and codes

The *stop sign* is a prohibitive sign. It is only 8-sided polygon with the word STOP written within the polygon. It means complete stop before entering, and proceed when safe to do so.

Figure 10-26: Stop signs

Informative signs are usually rectangular in shape and provide guidance information such as location of hospital, school area etc.

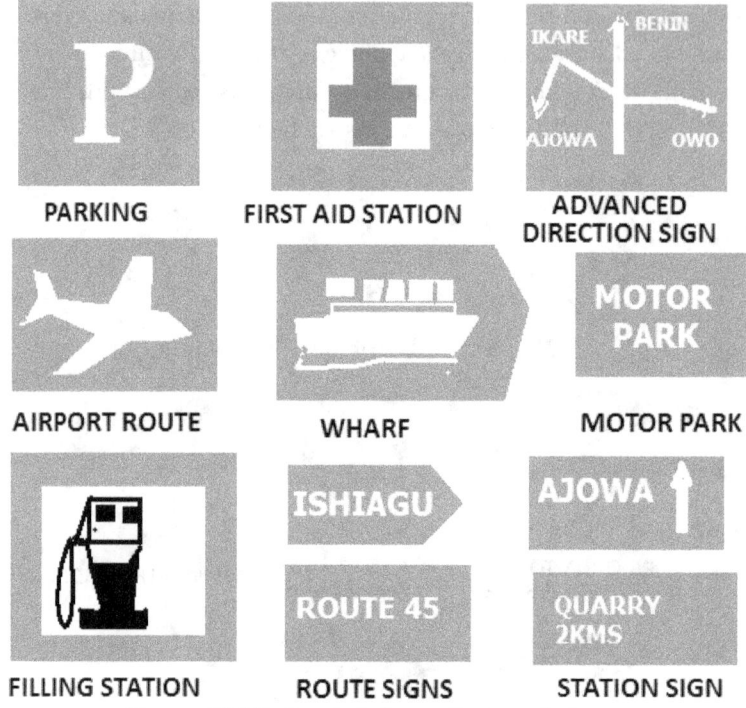

Figure 10-27: Some informative road signs

Warning signs are usually triangular in shape, with red perimeter. Warning signs with inverted triangle means GIVE way or YIELD.

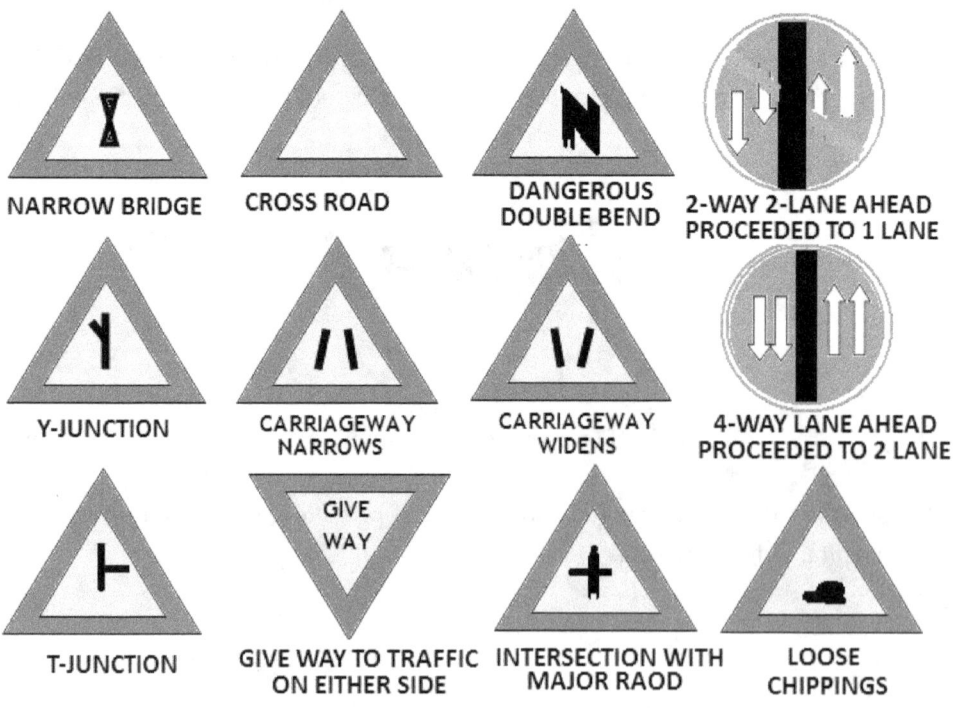

Figure 10-28: Some highway warning signs

FARM TRACTOR SYSTEMS

Slow moving vehicle (SMV) emblem

Traffic law requires that farm tractors with a maximum speed of 25 miles per hour should have a Slow-Moving-Vehicle (SMV) emblem on the rear of the tractor. When towing a trailer or other equipment that blocks the SMV emblem, another emblem must be attached to the rear of the towed equipment.

Figure 10-29: Slow moving vehicle emblem

Standards for shape, colour and placement of the SMV emblem established by the American Society of Agricultural Engineers, the American National Standards Institute, and the Society of Automotive Engineers have been adopted into law. The emblem must be an equilateral triangle at least 13.8 inches high (plus or minus 0.3 inches) and must be a fluorescent, red-orange material with a border of red retro-reflective material. The fluorescent material is visible in daylight and the reflective border shines when illuminated by headlights at night.

The slow moving vehicle (SMV) emblem must be mounted at the rear and as close to the center of the tractor or trailing equipment as possible and between 60 cm and 1.8 meters above the road surface. It must be mounted with the point up; the lower edge of the emblem must be at least 2 feet and not more than 6 feet above the ground. It must be clearly visible from a distance of 150 meters. SMV signs must be kept clean. Faded or damaged signs should be replaced.

Hand traffic control signals

The most effective kind of communication under variable conditions such as noise, weather, and human traffic situation is *Hand Signals*. Just as there are traffic control signals for vehicles operating on the highways, we equally have hand signals for field operations.

Hand traffic control signals could be used when traffic indicators are not fitted. Traffic control signals are made with hand and the following hand signals are recognized by the Nigerian Highway Code.

If I want to move into the left/turn left my hand is thrown out as shown below:

FARM TRACTOR SYSTEMS

Figure 10-30: Left turning hand signals

If I intend to move out to the right/turn right indication is as shown below:

Figure 10-31: Right turning hand signals

Figure 9-9 shows the sign of intention slow down or stop. This signal should be used when slowing down or stopping at zebra crossings.

Figure 10-32: Straight drive and slow down signals

Field operation control signals

Often, people must communicate with the operator while operating machines. But because of the noise generated by such machines particularly during tractor operation, they usually cannot be heard over the tractor. A set of universal signals (Figure 10-33-) have been encoded by safety organizations, which have been accepted as a standard for communicating instructions on the field. These signals were developed by the Indiana Farm Safety Council, refined by the National Institute for Farm Safety, and then adopted by the Society for Engineering in Agricultural, Food, and Biological Systems as Hand Signals for use in Agriculture; ASAE Standard ASAE S351. Such signals include:

1. Instruction to *start the engine*: Fold your hand into a bunch and rotate your arm in a circle in front of you.

Figure 10-34: Start the engine signal

Example: After some adjustments have been made to the engine, you need to signal the operator to start the engine by demonstrating this signal sign.

2. Instruction to *stop the engine*: Bend your fore arm to your breast level and move your arm sideways along the horizontal.

Figure 10-35: Stop the engine signal

3. Instruction to *increase speed*: Raise the hand to the shoulder, fist closed and pointing upward, thrust or stretch out your fist above your head full extent and down repeatedly.

Figure 10-36: Increase the engine speed signal

Example: Move the unit out now; the way is clear. We need to move on.

4. Instructions to *slow down the tractor*: Stretch out your right hand sideway and wave up and down along the vertical axis.

Figure 10-37: Slow the engine speed signal

Example: You are going too fast; slow down.

5. Instructions to *raise equipment/ implement*: Raise your arm above your head at right angle and pointing the forefinger up and then swinging the wrist in circular arc.

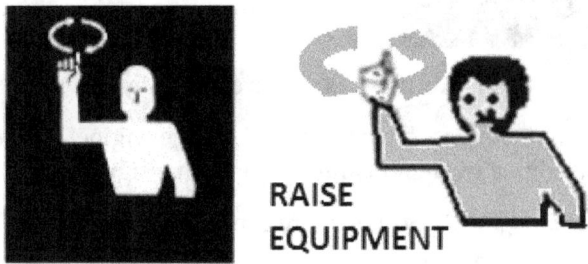

Figure 10-38: Lift up equipment signal

Example: Use this signal to have operator raise the attached implement after a check on the hydraulic system or brief stop.

6. Instructions to *lower equipment*: Point down your hand straight with the first finger pointing down and swing in circle.

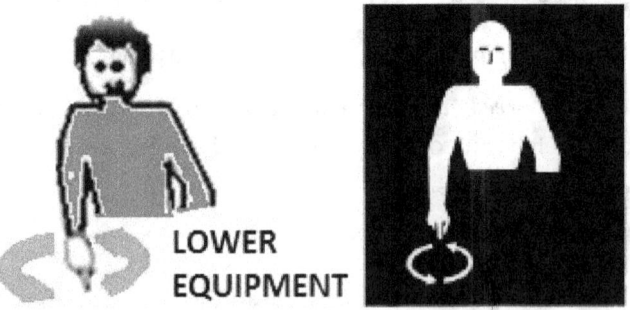

Figure 10-39: Lower the implement signal

Example: Use this signal to have operator lower implement for a quick check or refuel.

7. Instructions giving indication of *come to me*: Just raised up your hand above your head with the fingers stretched out and swing in a horizontal circle.

Figure 10-40: Signal for operator to come to instructor/mate

Example: While on the field and the tractor operator is too far for oral instruction: You will signal in this manner.

8. Instructions to *move toward me-follow me:* The instructor stretches his arm out in front of you, palm up, point toward the person, vehicle or unit and moves it toward you severally.

Figure 10-41: Signal for operator move toward instructor/mate

Example: Use this signal to motion a tractor operator to move toward you to position or move tractor out in a marshy area where side visibility is poor.

9. Instructing on *how far to go*: Raise the two arms (folded at right angle) about your head level and wave them outward. The fingers must be stretched out straight. When the palms are at ear level facing head and move inward, it shows the remaining distance to go.

Figure 10-42: Signal for extent to advance

Example: Use this signal to assist a tractor operator in backing a loaded wagon or hitching to a wagon.

10. Instructions on *stopping the operation*: Raise your right hand up straight with the fingers stretched out straight.

FARM TRACTOR SYSTEMS

Figure 10-43: Signal for machinery to stop operation

Example: The tractor and forage wagon are now positioned for unloading into the silage blower. You signal the operator to stop.

11. Instructions to *move out of field*: Face the desired direction of movement, hold the arm extended to the rear, raise your right hand sideway with the fingers folded and facing upward and drive it above your head through to the other side until it come to rest at the other side, your palm now facing down.

Figure 10-44: Signal for machinery to leave the field

Example: You have hitched the machine for the operator and connected the PTO; signal the person to move out for field work.

Types of roads

Roads may be classified as private drive pathways, two-lane highways, dual carriage ways, expressway and pathways.

1. *Private drive pathways*: These are roads owned, maintained or controlled by an individual, agency or organization.
2. *Two-lane highways*: These are the usual single carriageways which may be rural, urban, intra or intercity roads.
3. *Dual carriage ways*: these are roads having multiple lanes with traffic going in opposite direction
4. *Expressway*: this is a specially designed and restricted highway divided with barriers which made traffic in opposite directions completely separated from each other.

Categories of road users

Categories of road users include the motorists, pedestrian, cyclists, motorcyclists, children, animals, hawkers, traders etc.

Figure10-45: Some motorists on highway

Motorists: This class of road users must be aware of the rules for road use and be conscious of the drivers' protection against long distance driving, fatigue etc.; vehicle protection as well as passengers' protection. All these protections are contained in S.10 (4) (ee), FRSC act, 2007 and S.10 (4) (w), FRSC act, 2007 respectively.

The cyclist/motorcyclists: They ride on two wheeled vehicles. The riders protection is contained in S.10 (4) (ii), FRSC act, 2007

Figure 10-46: Motorcyclists on highway

The pedestrian: Generally, there are three classes of pedestrians using the road; the children, the elderly and the adults. Any of these could be involved in road accident so care must be taken avoid such occurrence. There are provisions made to prevent road accident with the pedestrians including traffic drill for children and elderly, pedestrian or zebra crossing- drivers have no right of way at such crossings once pedestrian have step on it so you must exercise caution, traffic control officers and traffic lights. Pedestrians are not to cross when the light show a red man while they have the right of way when it turn green.

The children: The drivers are urged to be careful near schools, churches, mosques, markets, and built-up areas. Children between ages 0-12 months should not ride in front of the vehicle

Animals: Animals in the vehicles should be kept under control, while animals walking on the road should be kept close to the edge of the road. Animals on herd should have a lead in front and another at the back alerting the on-coming vehicles and controlling the animals respectively.

Driver's/operator's license

Driver's license is a legal document that confers on a driver the right to drive or operate a machine. You will be qualified to own a license after satisfying all the requirement of a driving test.

Figure 10-47: Sample of driver's license

The following classes of driver's license are available to various category of driver.

- A   Motorcycle drivers
- B   Motor vehicle of less than 3 tonnes gross weight other than motorcycle, taxi, stage carriage articulated vehicles or omnibus
- C   Motor vehicles of less than 3 tonnes gross weight, other than motor cycle
- D   Motor vehicle other than motorcycle taxi, stage carriage or omnibus excluding an articulated vehicle or vehicle drawing a trailer
- E   Motor vehicle other than a motorcycle or articulated vehicle
- F   **Agricultural machines and tractors (for tractor/machinery operators)**
- G   Articulated vehicles
- H   Earth moving vehicle
- J   Special, for physically handicapped persons

Note: All tractor operators must hold an F-class license to operate a tractor.

Road mark identification

Lines and symbols on the road are meant to show the alignment of the road. They are reflective so that you may clearly and safely follow the road even at night. Road markings are basically of four major types: center lines, edge lines, cross walks and pavement messages.

Figure 10-48: Conditions for overtaking on highway

Center lines: These are lines in the center of the road to separate traffic in opposite directions. Broken lines are used in areas where there are no restrictions on overtaking. A solid line is painted alongside the broken line. You may not overtake if the solid line is on your side of the center line.

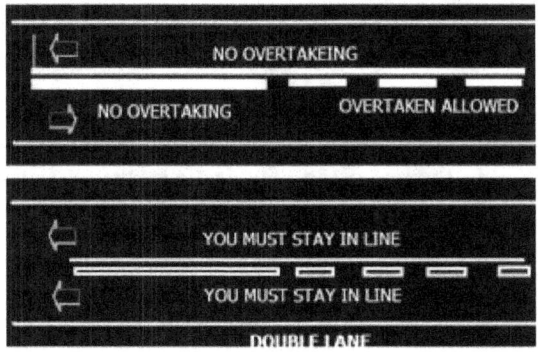

Figure 10-49: Conditions for no-overtaking on highway

Edge lines and lane lines: These are solid lines along the side of the road. They indicate where the edge is and can be used also as traffic guidance. An edge line that slants toward the center of the road forewarns that the road narrowed ahead.

Figure 10-50: Edge lines

FARM TRACTOR SYSTEMS

Cross walk: White solid lines across the road are usually used to denote pedestrian's crosswalks commonly at intersections.

Pavement messages: these are messages or symbols which are lettered or painted on the roads to warn of conditions ahead.

Zebra lines: These are used to indicate where pedestrians can cross the roads. You must stop for pedestrians that have stepped on the lines. In traffic queues leave pedestrian crossing clear.

Figure 10-51: Pedestrian/zebra crossing marks

Tractor safety features on highway

The maximum legal speed for driving a tractor on the road is 20 Km/hr. This is reduced to 12 mph when towing two trailers or driving wide self-propelled machines. If the tractor is equipped with a rollover protection structure (or ROPS), make sure to securely fasten the seat belt. This will keep you within the protected zone if the tractor overturns.

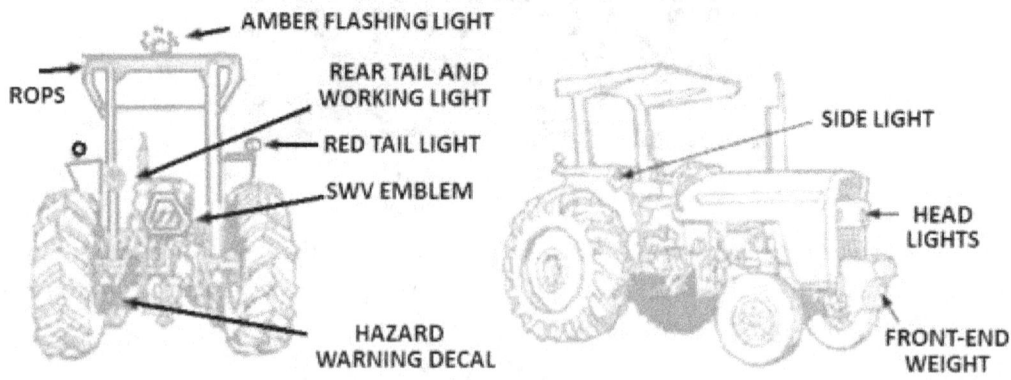

Figure 10-52: Highway safety features on tractor

Remember to turn on headlights and flashers to warn other drivers of your presence. When driving on the highway, *stay as far to the right as possible*, but avoid driving on uneven road surfaces at high speeds (i.e., driving with the left wheels on smooth pavement and right wheels on rough, loose shoulders). This could cause erratic steering, uneven braking, loss of control, and tractor overturns. If it is necessary to let cars pass you, slow down, pull to a secure shoulder, stop, and let them pass. Avoid driving on steep inclines and be careful when re-entering the highway.

Use turning signals and/or hand signals to warn other motorists of your intent to change lanes, slow down to stop, make a turn, or pull onto a highway. Make sure to give motorists advance warning by signaling at least 100 feet before you change speed or direction.

Tractor driving tests

Forward and reverse drive

When driving a tractor with or without attached implement, the operators may be tested on how well they can maneuver the machine by avoiding cone obstructions placed 20 feet apart as shown in Figure 10-53 without knocking the cones down. The tractor is expected to follow the forward arrow in the order indicated and equally follow the reverse arrow back.

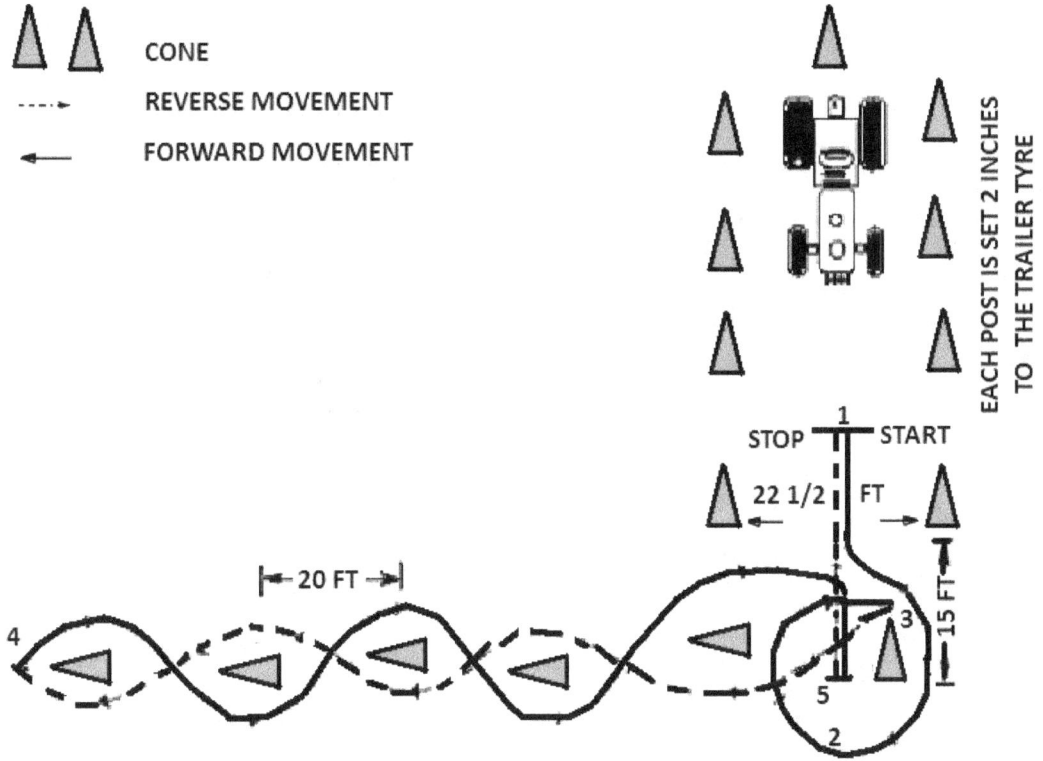

Figure 10-53: Layout for tractor driving test

Driving test instructions:

 Set out the driving course as indicated in Figure 9-42.

FARM TRACTOR SYSTEMS

- Observe all initial pre-operational procedures.
- With the tractor in parked in position 1, drive through course 2 to point 3.
- Engage the reverse gear and drive through the dotted arrow to point 4. At this point, engage the forward gear and drive through to point 5.
- Reverse to the stop position and park the tractor at the stop position.

Driving with trailer

1. Two trainers should be available – one for timing and recording marks and one for measuring and observing operation.
2. 1 stop watch and meter tape or 1 m ruler.

A two-wheeled trailer with 8 x 12 foot frame is specified. However, if this size trailer is not available, clearances will be adjusted for size available.

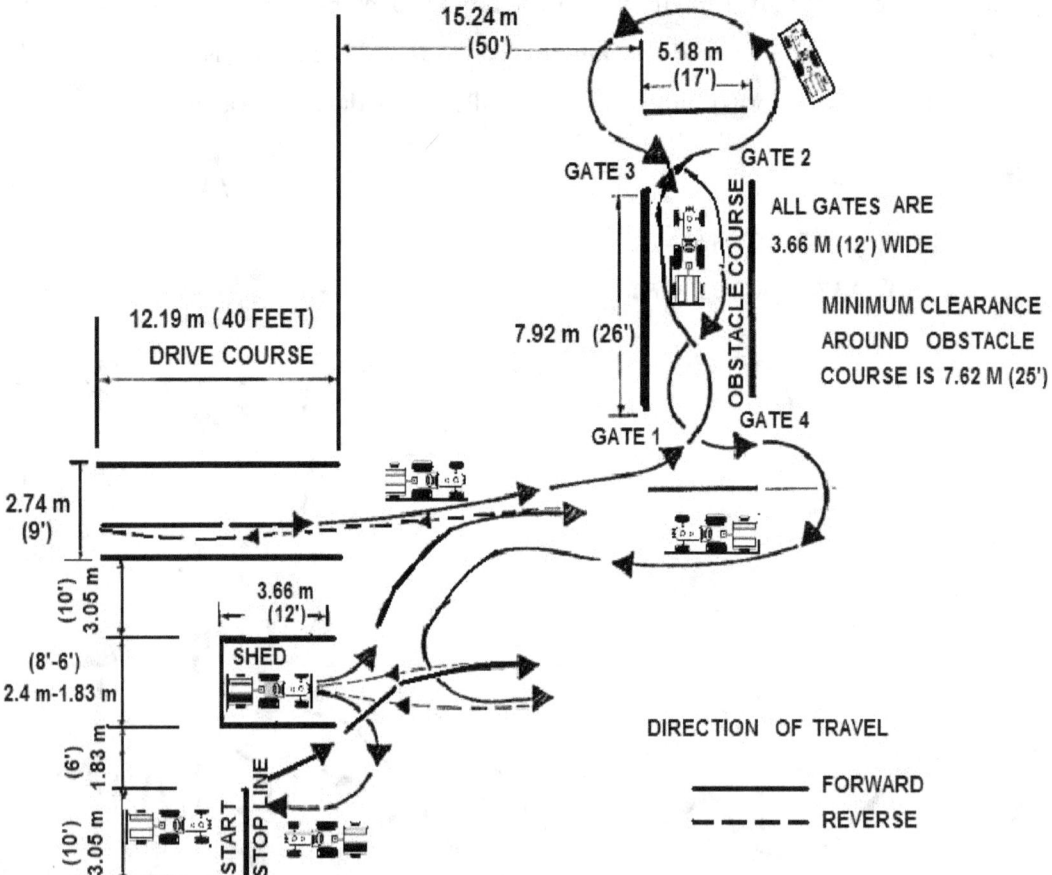

Figure 10-54: Layout for 2-wheel trailer driving test

Two different course layout patterns is shown, trainer/examiner could make a choice or better subject the trainee to both driving courses.

FARM TRACTOR SYSTEMS

Driving test instructions

- The tractor is positioned on the starting line (Figure 10-54). The trainee starts the tractor, leaves engine running with brakes set or transmission in park.
- Trainee signals official that he is ready to begin. Official then signals to start.
- Line up with two-wheel trailer and hitch safely. Pull trailer from shed, and line up with 40-foot tractor way. Back down 40-foot tractor way until signaled to stop. Pull back through the tractor way and proceed to Gate 1.
- Draw the implement through gate 1 and exit through gate 2, turn around the end and enter back through gate 3 and exit through gate 4.
- Line up implement and back into shed; unhitch.
- Return tractor to stopping line and stop engine (time stops when engine fan stops).

Driving test instructions

In another pattern (Figure 10-55and 56), the driving course layout could take the following pattern:

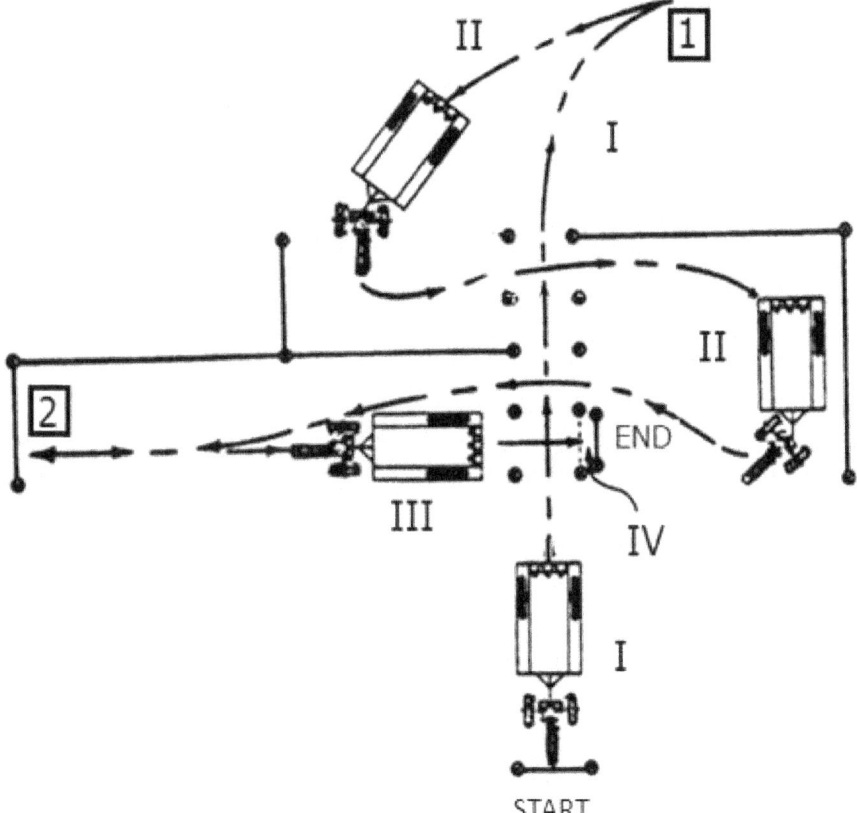

Figure 10-55: Pattern 1: 2-wheel driving test

- From the start point, reverse through the course to point 1
- Drive through the curved path to position 2

FARM TRACTOR SYSTEMS

- Reverse toward the end, and
- Stop at about 14 inches from the end.

Note: Obstacle course openings are adjusted to ally between 10 and 11 inches.

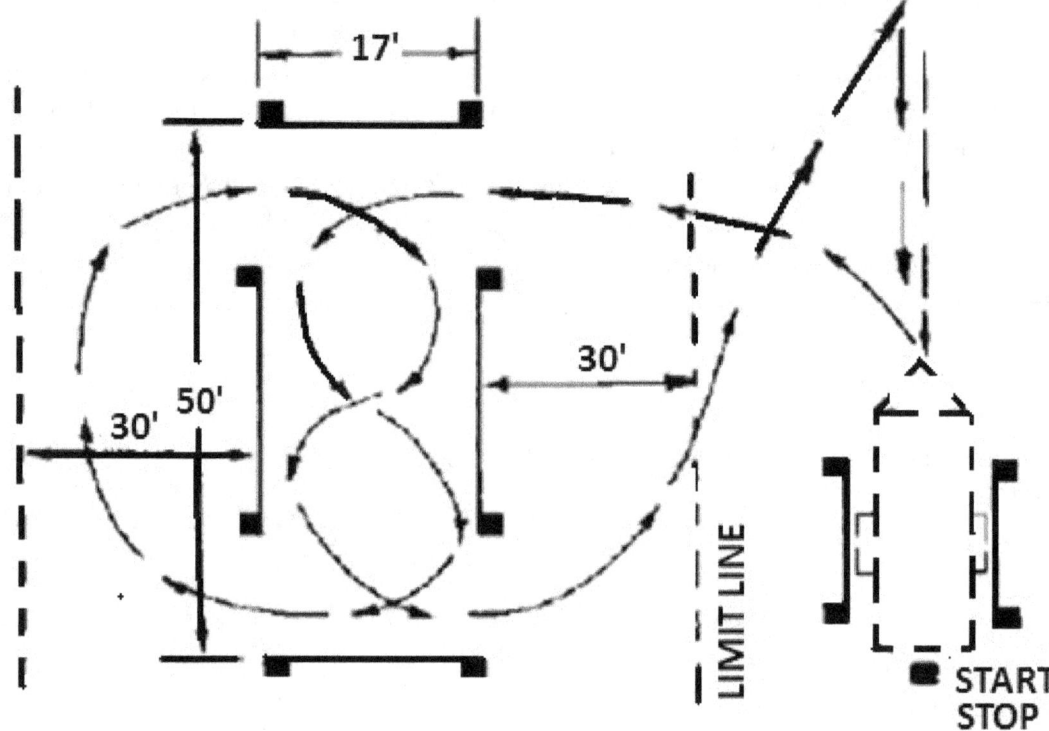

Figure 10-56: Pattern 2: 2-wheel driving test

Safety rules

1. When dismounting from tractor, both brakes must be set with transmission in neutral or in park.
2. Time begins on signal from trainer and stops when trainee returns trailer to start/stop line and engine fan stops.

Tractor field operations

Tractor is a farm vehicle used to produce power for the purposes of pulling, hauling and driving of farm machinery from one place to another within and outside the farm. The tractor drawbar is used to pull single hitch implements and such as trailer or water tanker. The three point hitch linkage is used for attaching mounted implements such as plough, harrow etc. the tractor also has provision for front end loader and attachments for earth scooping, land grading and manure loading. Power is also produced and made available at the PTO shaft for the purpose of turning stationary equipment.

FARM TRACTOR SYSTEMS

Figure 10-57: Tractor points of attachments

Guides to tractor operation

Tractors have been identified as one of the main causes of accidental deaths on farms. Over the years, many farmers, farm workers and others living on or visiting farms, have either been killed or seriously injured falling from moving tractors, being run over by tractors, or being crushed when a tractor rolls sideways or backwards.

Figure 10-58: Tractor rollover accident

In order to minimize or entirely avoid such incidence; the following guides must be followed while preparing the tractor for operation or during operation.

Spot the hazard

Regularly check for hazards relating to tractors, attached implements and field conditions. Hazard areas could include mechanical parts, operator training, other people, work procedures, unsafe jacking, climatic conditions, chemicals used, uneven terrain, and any other potential causes of an injury or a hazardous incident. Keep a record to ensure identified hazards are assessed and controlled.

Assess the risk

Once a potential hazard has been identified, assess the likelihood of an injury or hazardous incident occurring. For example, risk to children playing near a tractor will vary, depending on what the tractor operator is doing, how close they are to the tractor and whether the operator knows they are there. Consider ways of minimizing risk.

Figure 10-59: Checking for hazards

Make the changes

Appropriately make changes that could guarantee safety in tractor operations. Take steps in removing those risks identified that could cause potential danger.

Attaching implements to tractor

Implements are attached to tractor through the following devices;

1. Single point hitch to drawbar
2. 3-point hitch to drawbar
3. The 3-point linkage

FARM TRACTOR SYSTEMS

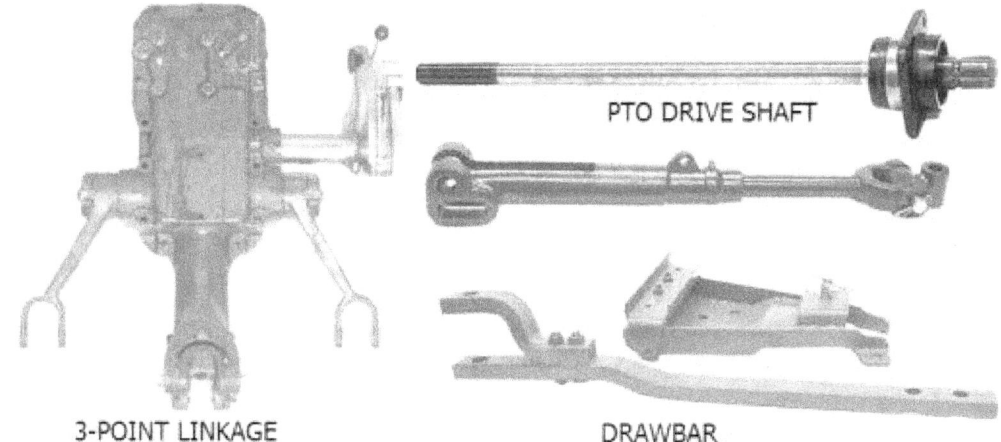

Figure 10-60: Points of implement attachment to tractor

Attaching rear mounted equipment to 3-point hitch

The 3-point hitch assembly consists of a rockshaft, two lift arms, two lift links, two draft links, and an upper link. The draft links and the upper links are the three points connected to the implement. Other essential components of the 3-point hitch include stabilizer

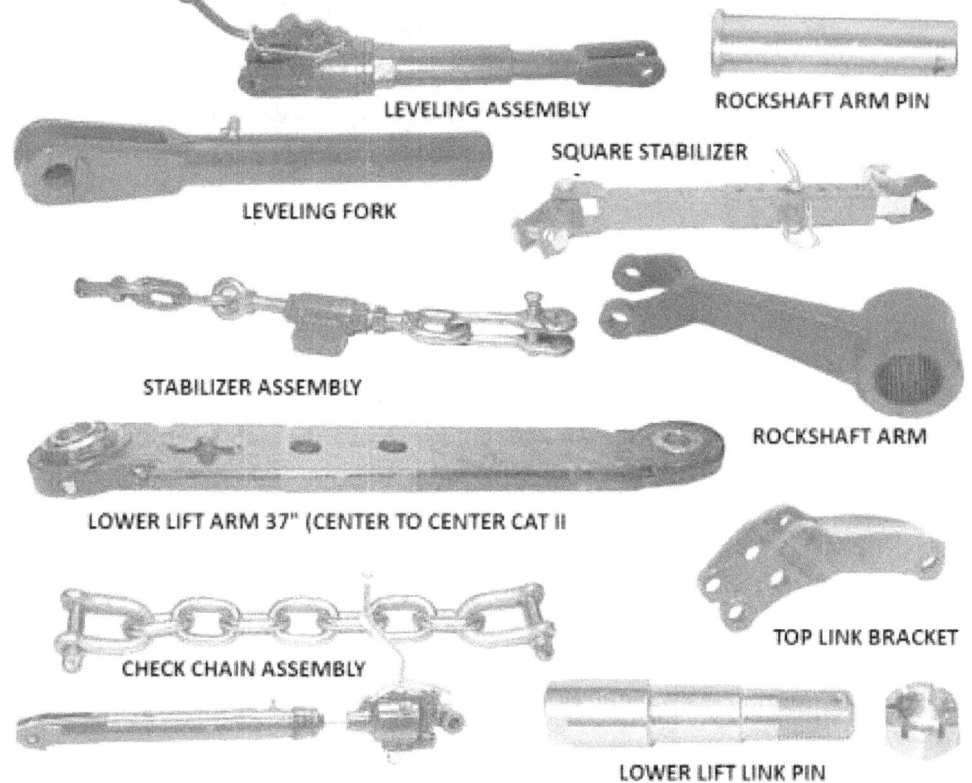

Figure 10-61: 3-hitch components

FARM TRACTOR SYSTEMS

Before attaching 3-point implements, the following requirements are necessary:

- Check the drawbar to be sure it will not interfere with the draft links or the implement frame when the equipment is raised or lowered.
- It may be necessary to install a tubular guard over the PTO shaft or remove the main PTO shield for adequate clearance.

To attach the implement to the 3-point hitch, follow these steps:

Step 1: Back the tractor so the draft links are in position for connection to the hitch pins.

Step 2: Remove the drawbar, or move the drawbar forward or to the side for clearance.

Step 3: Back the tractor so the pinholes of the tractor's draft arms are nearly aligned with the implement's lower hitch pins (Figure 10-62).

Figure 10-62: Backing tractor to attach implement.

Step 4: Raise or lower the lift arms with the hydraulic control to the height needed to connect the implement, ensure the engine is running to activate the hydraulic system.

If you have no MATE, stop the engine, securely park the tractor, and set the brakes.

Step 5: With flexible lift arm tractors, connect THE LEFT DRAFT LINK FIRST and secure with the hitching pins and security clips (Figure 10-63). The left link is attached first because the right lift link is usually equipped with a crank adjuster (Figure 10-63 circled) to position the right link.

Step 6: NEXT ATTACH THE RIGHT ARM in the same manner. If the right draft link is too low or high for connection, it can be raised or lowered by the crank adjustment.

Figure 10-63: Right lift link with a crank adjustment

Step 7: After connecting the draft links, CONNECT THE UPPER LINK (Figure 10-64(b)).

Step 8: Adjust the upper link to reach the mast by turning the outer housing.

On both the left and right sides of the implement, insert the draft arm attachment pin of the tractor into the pin holes of the implement's lower hitch assembly. Secure the hitch with the proper size hitch pins and security clip. Remount and restart the tractor, and slowly raise the tractor's draft arms with the hydraulic lift controls to closely align the upper hitch points.

Figure 10-64: Hitching the lower and top links

After adjusting the upper link of the tractor's 3-point hitch to align with the upper hitch point of the implement, secure the equipment with the proper size hitch pin and security clip. The circled area in Figure 10-64(b) indicates where the upper link can be adjusted for fit. The upper link may need to be lengthened or shortened to fit. The implement must be in a level position after the connection is made.

If the upper link is equipped with a latch assemble, depress the lever, fit the latch over the ball and release the lever. This locks the upper link into place. If it has a hole, insert a pin through the holes (Figure 10-64(b)) in the mast and turnbuckle-fasten the pin so it cannot bounce off the connecting point.

Hitching single point hitch to drawbars

Performance skill to learn: coupling trailer to tractor through the drawbar.

Work to be performed: The operator will demonstrate safe coupling of typical tractor to a trailer unit to student.

Performance criteria

   i. Operator should couple tractor-trailer units within 8 to 10 minutes.
  ii. Complete coupling in accord with safety requirements and approved practices.
 iii. Complete coupling with secure connections, including air lines and electrical cables.

Coupling procedure and assessment

1. Align tractor and trailer units and back tractor to position to touché the apron of trailer.
2. Secure trailer against movement back tractor slowly and straight into trailer kingpin at right level and with appropriate force, check coupling and pin engagement.

Figure 10-65: Drawbar coupling pin

3. Check connection for security by pulling tractor forward gently. If it is okay, release brake; if not, secure connection.
4. Check for improper connections and make necessary adjustments.

FARM TRACTOR SYSTEMS

Attaching 3-point hitch to drawbars

Precautions ⚠ Hitching implement wrongly could lead to tractor back flip. Hence never pull a load with the 3-point hitch drawbar more than 13-17 inches above the ground or the pulling forces will be higher than the tractor's center of gravity.

Stay braces (rigid or adjustable) could prevent the 3-point hitch drawbar from being lifted too high (Figure 10-66). If the hitch point is too high, rear overturn hazard may develop as the tractor moves forward.

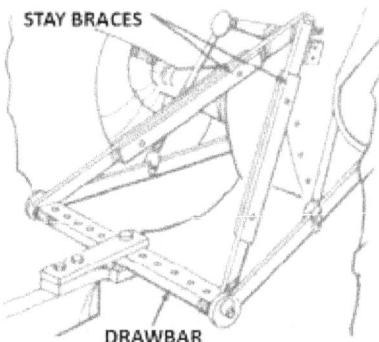

Figure 10-66: 3-Point hitch attachment to drawbar

Mounting slasher to tractor

Before mounting the slasher to the tractor, ensure that the stabilizer bars and adjustable top-links are fitted to the tractor. It is essential that the slasher should be able to ride easily over obstacles. There must therefore, be no downward pressure from the tractor's hydraulic system, and all "down pressure" pins or other means of applying downward pressure, should be removed. On many tractors the lift rods have an adjustable collar for exerting such pressure. This collar must be moved to its lowest point so that both lift rods may "float" and allow the slasher to lift-up if it hits an obstacle.

Mounting must take place with the tractor and slasher on level ground. All directions given are from the rear of the slasher. Ensure stabilizer bars or chains are fitted to tractor lift arms. Slacken off stabilizer chains or remove sway bars to allow the fitting of lift arms to the clevis hitches.

Back the tractor up to the slasher until it is positioned relative to the tractor PTO shaft and linkage points. Connect the lift arms in the following order:

STEP 1: Left hand lift arm to left hand clevis hitch.

STEP 2: Right hand lift arm to right hand clevis hitch.

STEP 3: Adjust both lift arms until they are firm against the clevis hitch shoulders. Then secure each arm with pins.

STEP 4: Adjust the tractor top link to approximate length and attach to the top hitch on the headstock section.

STEP 5: Secure with pin.

STEP 6: Connect the drive shaft to the tractor's power take off, ensuring that the release pins on the clutch are located in the PTO groove and the safety cover does not interfere with the tractor PTO guard.

It is important to test that the two sections of the driveshaft do not, at any position in the lifting arc, close completely up, or extend such that there is less than 15m of inner tube engaged.

STEP 7: Raise and lower the slasher to ensure that the drive shaft does not jam or become disengaged. If needed, the inner & outer tube, as well as the plastic cover of the drive shaft, may need shortening. If so, use a hacksaw to cut both the inner and outer tubes, to equal length. Do not overcut. Cut off only a little at a time. Similarly, cut plastic cover to appropriate length. Refit and re-test lifting arc.

STEP 8: Level the slasher by adjusting the right hand tractor lift rod so that the blade beam clearance from ground is equal across the width of the slasher when the weight of the slasher is supported on the three point linkage.

STEP 9: Lower the slasher onto the ground and adjust tractor top link until the front rollers just clear the ground.

Attaching remote hydraulic cylinder

Remote hydraulic implements such as tipping trailer, front-end loaders, post-hole-diggers, and ditchers extend the usefulness of the agricultural tractor. These implements may be attached directly to the tractor or some distance away, connected by a hydraulic line. Usually, the tractors' own hydraulic PTO powers the remote units. Some implements however, are equipped with their own hydraulic pump, which is driven from the tractor's PTO. These systems are generally powerful.

Caution ⚠

Before attempting to attach a remote cylinder, the following steps are required:

1. Be familiar with the mode of operation of the hydraulic cylinder. Determine whether it is single-acting or double acting by examining the connections on the hydraulic pump.
2. Next determine if the tractor supplies 1-way, 2-way or both types of pressure.

Figure 10-67: Trailer remote hydraulic cylinder

3. Before connecting hoses, remove the dust plugs from the ports and make sure the ports are free of dirt and foreign materials.
4. Clean the couplers.

Relieve the pressure from the tractor lines; high pressure in the lines can make connecting the hoses difficult.

Locking devices for PTO

Four types of locking devices are used to hold the implement power shaft onto the PTO stub shaft as applicable.

1. *Spring loaded lock pin*: Slide the power shaft onto the stub shaft until resistance is felt. Then depress the pin and move the shaft about one-half inch and then release the pin.

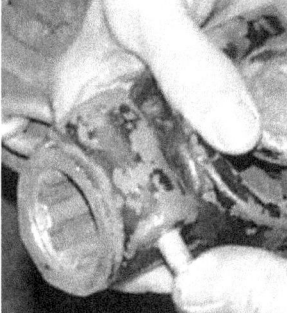

Figure 10-68: Spring loaded lock pin

Push the shaft forward and the locking mechanism will come in contact with a slight grove in the stub shaft. Check to make sure that the locking mechanism has engaged before operating the machine.

2. *Bolt and nut*: This allows the split front end of the drive shaft to be placed over the stub shaft. The bolt is inserted through the side of the power shaft at one end and through the machined groove in the stub shaft at the other end. When lightened, the bolt and nut exert a clamping force to the shaft and to the stub shaft.
3. *Locking pin:* This is inserted through the center cross-drilled hole in the power shaft and stub shaft. When connecting these shafts, align the holes with each other as well as the grooves' and splines'. Then insert the largest diameter pin the hole will accept.
4. *Self-locking coupler:* This is similar in operation to spring-loaded lock. Pull the locking collar back to fit the power shaft onto the stub shaft. Release the collar to lock onto the stub shaft.

Caution ⚠ Always shut off the tractor engine and disengage the PTO before attempting to connect, disconnect or service PTO-driven equipment.

Before connecting the implement power shaft,

1. Check the coupling to make sure it has the same number of splines as the stub shaft it is being connected to.
2. Next, clean the grease and debris off the insides of the power shaft coupling. If the tractor stub shaft or the power shaft splines are rusty, oil them to aid connection.
3. Align the grooves in the stub shaft with the splines of the power shaft, and slide the power shaft of coupling over the stub shaft.

Attaching PTO shaft to 3-point hitch implement

To attach the PTO shaft to a 3-point hitch implement, follow these steps.

1. Connect the 3-point hitch of the implement using the approved steps to align the hitch and to park the tractor securely.
2. Attach the implement driveline shaft to the PTO stub shaft of the tractor.

Here are some suggestions to make connecting the PTO easier.

(a) Align the implement PTO shaft splines with the splines of the stub shaft of the tractor.
(b) Press the detent lock (skewed arrow in Figure 9-62) inward as you slide the implement shaft onto the tractor PTO stub shaft.

Figure 10-69: Detent lock of the PTO shaft

This lock engages the groove in the stub shaft to secure the PTO driveline shaft to the stub shaft. Slide the implement shaft forward far enough to make sure the detent lock has snapped into the lock position. To attach the PTO shaft, be sure the tractor engine is shut off and is securely parked.

Tractor gear selection for field operations

In selecting gear for field operation, the forward speed is combined with either HIGH or LOW gear for all field operations such as slashing, ploughing, harrowing etc depending on the condition of the tractor as follows.

Slashing (Both lawn and jungle): Use gear 3 in combination with LOW gear

Ploughing: Use gear 1 in combination with HIGH gear for all tractors

Harrowing: Use gear 1 in combination with HIGH gear for most MF/Ford tractors or 4 in combination with LOW gear.

: Use gear 2 in combination with HIGH gear for most Fiat tractors

Ridging: Use gear 1 in combination with HIGH gear for expanded track tractors

: Use gear 4 in combination with LOW gear for non-expanded track tractors

Carting/towing: Use gear 1-2-3-4 in combination with High gear for all tractors.

Steering the tractor: Keep both hands on the steering wheel unless you are operating another control.

FARM TRACTOR SYSTEMS

Determination of shaft speed and speed ratio

In determining the output shaft speed and speed ratio of first gear, the following sample calculation is used. Given the power transmission line shown in Figure10-63, it is required to determine the output shaft speed and speed ratio of first gear if the input shaft has a speed of 1000 rpm.

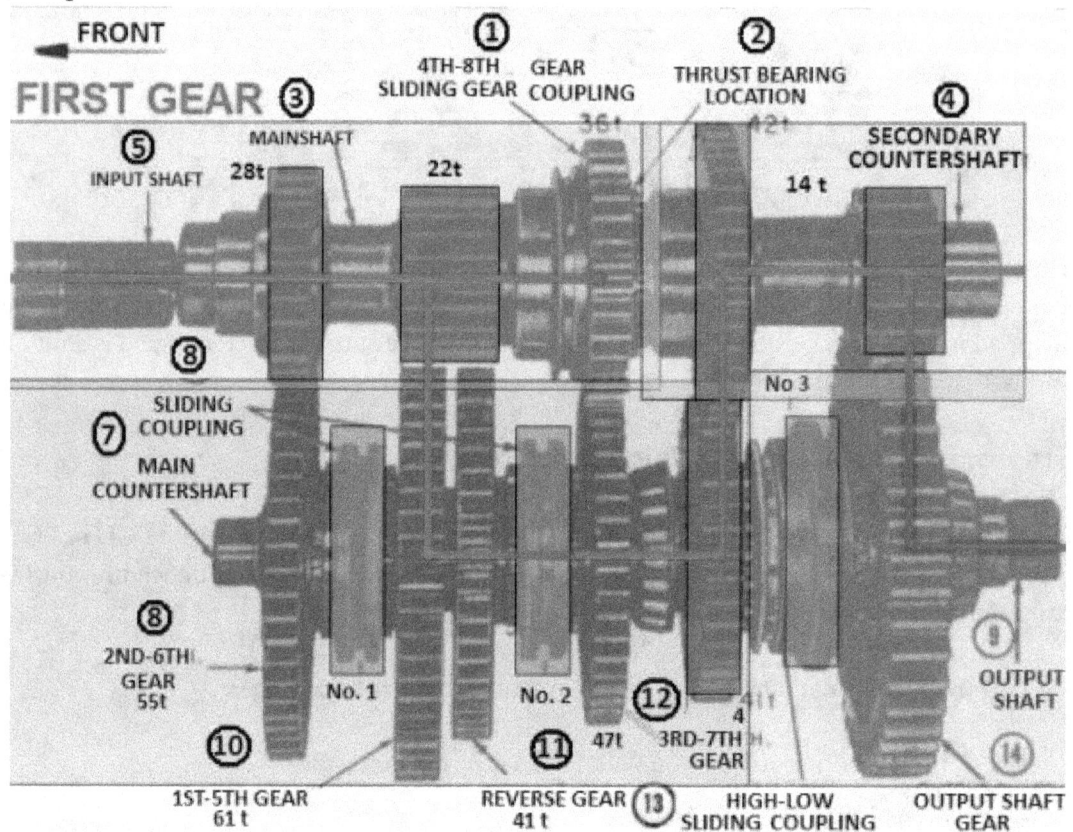

Figure 10-70: Line transmission on braking tractor

Solution: The main countershaft speed will be estimated by the formula:

$$\Rightarrow Main\ countershaft\ speed = \left[\frac{No.\,of\ teeth\ in\ maincountershaft}{No.\,of\ teeth\ in\ meshing\ gear}\right] x\ First\ gear\ speed$$

From the figure,

First gear speed = 1000rpm

No. of teeth in main countershaft = 22 teeth

No. of teeth in meshing gear = 61 teeth

$$\Rightarrow Main\ countershaft\ speed = \left[\frac{22}{61}\right] \times 1000 = 360\ rpm$$

$$Speed\ ratio = \left[\frac{Input\ speed\ or\ (output\ teeth)}{Output\ speed\ or\ (input\ teeth)}\right] = \frac{1000\ or\ (61)}{360\ or\ (22)} = 2.78$$

Output speed of the secondary countershaft is determined from: $360\ rpm \times \frac{41}{42} = 351\ rpm$

Output shaft speed is determined thus: $351\ rpm \times \frac{14}{45} = 109\ rpm$

Output shaft speed ratio is given by

$$Speed\ ratio = \left[\frac{Input\ speed}{Output\ shaft\ spped}\right] = \frac{1000}{109} = 9.2$$

In similar way, the product of output teeth/input teeth $= \left(\frac{61}{22}\right)\left(\frac{42}{41}\right) = 9.2$

The tractor brake system

The tractor has two sets of brakes:

1) The independent braking systems called the differentials and
2) The master brake.

Figure 10-71: The differential brake pedals

The differential (in case of tractors and heavy duty vehicles) allows each wheel to travel at a different speed and still propel its own load. When the machine is moving straight ahead, both wheels are free to rotate.

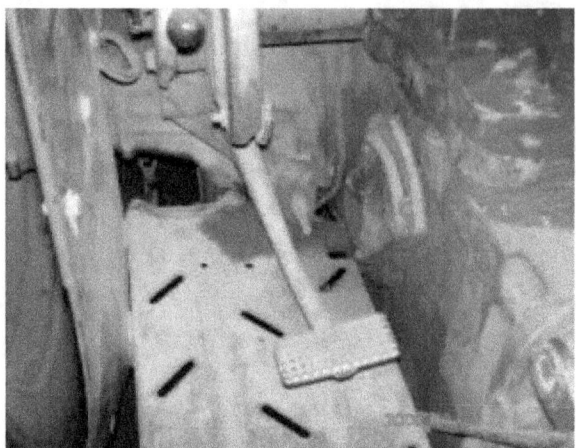

Figure 10-72: The differential lock pedals

When the machine turns a sharp corner, only one wheel is free to rotate. The master brake allows the two differential wheels to be locked together for highway movement. Differential action can be a disadvantage when a farm tractor is ploughing, for example one wheel may lose resistance on sticky ground and start spinning. To prevent a loss of power, a differential lock is often used to lock the two axles together until the slippery spot is passed, then they are released

The differential lock pedal

The differential lock pedal located at the lower side of the operator seat could be activated to increase the torque in one rear tyre to get the tractor out of bug if one of the rear tyres is sinking Figure 10-73.

Draft control on the field

In selecting and performing heavy duty cultivation works such as ploughing or subsoiling, the draft control must be adjusted as follows:

Lever position settings: Position Control Lever (Figure 10-73) should be fully up in TRANSPORT position 3, while the Response Control Lever should be in the SLOW position 2. Operate the lift hydraulic system using the Draft Control lever 1.

FARM TRACTOR SYSTEMS

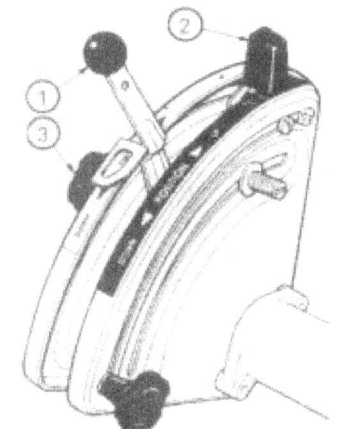

Figure10-73: Draft control setting

Entering the field: Move the draft control lever to the down position to lower the implement. The implement will start to penetrate as the tractor moves forward. The further the lever is moved to the down position, the deeper the penetration.

While at work on the field: When the desired working depth has been reached, move the adjustable stop back to lever and lock into position with the hand knob. The draft control lever can be slightly moved either side of the stop to adjust for varying soil conditions. Set the response control toward SLOW but in a position to allow fast entry into work. If the implement moves up and down erratically, turn the response control toward SLOW.

When leaving the field: On reaching the headland, you raise the implement by moving the Draft Control Lever back to the UP position. Re-enter the field by moving the Draft Control Lever forward to the Adjustable Stop. On some light draft implements (cultivators), it may be necessary to move the draft lever beyond the working depth stop setting to engage implement into the soil. Once the draft forces are created, the lever may then be positioned to the previous stop knob position setting.

Position control-inner lever-red sector

Selecting a type of work on the field operation which requires the implement to be kept at a constant height above the ground e.g. mowers, spinner broadcasters, leveling blades and other implements producing little or no draft.

Lever position: control must be adjusted as follows: Let the Draft Control (Figure 10-74) fully be in the UP position and the Response control in the SLOW position. Operate the hydraulics by using the Position Control Lever.

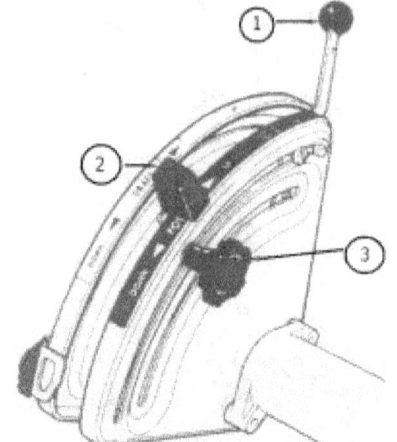

Figure 10-74: Position control setting

Starting work on the field: Move the Position Control Lever DOWN until the required implement height above the ground is obtained. Set the adjustable stop in line with the position control lever. Set the response control toward *slow* but in a position to allow fast entry into work.

While at work: No further adjustment is necessary.

When leaving the field: Move the position control lever up to the *transport* position, if required. Re-enter the field by moving the position control lever forward to the adjustable stop.

Transport position

The Transport position is used for transporting implements on three-point-hitch in the fully raised position.

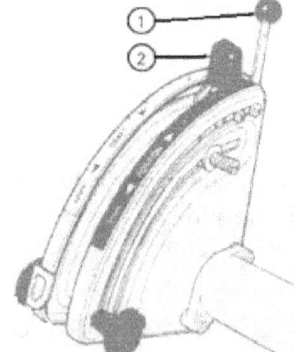

Figure 10-75: Draft control position in transport

Lever position: Draft control lever should be in the fully UP position, Position control lever adjusted to the TRANSPORT position against the stop. This will lift the linkage and the implement and hold it in the raised position.

Constant pumping position

The constant pumping position provides additional height for the lift arms and unlatches an automatic trailer hitch. NEVER position lever into *constant pumping position* or hydraulic oil heating will occur due to continual relief valve opening.

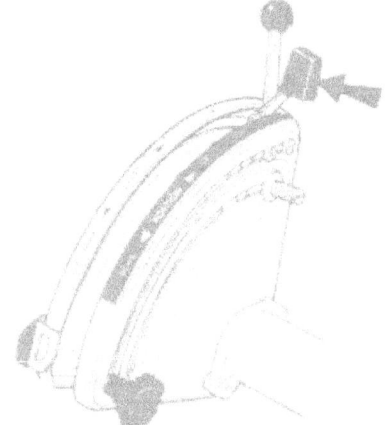

Figure 10-76: Draft position in constant pumping

Field operation patterns and methods

Ploughing operation

Ploughing operation should be performed working from the left hand side to the right hand side. This is so because of the direction of the concave of the blades. On a typical plough (for instance 3-bottom plough), the discs were arranged with the curvature facing left hand side of the tractor. When the implement is mounted and taken to the field, the tractor create furrow to the left while soil is thrown to the right. Thus in order to effectively plough the field, the tractor must always work through the left hand side throughout the field.

Figure 10-77: Creating bond on plough field

The best ploughing practice is to start work from the center of the field and end at the boundaries in order to avoid soil compaction as it is always the case when you work from the edges and end at the center. When you open the first furrow at the center of the field, ploughed soil is placed on unploughed land strip which must be ploughed. To achieve this, the tractor should work back through the unploughed portion of the field. Turn at thee headland and put the front right tyre on the furrow created and work back through the unploughed soil thus creating what is called a *bond*. The operation can then proceed from left to right irrespective of the pattern or method used.

Plough patterns

Four types of ploughing patterns have been identified as common methods used on the field. These patterns and their advantages are discussed below.

1. *Casting or headland pattern*: In this pattern of ploughing, spaces were provided at the boundaries (called headland) of the field where tractor can make a turn and work back from the initial furrow.

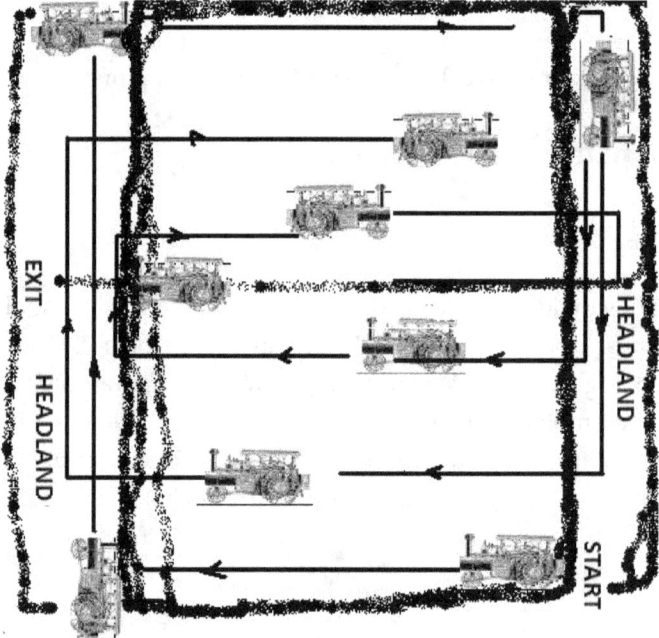

Figure 10-78: Casting or headland pattern

You start working from the edge of the field to the end of the lane, turn through the headland to the right edge of the field thus creating furrows at the two edges of the field. When you get to the end of the right lane, travel to the left edge and continue until you get to the center of the field where two furrows will be created. You can now close the furrows, the headlands and exit from the last headland.

2. *Gathering or alternation method:* In this method of ploughing, the field is divided into two and the tractor work from the center, creating a bond at the center and continue to work from left to until the field is finished at the edges of the field. In this method, headland is

required. It has the advantage of shorter time than the cast method. Time for turning and travelling at the headlands is avoided. Compaction of ploughed land is non-existence.

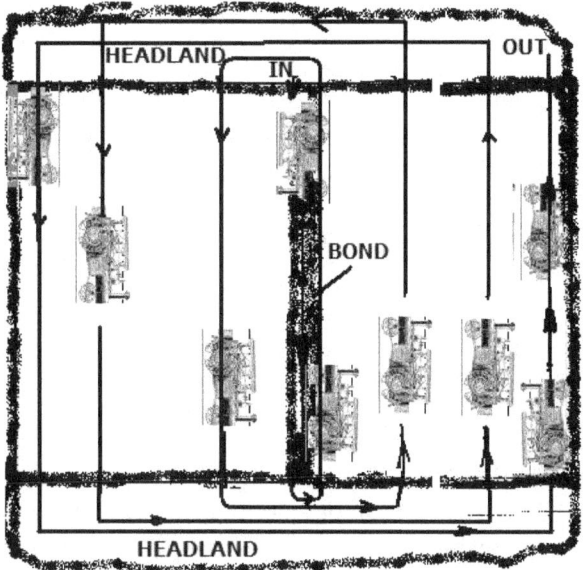

Figure 10-79: Gathering or alternation pattern

3. *Square or circuitous method*: In this method, the tractor enters the field at one end, working through left and *continuously round* the boundary in circuits until you end at the center where you close up and take exit from one point.

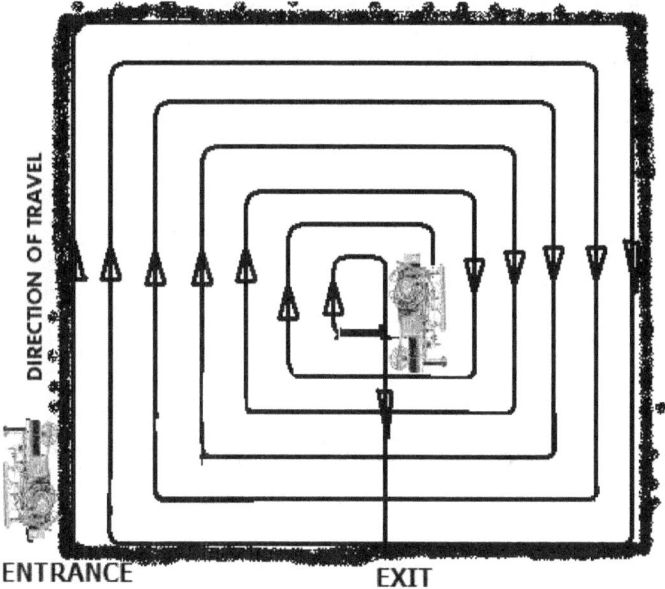

Figure 10-80: Circuitous method

4. *One way method*: In this method, the tractor travels up the longest length of the field; return to the same track and continuously till you get to the end of the field. This method is time consuming and fuel wastage.

FARM TRACTOR SYSTEMS

Methods of opening field

There are three types of field opening methods for ploughing operation as explained below:

a. *Arable opening method*: This is when the land is properly ploughed with nowhere left unturned.
b. *Grassland opening method*: In this method, the land is ploughed on top of the unplouhged area.
c. *Plot by plot opening method*: This method help you plough land plot by plot or combine two plots by gathering method.

Field harrow patterns

A typical disc harrow has two sets of discs arranged in gangs (with sets of discs on each gang) opposite one another. The first set close to the tractor is scalloped and each disc curvature faces right and the solid disc faces left. In operation, harrowing is done from right to left side of the field. When the harrow get to the end of the row, it turned at the headland and work back with the right front tyre running along the last furrow created by the solid disc.

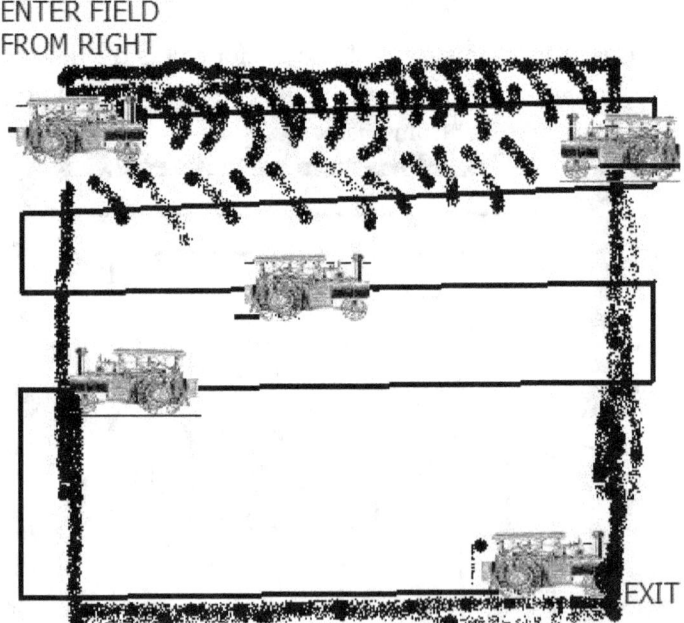

Figure 10-81: Harrowing pattern

Ridging operation

Ridging is a secondary operation done after the initial primary operations, pulverization and harrowing. Fields are often ridged for row crop planting. In ridging with double gang ridger, two beds/rows were formed in each pass.

Figure 10-82: Expanded tractor for ridging

The possibility of not having fully formed ridges at the first pass will necessitate a second pass. In such cases, the tractor track width should be wide enough to prevent the front wheel from climbing already formed ridge. To avoid such instances, the tractor is often expanded for such ridging operations as seen in Figure 10-82.

Time losses in field operation

The following activities have been found to account for the majority of time lost in the field due to size (area) of field, t_{lp}:

a. Time loss in Turning at the headlands and idle travel;
b. Materials handling: seed; fertilizer; chemicals; water; harvested material;
c. Machine adjustment;
d. Lubrication and refueling (besides daily service);
e. Waiting for other machines.
a. Time loss in resting and feeding
b. Time loss in checking faults and correction
c. Time loss in marking the field

Time loss independent of area or size of field

a. Time loss due to coupling of the implement
b. Cleaning clogged equipment;
c. Time loss in opening and closing of the gates
d. Time loss in maintaining the implements
e. Time loss in fueling and lubricating

f. Time loss in moving to the site

Estimating tractor operating costs and fuel consumption

Agricultural producers and machinery operators are currently facing the highest nominal price of diesel they have ever seen in history. However, these higher costs will likely reflect a direct reduction in net income in the short run because producers are limited as to the changes they can economically make. Fuel, oil and lubrication costs vary with the annual use of the machine and its maintenance schedule. Lubrication costs add approximately 15% to fuel costs. The best source of information for fuel use is past records. If these records are unavailable, calculate annual fuel consumption using the following method:

Average petrol consumption (liters/hour)

$$= (0.229) \times maximum\ PTO\ horsepower\ per\ hour \ldots\ldots\ldots\ldots.10.1$$

Diesel units will use approximately 73% less fuel than gasoline engines.

Average diesel fuel consumption (liters/hour)

$$= (0.229) \times maximum\ PTO horsepower/hr \times (0.73) \ldots\ldots\ldots\ldots\ldots\ldots.10.2$$

Or

$$= (0.167) \times maximum\ PTO\ horsepower \ldots\ldots\ldots\ldots\ldots\ldots\ldots 10.3$$

The maximum PTO horsepower per hour can be obtained from the *Nebraska tractor test data* published by the Nebraska Tractor Test Laboratory, University of Nebraska. If the maximum PTO horsepower for a particular tractor is not known, the advertised PTO horsepower per hour or the Nebraska Tractor Test Data for a tractor with similar displacement can be used.

Fuel and lubrication costs

$$= liters\ of\ fuel\frac{used}{hr} \times hours\ of\frac{use}{yr} \times fuel\frac{cost}{L} \times 1.15 \ldots\ldots\ldots\ldots.10.4$$

Fuel requirement for field operations

Approximately 1/3 of the energy used in agricultural production is used during field operations. To assist in providing a basis for making good fuel management decisions in this area, an on-farm fuel study is required to be conducted during cropping seasons.

Fuel consumption can greatly vary from farm to farm. Therefore, the information here should be used as an estimate. More accurate figures can be obtained by keeping individual records. The table below contains estimates of the average quantity of diesel fuel required for some field operations.

Field operation	Diesel gallons per acre	Field operation	Diesel gallons per acre
Fertilization		*Weed control*	
Spreading dry fertilizer	0.15	Sprayer, trailer type	0.10
Anhydrous ammonia	0.55	Rotary hoe	0.20
Tillage		Row-crop cultivator 0.40	0.40
MB plough	1.70	**Harvesting**	
Subsoiler/ripper	1.70	Mower	0.30
Disk-chisel plough	1.30	Mower-conditioner, PTO	0.55
Offset disk	0.85	Rake	0.25
Planting		Corn grain	0.20
No-till planter	0.45		
Grain drill	0.15		
Broadcast seeder	0.70		

The estimates include only the fuel required for actual field work and do not account for machine preparation or travel to and from the field. Because fuel consumption values for any particular operation vary among tractors, soil type, etc., actual fuel requirements may be as much as 35 percent higher or lower than the values listed in the table. Engine efficiencies increase over time. Older tractors manufactured before 1990 or in poor repair may use 10 percent or more additional fuel. Older gasoline powered equipment uses about 50 percent more gallons of gasoline than the table values indicated.

The values for row-crop operations were calculated for 30-inch rows. They should be adjusted for other row widths. All values were calculated assuming efficient materials handling in the field, proper tractor ballasting to keep wheel slippage below 15 percent, properly tuned and adjusted tractor engines, and partial load tractor operation efficiency by shifting up a gear and throttling the engine back.

Saving fuel used in field operation

FARM TRACTOR SYSTEMS

The first and most important step in saving fuel used for field operations is to ask these questions;

- "Is this operation really necessary?
- Will I gain any benefits in excess of the costs?"

When tractor power is used for fieldwork the following strategies can save fuel.

1. Keeping a current maintenance schedule. Keeping your maintenance schedule current as suggested in the tractor owner/operators' manual will improve fuel economy.
2. Keeping proper ballasting and tyre inflation pressures, and
3. Selecting a fuel saving gear and throttle setting.

Although ballasting is recognized as important, tractor operator's often do not check total tractor weight or front- and rear-axle ballasting. Proper ballasting enables the tractor to efficiently transfer power to the drawbar and avoids wasting energy. Total tractor weight required depends on tractor style (true four-wheel drive, mechanical-front-drive, or two wheel drive) and field speed.

Having this weight properly split between the front- and rear-axle also affects efficiency. Proper weight split is affected by tractor style and whether the attached implement is pulled or mounted. Using cast-iron weights allows ballast to be removed for fuel savings in lighter drawbar work (e.g. spraying).

Tyre inflation pressure should be correctly adjusted for the load the individual tyre is carrying. Consult the tractor operator's manual or check with the tractor or tyre dealer as correct inflation pressure for a given weight depends on tyre size, use as a single or dual, and if the tyre will be used at high speed (e.g. greater than 25 mi/h). Most operators are aware of the damage under-inflation can cause to tyres. Over inflation contributes to excessive wheel slippage and fuel use.

In summary, energy saving can be significant when best conservation practices are used, in tillage, planting, harvest, residue management and distribution, nitrogen application, and proper operation and maintenance of farm equipment. Energy use is a significant component of crop production and it should be considered as a key production input. As stated earlier, questions should be asked before any field operation decision, "Is this field operation necessary? Do benefits exceed losses?"

FARM TRACTOR SYSTEMS

Tractor power and field calculations

Engine power losses occur when transmitting engine power through the drive wheels, the PTO shaft, and the hydraulic system. If Flywheel power is known, multiply it by 0.9 to estimate PTO power.

Tractive efficiency (TE) is the ratio of drawbar power to axle power and can be estimated when slip is known. TE (and hence tyre efficiency) of a wheel can be predicted using a series of equations that take into account tyre dimensions, soil conditions, slip, etc. Tractor performance is calculated by summing the individual wheel performances. Maximum TE is obtained with slip ranges of:
4-8% for concrete;
8-10% for firm soil;
11-13% for tilled soil;
14-16% for soft soils and sands

Implement (machine) power requirements

Draft (D) is the total force parallel to the direction of travel that is required to pull the implement. Typical draft requirements can be calculated as:

$$D = f * i [a + b + c(s)^2] w * t \quad \ldots\ldots\ldots\ldots\ldots\ldots\ldots\ldots.10.5$$

Where:
  D = implement draft, N (lbf);
  F = a dimensionless soil texture adjustment parameter;
  i = 1 for fine, 2 for medium and 3 for coarse textured soils;
  A, B and C are machine-specific parameters given in a table;
  S = field speed, km/h (mile/h).
  W = machine width, m (ft) or number of rows or tools;
  T = tillage depth, cm (in.) for major tools, or equals one for minor tillage tools and seeding implements.

Drawbar power for tractor-powered implements (and propulsion power for self-propelled implements) is computed as:

$$P_{db} = \frac{d \times s}{3.6} \quad \ldots\ldots\ldots\ldots\ldots\ldots\ldots\ldots..10.6$$

Where:
  P_{db} = drawbar power required for the implement, kW; D = implement draft, kN;
  s = travel speed, km/h;

Or,

$$P_{db} = \frac{d \times s}{375} \quad \ldots\ldots\ldots\ldots\ldots\ldots 10.7$$

Where:
P_{db} = drawbar power required for the implement, hp;
D = implement draft, lb;
s = travel speed, mph.

Power-takeoff (PTO) power is power required by the implement from the PTO shaft of the tractor or engine. Typical PTO power requirements can be determined using rotary power requirement parameters available from a number of sources. Implement power take-off power can be calculated as:

$$P_{PTO} = a + bw + cf \quad \ldots\ldots\ldots\ldots\ldots 10.8$$

Where:
P_{PTO} = power-takeoff power required by the implement kW (hp);
w = implement working width, m (ft);
F = material feed rate, t/h (ton/h) wet basis;
a, b, and c are machine specific parameters

Hydraulic power is the fluid power required by the implement from the hydraulic system of the tractor or engine. Implements hydraulic power can be computed as

$$P_{hyd} = p \cdot \frac{F}{1000} \quad \ldots\ldots\ldots\ldots\ldots 10.9$$

Where:

P_{hyd} = hydraulic power required by the implement, kW;
F = fluid flow, L/s;
p = fluid pressure, kPa;

Or,

$$P_{hyd} = p \cdot \frac{F}{1714} \quad \ldots\ldots\ldots\ldots\ldots 10.10$$

Where:

P_{hyd} = hydraulic power required by the implement, hp;
F = fluid flow gal/min;

p = fluid pressure, psi.

Electric power is required to operate components of some implements. To compute implement electric power, use the formula:

$$P_d = i \cdot \frac{E}{1000} \quad \ldots\ldots\ldots\ldots\ldots 10.11$$

Where:
   P_{el} = electric power required by the implement, kW;
   I = electric current, Amp;
   E = electric potential, V;

Or,

$$P_d = i \cdot \frac{E}{746} \quad \ldots\ldots\ldots\ldots\ldots 10.12$$

Where:
   P_{el} = electric power required by the implement, hp;
   I = electric current, Amp;
   E = electric potential, V;

Total power requirement for operating implements (drawn or self-propelled) is the sum of implement power components converted to equivalent PTO power. Total implement power requirement can be computed as:

$$Pt = \left(\frac{P_{db}}{e_m e_s}\right) + P_{PTO} + P_{hyd} + P_d \quad \ldots\ldots\ldots\ldots 10.13$$

Where:
   PT = total implement power requirement, kW (hp);
   E_t = traction efficiency (expressed as a decimal);
   P_{db} = drawbar power required for the implement, kW (hp);
   P_{hyd} = hydraulic power required by the implement, kW (hp);
   P_{PTO} = power-takeoff power required by the implement, kW (hp);
   P_{el} = electric power required by the implement, kW (hp);
   E_m = mechanical efficiency of the transmission and power train. This coefficient is typically 0.96 for tractors with gear transmissions.

Note: Additional power is required to accelerate and overcome changes in topography, soil and crop conditions. Additional power is also required for operator-related equipment such as hydraulic control systems, air conditioning, etc.

FARM TRACTOR SYSTEMS

Practical exercises

1. Carry out ONE of the following instructions:

   a. Make necessary adjustments to ready a piece of farm equipment or machinery for field operation.
   b. Choose a piece of farm machinery or equipment. Check all nuts, bolts, and screws. Tighten any that are loose. Replace those that are missing, worn, or damaged.
   c. Repair broken or worn farm machinery or equipment.
2. Make a list of safety precautions for adjustments or repairs you make for question 1.
3. Observe the tractor operator preparing for field operation and report on procedure for starting and operating the tractor. What are the safety precautions and measures taken in operating farm vehicle?
4. Prepare farm machine for end of season storage. List the procedure, precautions and activities to be carried out
5. Carry out field operation using common types of tillage equipment e.g. ploughs, harrow, ridger etc. Follow a coupled tractor to the field and observe the mode of operation on the field, note depth of cut, width of row etc. and report on the type of field pattern used and your field experience.
6. Do the following

   a. Identify the universal joint with the PTO of a tractor
   b. Couple machinery to the PTO of a tractor
   c. Operate mounted machinery with PTO coupled to tractor
   d. Prepare workbook/field report of activities and sketches

References

Ag Decision Maker, 2009.Farm Machinery. Cooperative Extension Service, Iowa State University of Science and Technology, Ames, Iowa. Selection File A3-28 PM 952 Revised October 2009.

Alec Osborn (2006).The Fascination of Engines Formula 1 to Farm Machinery I Mech E Presidential Address 2006 pp 1-25

Anozodo U.G.N. et al (1989). Perspective plan for agriculture development in Nigeria (1989-2004).Agricultural Mechanization study report, Federal Agricultural coordinating Unit (FACU), Ibadan.

Arvin Meritor - an ArvinMeritor Brand the Future Generation of Drum Brake Linings. Vehicle Aftermarket Park Lane, Great Alne, Alcester, Warwickshire, B49 6HT, England

Bankim Shikari (2004). Automation In Condition Based Maintenance Using Vibration Analysis. Dept. Of Mechanical Engineering MANIT, Maulana Azad National Institute of Technology Bhopal, India, Pin: 462007

Bello R. S., (2009). Farm Power and Machinery Workbook. Pub. Climax Printers #26/30 College Rd. Ogui New Layout, Enugu Nigeria. ISBN: 978-376-671-8.

Bello R. S., (2007). Fundamental Principles of Agricultural Engineering Practice. Pub. Climax Printers #26/30 College Rd. Ogui New Layout, Enugu. ISBN: 978-080-015-8.

Bello R. S., (2006). Guide to Agricultural Machinery Maintenance, and Operation Pub; Fasmen Communications 79/94 Owerri Road, Okigwe Nigeria. ISBN: 978 - 2986 - 90 – 9).

Briggs Stratton. Small Gas Engine Service & Maintenance. Service and Repair Instruction Book

Brinkmann, W. (1985). Precision drilling of sugar beets, fodder beets and corn (in German). In: Landtechnik, Heraugegeben von Horst Eichhorn: Stuttgart, pp. 279–297.

Claassen S. L. 1995. Management of Tillage Equipment on Research Farms. IITA Research Guide 10.Training Program, International Institute of Tropical Agriculture (IITA), Ibadan, Nigeria.33 pages.

Clare Bishop-Sambrook, 2003. Labour Saving Technologies And Practices For Farming And Household Activities In Eastern And Southern Africa. Labour Constraints and the Impact of HIV/AIDS on Rural Livelihoods in Bondo and Busia Districts Western Kenya accessed June 25 2006.

Daken, 2000.Slashers Owner's Manual / Parts Lists. Daken Pty Ltd ACN 004 476 Web Site: www.daken.com.au 00-SDESLSMAN.p 65.

David W. Smith, 2004. Safe Tractor Operation: Driving on Highways. Texas Cooperative Extension at http://texasextension.tamu.edu

Donald E. Borgman, Everette Hainline, Melvin E. long (1981).Tractors- Fundamental of machine operation. John Deere service training, Dept. F, John Deere Road, Moline, Illinois 61265.

Douglas C. Couper (1996). Tractor design and operation for research stations. IITA Research Guide 4 Training Program, International Institute of Tropical Agriculture (IITA), Ibadan, Nigeria

Eather, J. J, and Fragar, L. J., 2005. Health and Safety in the Farm Workshop: a Practical Guide. Australian Centre for Agricultural Health and Safety

Eric Hallman Power Take-Off (PTO) Safety. Cornell Agricultural Health & Safety Program. http (accessed June 13, 2007).

Fleet guard Nelson, New Zealand. A Practical Guide for Farm Equipment Maintenance Complete Engine System Protection for Farmers. Web Site: www.fleetguard.com (accessed October 19, 2009).

Frank Buckingham (1981). Machinery maintenance. John Deere service training Dept. F. John Deere road, Moline Illinois 61265 USA

FRSC, 2008.Nigeria Highway Code. Federal Road Safety Commission FRN

John B. Liljedahl, Paul K. Turnquist, David W. smith, Makoto Hoki (1989). Tractors and their power units, 4th Ed. Published by Von Nostrand Reinhold. NY.

HOSTA Task Sheet 5.3, 2004.Using 3-Point Hitch Implements. National Safe Tractor and Machinery Operation Program (HOSTA). The Pennsylvania State University 2004 Cooperation provided by the Ohio State University and National Safety Council.

HOSTA Task Sheet 5.4.1, 2004.Using Power Take-Off Implements. National Safe Tractor and Machinery Operation Program (HOSTA). The Pennsylvania State University 2004 Cooperation provided by the Ohio State University and National Safety Council.

HSE AS24 (rev), 2004.Power take-offs and power take-off drive shafts as Agricultural Safety. www.hse.gov.uk.) (accessed October 19, 2009).

Kevin Dhuyvetter and Terry Kastens, 2005.Impact of Rising Diesel Prices on Machinery and Whole-Farm Costs (accessed October 19, 2009).

MSI Motor Service International GmbH, 2000.http//: www.msi-motor-service.com

MU guide, 1997.Tractor Tyre and Ballast Management. Agricultural Equipment Published by University Extension, University of Missouri-Columbia. Compiled by William W. Casady Pp 1-4.

Odighoh E.U (1991). Research, Development and Manufacture of Agricultural Machinery in Nigeria: A National imperative, NSE 1992 food and agricultural lecture. NCAM, Ilorin July 11 1994Winchell W. (2001).

Mark Hanna (2000). Planter Clean-out Tips When Changing Seed Varieties Cooperative Extension Service, Iowa State University of Science and Technology, Ames, Iowa. PM 1847 Revised December 2000.

John E. Morrison, Jr. and Ronald R. Allen Planter and Drill Requirements for Soils with Surface Residues.

Marcus Bengtsson (2004). Condition Based Maintenance System Technology – Where is Development Heading? Euromaintenance 2004 – Proceedings of the 17th European Maintenance Congress, 11th – 13th of May, 2004, AMS (Spanish Maintenance Society), Barcelona, Spain, B-19.580-2004.

Mark Trudeau, 2005. Part One: Considerations for the First-Time Tractor Buyer. Whitetail News / Vol. 16, No. 3 www.whitetailinstitute.com

Mitchell, J. S., (1998). "Five to ten year vision for CBM, ATP Fall Meeting – Condition Based Maintenance Workshop", USA, Atlanta, GA, (PowerPoint Presentation), http://www.atp.nist.gov/files/3 (2003-10-08).

Neilsen M.L., Lenhert D.H., Mizuno M., Singh G., Staver J., Zhang N., Kramer K., Rust W.J., Stoll Q., and Uddin M.S., 2004. Encouraging Interest In Engineering Through Embedded System Design Proceedings of the 2004 American Society for Engineering

Education Annual Conference & Exposition. Copyright © 2004, American Society for Engineering Education

Renius, K. Th. 1999. Tractors. Two Axle Tractors. In: CIGR Handbook of Agricultural Engineering. Vol. III, Plant Production Engineering (edited by B.A. Stout and B. Cheze):115-184.

Renius K. Th. and Rainer. 2005. Continuously Variable Tractor Transmissions. ASAE Distinguished Lecture No. 29, Agricultural Equipment Technology Conference, 14-16 February 2005, Louisville, Kentucky, USA. Copyright 2005 American Society of Agricultural Engineers pp. 1-37.*ASAE Publication Number 913C0305*.

Road Traffic, 2008 No. 1980. The Tractor etc (EC Type-Approval) (Amendment) Regulations 2008. Printed and published in the UK by the Stationery Office Limited.

Ricky Smith (2000). Best Maintenance Repair Practices. Technical Training Division Life Cycle Engineering, Inc.

Terry Kastens, 1997. Farm Machinery Operation Cost Calculations. Kansas State University Agricultural Experiment Station and Cooperative Extension Service.

Terry Kastens and Kevin Dhuyvetter, 2008.Evaluating Self-Propelled Sprayer Ownership with the Own Sprayer Spreadsheet. *K-state department of* Agricultural Economics. Available at www.agmanager.info(accessed October 19, 2009).

Tom Karsky and A. K. Jaussi, 1998.Highway Transport of Agricultural Equipment: Preventing Public Road Accidents. Pub. By the University of Idaho Cooperative Extension System, the Oregon State University Extension Service, Washington State University Cooperative Extension, and the U.S. Department of Agriculture cooperating.

Kolberg, Robert L., and Lori J. Wiles. 2002. Effect of steam application on cropland weeds. Weed Technology. Vol. 16, No. 1. p. 43–49.

Flame Engineering, Inc. OnLine Agricultural Flaming Guideflame@flameengineering.com

SAF, 2006.Maintenance and Repair Manual for SAF Disc Brakes. SK RB 9022 K/SK RB 9019 K with KNORR brake caliper SK 1000 ET 120. Edition 01/2006

SIME-Stromag, 2004.Disc Brakes vs. Drum Brakeshttp://www.sime-stromag.com accessed on August 13 2008.

Steve Diver, 2002.Flame Weeding for Vegetable Crops NCAT Agriculture Specialist. Editor Richard Earles, Slot 211

Valtra T series, 2002.New Large Tractors from Valtra.http://www.valtra.co.uk/news/534.asp. Date modified 24.11.2012

Von H. Jarrett, 1995.Getting the most out of your Tractors. Farm Machinery Fact Sheet FM-14.Cooperative Extension Service, Utah State University, Logan, Utah.

Von H. Jarrett, 1995.Buying a Used Farm Machine. Farm Machinery Fact Sheet FM-02.Cooperative Extension Service, Utah State University, Logan, Utah.

FARM TRACTOR SYSTEMS

Glossary of Terms

The following terms were used frequently in connection with maintenance practices and thus apply to this book. They are sometimes referred to as maintenance functions:

Actions: These include processes such as welding, grinding, riveting, straightening, facing, machining, and/or resurfacing.

Adjust: To maintain or regulate, within the prescribed limits, by bringing into proper or exact position, or by setting the operating characteristics to specified parameters.

Align: To adjust specified variable elements of an item to bring about optimum and desired performance e.g. belt alignment, tyre alignment etc.

Back furrow: Two adjacent furrows with soil

Blow-by: The term blow-by denotes the escape of compression and combustion gases past the pistons and piston rings into the crankcase.

Borders or boundary: These are the ends of the field defining the area of the field. There are always 4 borders to a field, though it could be bounded by other fields already prepared.

Calibrate: This consists of comparisons of two instruments, one of which is a certified standard of known accuracy, to detect and adjust any discrepancy in the accuracy of the other instrument being compared.

Dead furrow: These are two adjacent furrows without soil

Disassembly/Assembly: This is the step-by-step breakdown (taking apart) of a spare/functional group to the level of its least component.

Fault location/troubleshooting: The process of investigating and detecting the cause of equipment malfunctioning; the act of Isolating a fault within a system or unit Under Test (UT).

Headland: These are the pieces of land adjacent the boundary which are provided for tractor turning just out of the field.

Inspection: This is a process of determination of the serviceability of an item or component by comparing its physical, mechanical, and/or electrical characteristics with established standards through examination (e.g. by sight, sound. or feel)

Install: Installation is the act of placing, seating, or fixing into position a spare, repair part, or module (component or assembly) in a manner to allow the proper functioning of an equipment or system.

Machinery mechanics: These are personnel that carry out adjustments, maintain, and repair machinery such as engines, conveyors, tractors etc. they are sometimes called machinery repairers or machinery maintenance mechanics. Mechanics must be able to stand, stoop, lean, lift, squat, and have full use of both hands.

Millwrights: These are personnel that specialize in installation, dismantling, repairs and replacement of machinery and heavy equipment

Open furrow: This is the furrow created by one of the disc as the tractor travels up the field.

Overhaul: That maintenance effort (service/action) prescribed to restore an item to completely serviceable/operational condition as required by maintenance standards in appropriate technical publications. Overhaul does not normally return an item to new condition.

Rebuild: This consists of those services/actions necessary for the restoration of unserviceable equipment to a near-new condition in accordance with original manufacturing standards. The rebuild operation includes the act of returning to zero those age measurements (e.g. hour/miles) considered in classifying equipment/components.

Remove: To remove an item means the taking out and installation of the same item with the same one when performing service or other maintenance functions.

Repair: The application of maintenance services, including fault location / troubleshooting, removal/installation, and disassembly/assembly procedures and maintenance actions to identify troubles and restore serviceability to an item by correcting specific damage, fault, malfunction, or failure in apart, subassembly, module (component or assembly), or system.

Replace: This means the removal of an unserviceable item and installation of a serviceable counterpart in its place.

Round: This is one up and down movement of tractor through the length of the field i.e. 1round = 2 trips.

Service: This is an operation required periodically to keep an item in proper operating condition, e.g., to clean (this includes decontaminate, when required), to preserve, to drain, to paint, or to replenish fuel, lubricants, chemical fluids, or gases.

Test: This is to verify serviceability by measuring the mechanical, pneumatic, hydraulic, or electrical characteristics of an item and comparing those characteristics with prescribed standards.

Tractor trip: This is one movement of tractor up the field length.

Work environment: This is the area in which maintenance work is carried out. This area is usually noisy, but well ventilated and lighted. Maintenance work is sometimes greasy and dirty.

Appendix

Appendix I: Tractor systems troubleshooting chart

Problem	Causes	Remedy
Engine		
Engine is hard to start or will not start	1. No fuel or improper fuel 2. Water or dirt in fuel or dirty filters 3. Air in fuel system (diesel) 4. Improper timing 5. Defective coil or condenser (spark ignition) 6. Pitted or burned distributor point 7. Cracked distributor cap or eroded rotor 8. Distributor wires loose or installed in wrong order 9. Fouled or defective spark plug 10. Poor injection nozzle operation (diesel) 11. Liquid fuel in line lp-gas) 12. Engine flooded	1. Fill tank. If wrong fuel, drain and refilled with proper fuel. 2. Check out fuel supply. Replace or clean filters 3. Bleed air from system 4. Check distributor (spark ignition). Check injection pump 5. Replace coil or condenser 6. Clean or replace points 7. Replace cap or rotor 8. Push wire into sockets. Install wires in correct firing order 9. Clean and re-gap plugs or replace them 10. Have service shop clean or repair nozzles 11. Always turn on vapour valve when starting engine 12. Wait several minutes before attempting to start engine. Do not choke engine again
Engine starts but will not run properly	1. Fuel problem-dirt, air restrictions, or clogged filters 2. Carburetor needs adjustment 3. Defective coil or condenser 4. Defective ignition resistor or key switch 5. Pitted or burned distributor points 6. Fouled or defective spark plugs 7. Cracked distributor cap or eroded rotor	1. Check fuel supply, bleed system (diesel), check for restrictions, clean or replace fuel and air filters and screens. 2. Adjust idle and load air-fuel mixtures 3. Replace as necessary 4. Replace resistor or switch 5. Clean or replace points 6. Clean and re-gap plugs or replace them 7. Replace cap or rotor
Engine detonates (gasoline)	1. Wrong type of fuel	1. Use proper octane fuel
Engine pre-ignites (gasoline)	1. Distributor timed too early 2. Distributor advance mechanism stuck 3. Faulty spark plug heat range too high	1. retime distributor 2. Free mechanism 3. Install new plugs that have proper heat range

FARM TRACTOR SYSTEMS

Problem	Causes	Remedy
Engine back fires	1. spark plugs cables installed wrong 2. Carburetor mixture too lean	1. Install in correct firing order 2. Adjust carburetor
Engine knocks	1. Improper distributor timing 2. Improper injection timing 3. Wrong engine bearings or bushings 4. Loose bearing caps 5. Foreign matter in the cylinder	1. Time distributor 2. Check injector pump 3. Replace bearings 4. Tightened to proper torque 5. Remove material and repair engine
Engine overheats	1. Defective radiator cap 2. Radiator core plugged with dirt and debris 3. Defective thermostat 4. Loss of coolant 5. Loose fan belt 6. Cooling system has scale deposit build up 7. Overload engine 8. Incorrect engine timing 9. Engine low in oil 10. Wrong type of fuel	1. Replace cap 2. Clean radiator core 3. Replace thermostat 4. Check for leakage and correct 5. Adjust tension 6. Use cooling system cleaner to remove scale 7. Reduce load or shift into a lower gear 8. Time distributor (si). Set injector (ci) 9. add oil to proper oil level 10. Use recommended fuel
Lack of power	1. air cleaner dirty or obstructed 2. restriction in fuel lines, filters or carburetor 3. wrong type of fuel 4. frost at fuel-lock strainer (lp-gas) 5. governor not operating properly	1. Clean or replace fuel cleaner. Remove obstruction 2. Clean clogged parts 3. Use recommended type 4. Clean strainer 5. Have service shop adjust or repair governor
High oil consumption /oil pressure too low	1. Crankcase oil too light 2. Worn piston and rings 3. Worn valve guides or stem oil seals 4. External oil leaks 5. Oil pressure too high	1. Use recommended weight oil 2. Recondition the piston through servicing 3. Replace valve guide and stems 4. Eliminate leaks 5. Adjust pressure/check system and relieve restrictions
Engine exhausts black or gray smoke	1. Improper type of fuel 2. Clogged or dirty air cleaner 3. Defective muffler 4. Engine overloaded 5. Fuel injection system faulty	1. Drain fuel tank and fill with recommended fuel 2. Clean or replace air cleaner 3. Replace muffler 4. Reduce load or shift into a lower gear 5. Check the fault and repair

FARM TRACTOR SYSTEMS

Problem	Causes	Remedy
Exhaust smoke white	1. Improper carburetor adjustment 2. Improper type of fuel 3. Low engine temperature 4. Defective thermostat 5. Engine out of time	1. Drain fuel tank and fill with recommended fuel 2. Allow engine to warm up to normal temperatures before loading 3. Replace thermostat 4. Time distributor or injector
Electrical system		
Low battery output	1. Low electrolyte level 2. Defective battery cell/ cracked battery case/ 3. Low battery capacity	1. Add electrolyte to proper level 2. Replace battery 3. Replace battery with right capacity battery
Low battery charge	1. Excessive loads from added accessories 2. Excessive engine idling 3. Faulting charging operation	1. remove excess loads or replace alternator with larger capacity 2. Allow engine to idle only when necessary 3. Perform low charging circuit output operation
Low charging circuit output	1. Slipping drive belts 2. Excessively worn or sticking brushes in alternator 3. Defective alternator or generator	1. adjust belt tension 2. Replace brushes 3. Replace unit or check alternator
Starter motor will not operate	1. Low battery charge 2. High resistance in circuit 3. Starter safety switch open 4. Defective starter / starter switch	1. Charge battery 2. Clean and tighten connection 3. Move shift lever to neutral or park position 4. Replace switch or service the starter
Misfiring of engine	1. Bad plug wiring 2. Worn spark plug electrodes or fouled plugs 3. Incorrect distributor timing 4. Insufficient voltage available to spark plug	1. Replace plug wires 2. Clean plugs and re-gap 3. Retime distributor 4. Perform low voltage at spark plug
Low voltage ta spark plug	1. Worn or improperly spaced distributor points 2. Defective condenser 3. Defective spark plug cables	1. Adjust point gap 2. Replace condenser 3. Replace cable
Oil pressure indicator fails to light	1. Burned out bulb 2. Open circuit 3. Defective lamp body 4. Faulty oil pressure switch	1. Replace bulb 2. Clean and tightened connections 3. Replace lamp body 4. Replace switch

FARM TRACTOR SYSTEMS

Problem	Causes	Remedy
Clutch system		
Clutch slips, Clutch grabs, Clutch squeaks, Rattles and vibrates	1. Too little clutch pedal free travel 2. Worn clutch disk 3. Clutch release bearing dry 4. Clutch actuating mechanism dry	1. Adjust clutch pedal free travel 2. Repair disk 3. Lubricate bearing and linkages 4. Service and adjust
Transmission system		
Transmission noisy, Transmission hard to shift, Transmission sticks in gear	1. Transmission oil low/worn or broken gears 2. - 3. Clutch not releasing/shift linkage binding/worn linkage	1. Fill with proper lubricant 2. Repair and adjust 3. Adjust clutch free travel/free linkage/repair
Differential system		
Turning difficult, Mechanical lock free, Hydraulic lock free	1. Differential lock stick/brakes dragging or diff. Lock stuck 2. Linkage not in adjustment 3. Valve malfunction or linkage not adjusted	Repair lock, brakes and shift linkage
System fail to operate or operate erratically	1. Little or no oil/wrong oil viscosity/oil filter plugged/oil leakage	1. Fill to proper level/refill with proper oil/replace filter/tighten fittings & lines
Hydraulic system		
System operates slowly, System overheating	1. Oil viscosity too heavy/engine speed too low /air in system 2. Operator holding controls in power position too long	1. Refill with correct oil/operate engine at recommended speed/bleed and tighten 2. Return control to neutral when not in use
Foaming oil, Noisy pump, Component leaking	1. Low oil level/water in oil/wrong kind of oil 2. Air leak/low oil level/air in oil 3. Failure of major parts	1. Fill to proper level/ drain and replace oil/drain and fill with rec. Oil 2. Tighten lines and fittings/fill with proper oil/tighten fittings & check leaks 3. Repair and service in workshop

FARM TRACTOR SYSTEMS

Problem	Causes	Remedy
Brake systems		
Brakes not holding Brakes not releasing	1. Glazed, greasy or worn pads/linings 2. Cables or linkages binding/foreign materials lodged in mechanism	1. Replace lining 2. Adjust linkage and cables/remove materials
Bouncing/spongy pedal No brake	1. Hydraulic brakes: air in the system 2. Hydraulic brakes: air in the system; 3. Manual :linings worn or linkage out of adjustment; 4. Power brakes: accumulators discharged	1. Bleed the systems 2. Bleed the system 3. Replace lining or adjust linkage 4. Repair accumulator on workshop
Machine pulls to one side	1. Brakes adjusted unevenly	1. Adjust brakes
Hydraulic brakes erratic	1. Contaminated fluid	1. Drain, clean and refilled with proper fluid, then bleed the system

Notes

Titles in author's list

Agriculture & mechanization series

- Farm power and machinery operations
- Agricultural machinery & mechanization
- Agricultural engineering: principles and practice (Vol 1)
- Agricultural engineering: principles and practice (Vol 2)
- Farm tractor systems: operations and maintenance
- Timeline of agricultural mechanization

Horticultural series

- Horticultural machinery: equipment and safety
- Fruits and vegetable technologies: management options

Workplace safety and machine technology series

- Agricultural machinery hazards & safety practices
- Workplace hazards risks & control
- Workshop technology & practice
- Technical drawing presentation and practice

Students' handbook series

- Study companion
- Path to exam success

Edited works on sustainable agriculture and environment series

- Sustainable agriculture: prospects and challenges
- Sustainable environmental management: issues and projections

More information available @:

http://www.amazon.com/Segun-R.-Bello/e/B008AL6RI0

www.ingramcontent.com/pod-product-compliance
Lightning Source LLC
Chambersburg PA
CBHW081233180526
45171CB00005B/414